'Russia and the other transition economies have taught us that policy change involves a struggle over assets and that market-oriented reform can be hijacked to sustain, or even generate, new structures of power and corruption. In this powerful and historically rich book, Robison and Hadiz make the same point for the Indonesian oligarchy, showing its tremendous resilience in the face of a variety of external and internal pressures, from the IMF to the democratic opposition.'

Stephan Haggard, *University of California San Diego*, USA

'[This] is the first study to spell out the nature of the relationship between a national oligarchy and the global market place. As such it is not only an important empirical study, it also represents a major critique of the neo-classical understanding of development that prevailed throughout the late 20th century.'

Richard Higgott, Editor, *The Pacific Review*

'Vedi Hadiz and Richard Robison argue that what is taking place in Indonesia is the reorganization of the power relations incubated within the Soeharto regime rather than their fundamental transformation, and that democratizing was instituted for the survival of the interests nurtured under Soeharto's rule as the institutional structures of the New Order became unviable. A must read for anyone interested in Indonesia and "third world" transformations.'

Takashi Shiraishi, *Kyoto University*, Japan

REORGANISING POWER IN INDONESIA

Reorganising Power in Indonesia is a new and distinctive analysis of the dramatic fall of Soeharto, the last of the great Cold-War capitalist dictators, and of the struggles that are reshaping the institutions and systems of power and wealth in Indonesia.

But this is more than a pathology of power and conflict in Indonesia. It is, at the same time, a broader political economy of regime change engaging the major theoretical debates about how institutions and states are changed, and how systems of social power survive, fail, or are transformed. The book challenges neo-liberal accounts where centre stage is taken by rational individuals making choices about policy, or by technocrats able to insulate themselves from predatory raiders within a state emptied of politics. Instead, it is argued that policies and institutions are forged in bitter social conflicts about power and its distribution. Thus, in the case of Indonesia, the dramatic events of the past two decades are understood essentially in terms of the rise of a complex politico-business oligarchy and the ongoing reorganisation of its power through successive crises, colonising and expropriating new political and market institutions. With the collapse of authoritarian rule, the authors propose that the way was left open for this oligarchy to reconstitute its power – to reinvent itself – via new accommodations with populist and predatory interests within broader society and within the new institutions of newly democratic Indonesia.

These are questions made even more critical today as the US struggles, once again, in Iraq and elsewhere, to comprehend why expectations of market reform, democratic transition and social change are often overtaken by a metamorphosis of reactionary interest or by a descent into chaotic and unconstrained systems of oligarchy.

This is a book not only for researchers and students but for anyone interested in political economy, political sociology, development studies, and Southeast Asian politics and society.

Richard Robison is Professor of Political economy at the Institute of Social Studies in The Hague, The Netherlands and was formerly Director of the Australian Research Council's Special Centre for the Study of Political and Economic Change in Asia at Murdoch University, Australia.

Vedi R. Hadiz is Assistant Professor in the Department of Sociology at the National University of Singapore and was previously Research Fellow at the Asia Research Center, Murdoch University, Australia.

ROUTLEDGECURZON/CITY UNIVERSITY OF HONG KONG SOUTHEAST ASIAN STUDIES

Edited by Kevin Hewison and Vivienne Wee

LABOUR POLITICS AND THE STATE IN INDUSTRIALIZING THAILAND

Andrew Brown

ASIAN REGIONAL GOVERNANCE: CRISIS AND CHANGE

Edited by Kanishka Jayasuriya

REORGANISING POWER IN INDONESIA: THE POLITICS OF OLIGARCHY IN AN AGE OF MARKETS

Richard Robison and Vedi R. Hadiz

REORGANISING POWER IN INDONESIA

The politics of oligarchy in an age of markets

Richard Robison and Vedi R. Hadiz

LONDON AND NEW YORK

First published 2004
by RoutledgeCurzon
Published 2015 by Routledge
2 Park Square, Milton Park, Abingdon,
Oxon, OX14 4RN

Simultaneously published in the USA and Canada
by Routledge
711 Third Avenue, New York, NY 10017

Routledge is an imprint of the Taylor & Francis Group

Typeset in Baskerville by
Taylor & Francis Books
Ltd

British Library Cataloguing in Publication Data
A catalogue record for this book is available from the British Library

Library of Congress Cataloging in Publication Data
A catalog record for this book has been requested

ISBN 978-0-415-33252-1 (hbk)
ISBN 978-0-415-33253-8 (pbk)

CONTENTS

Series editors' preface ix
Preface xii
Acknowledgements xv
Abbreviations and Glossary xvii

PART I
Historical and theoretical frameworks 1

Introduction: Economic crisis and the paradoxes of transition 3

1 Theories of change and the case of Indonesia 18

2 The genesis of oligarchy: Soeharto's New Order 1965–1982 40

PART II
The triumph of oligarchy 1982–1997 69

3 Hijacking the markets 71

4 Capturing the political regime 103

5 Disorganising civil society 120

PART III
The oligarchy in crisis 1997–1998 145

6 Economic catastrophe 147

7 Political unravelling 164

Part IV
Oligarchy reconstituted 185

8 Reorganising economic power 187

9 Reorganising political power 223

10 Can oligarchy survive? 253

Bibliography 267
Index 290

SERIES EDITORS' PREFACE

The establishment of the Southeast Asia Research Centre at the City University of Hong Kong in 2000 reflected an increased interest in Southeast Asia following two watershed changes. The first was the end of colonialism in Hong Kong, as the territory became a Special Administrative Region of China in 1997. This coincided with the second event, the Asian Economic Crisis, which struck down some of the major economies of the region, with some important political consequences.

This book series reflects the Centre's research agenda and seeks to advance the understanding of the political, economic and social forces that are shaping contemporary Southeast Asia. This series aims to produce books that are examples of the Centre's emphasis on multi-disciplinary, comparative and holistic research. It also recognises that the political and economic development of Southeast Asia has often been turbulent, and that the contemporary era is no different.

As the region emerged from decolonisation and war, rapid economic development reconfigured the societies of Southeast Asia. From the mid-1970s, a number of Southeast Asian economies enjoyed periods of significant economic growth. The economies of Singapore, Malaysia, Thailand and Indonesia benefited from a more generalised development in East Asia, and made rapid advances, becoming some of the most dynamic economies and societies in the world. Huge flows of foreign capital and the development of relatively powerful domestic capitalist classes rapidly transformed these economies and their societies. The international financial institutions celebrated the region's economic success, urged a continued unfettering of markets, and extolled the benefits of enhanced globalisation.

But the negative social outcomes of the 1997 economic crash posed new challenges for the region's development models, and demanded a questioning of the processes associated with capitalist globalisation. Further, the economic crash confronted the region's political regimes with significant challenges. The most notable of these was the collapse of Soeharto's New Order in Indonesia. This confluence of economic and political turmoil stimulated a reassessment of the impacts of globalisation and associated ideas about regionalisation.

Nowhere has this reassessment been more vividly revealed than in the economic rise of China and the challenges and opportunities this poses for Southeast Asia.

Understanding how Southeast Asians are negotiating the broad and multiple challenges – economic, political, social, religious and cultural – posed by globalisation, and how they are reinventing their societies are critical tasks. This is a central concern of the Southeast Asia Research Centre's research agenda. A second research focus is the divisions of class, ethnicity, gender, culture and religion that appear as faultlines underlying Southeast Asia's post-colonial nations. Such rifts shape diverse patterns of conflict in the region. A third area of research interest involves regional interactions, including those between states, within transnational civil society, business, labour, and migration flows in and beyond the region. Finally, attention is given to the ways in which Southeast Asian political economies are being reinvented following the Asian Crisis, examining new patterns of politics, accumulation and allocation in the region.

It is this last and important issue that is the focus of this book by Professor Richard Robison and Dr Vedi Hadiz. The book provides a carefully crafted and intelligent analysis of the sources of the deep and complex conflicts that have determined the trajectory of Indonesian capitalism over the last four decades. Robison and Hadiz use a broad political economy approach and focus on domestic social conflict as they map the development of Indonesian capitalism and its integration with global markets. Of course, the 1997 Crisis unleashed a period of turmoil that has made political, economic and social conditions in Indonesia more complex than ever. This book provides a systematic analysis of the reconfiguration of power that is shaping contemporary Indonesian society.

The authors' theoretical framework addresses broad issues that have wider implications and significance than for Indonesia alone. In fact, this volume confronts fundamental questions raised by recent economic and political transformations where the development of liberal markets and the collapse of authoritarian regimes have not produced the results many neo-liberal reformers had hoped would be the case. Robison and Hadiz show that liberal proposals and agendas may be hijacked in ways that facilitate the reconstitution and reorganisation of predatory forms of power that have distinctly illiberal political outcomes.

That neo-liberal agendas have not always developed as expected presents neo-liberals with a paradox. Whereas these reformers may have thought and hoped that old, predatory elites would be destroyed in Indonesia, the result has been their apparent entrenchment. Robison and Hadiz make it clear that embedded power relations may be reorganised to survive and prosper within new institutional frameworks. It is evident that any new policies and institutions that emerge do so only within the context of existing power structures and ongoing social conflicts, and continue to be shaped by these forces. This observation is timely for those who might think that it is possible to re-engineer so-called 'failed states' through new market-friendly institutions of governance.

Kevin Hewison (Director) and Vivienne Wee (Associate Director)
Southeast Asia Research Centre, City University of Hong Kong

PREFACE

Debates about the nature of power and the volatile collisions that have accompanied its transformation in Indonesia over the past four decades have been characterised by a remarkable degree of acrimony. In part, this is because many of the battlelines were drawn up at the height of the Cold War, laying the foundations for highly passionate attitudes towards the Soeharto regime – the ascendance of which was accompanied by the massacre of hundreds of thousands of real and imagined communists. The problem, however, is deeper and more critical not least because market-oriented liberals found themselves for over three decades immersed in an uncomfortable and ongoing embrace with a repressive and authoritarian political regime. They had to defend the very principles of neo-liberal ideology from attack as the advance of capitalism seemed to enhance rather than undermine an arbitrary and predatory system of power. Moreover, rather than dismantling the centralised authoritarian rule that underpinned Indonesia's state capitalism, the growth of powerful private interests increasingly became an integral part of it.

This study is therefore drawn into larger and more general debates about the relationships between globalisation, markets and systems of state power, investigating in the Indonesian context circumstances where global markets might actually throw out a lifeline to authoritarianism and enable an oligarchic ascendancy to prevail. Furthermore, in the volatile circumstances following the demise of the New Order, it challenges widespread neo-liberal assumptions about the processes of convergence. While the Indonesian debates recall in many ways earlier ones regarding market and democratic 'transitions' in former Soviet-bloc countries, they have also been increasingly drawn into newer controversies about the salvaging of so-called 'failed states', and about institution-building, especially given the ongoing experiences of Afghanistan after the Taliban and Iraq after Saddam Hussein.

It is significant that the case of Indonesia had long presented neo-liberal orthodoxy with difficult paradoxes. The protracted metamorphosis of power in which authoritarian rule and state capitalism were colonised and harnessed to the interests of a pervasive politico-business oligarchy in the 1980s and 1990s took place as Indonesia's integration with global financial and capital markets deepened and became more intense. Even the more direct and instrumental leverage over domestic policy agendas enjoyed by such bastions of neo-liberal orthodoxy as the IMF after the 1997 crisis and the collapse of authoritarian rule in 1998 did not produce the expected liberal transition. What instead emerged was an extraordinary scramble for power and wealth in an apparently chaotic system of parliaments and parties while old relationships between the state and business proved resilient as a procession of bankrupted (or should have been bankrupted) tycoons struggled to keep their corporate empires intact under assault from domestic reformers and foreign creditors.

The sheer density of events in Indonesia since then, the uncertainty of back room intrigues and shifting alliances, the confusing struggles to maintain or to dismantle economic empires, led some analysts towards highly descriptive attempts to keep up with the day-to-day complexities of economic and political life where the problem became one of tracking and cataloguing the chaos and the manoeuvres of individuals. Indeed, the puzzling events in Indonesia reinforced, for some observers, the view that there are no grand frameworks within which the apparently chaotic dynamics of change may be given an elegant simplicity; that history proceeds by accidents and coincidences or by the capacities of leaders to form strategies and weld alliances together. Here, structure is subordinated to agency and almost anything can happen, thereby justifying the close focus on the apparently unconnected minutiae of political and economic life.

But this descent into empiricism has been far outweighed by the way the crisis and its aftermath has also marked a return to grander theories of change and speculation about the way global markets might influence systems of administrative authority and governance, or how the rise of civil society and middle classes shapes the evolution of political democracy. Thus, this study is drawn into larger debates that spread beyond the Indonesian case about how market capitalism and democratic politics are forged and how the increasing influence of global factors intrudes on the equation. Theoretical debates have been fiercely engaged within the neo-liberal camp itself, where views of change as ultimately a process of abstracted and frictionless triumph of efficiency were challenged by institutional political economists for whom the problem was couched increasingly in the struggle to protect markets from politics, and where the central task became one of creating powerful institutions able to insulate rational technocratic decision-makers from the predatory raids of rent-seeking coalitions. As the problem was seen increasingly as one of salvaging a failed state, a romantic nostalgia for the Soeharto years was to surface.

However, there is something strangely abstracted and timeless in this view of change as a process of institutional engineering driven by far-sighted technocrats and politicians operating above the maelstrom of vested interest. The issue of power seems curiously neglected and abandoned for programmes that focus on capacity building and social capital, where institutions themselves are assumed to define the very possibilities of political and social options. Yet, it is a salient feature of the Indonesian experience that the oligarchy and its beneficiaries, the products of the Soeharto era, have reconstituted their social and political power within a new democracy, characterised by the rise of parties and parliaments to cement new predatory alliances. This suggests that old relations of power may survive, and even find new life, within a range of institutional frameworks very different from those in which they had originally emerged.

Thus, in this study, we address these apparent paradoxes not as problems of capacity or institutional design but as the products of bitter social conflicts about power and the way it is distributed. The central question, we propose, is how social interest is politically organised and how these complex coalitions reorganise their power and secure their collective interests in response to economic and political crises and the vast shifts in global and domestic power that accompany them. Specifically, this is a study of how a complex politico-business oligarchy emerged from within a system of authoritarian rule, reorganising its power through successive crises, colonising and expropriating new political and market institutions. It is a study of how an oligarchy fractured and weakened by crisis has reorganised its power by holding out against the 'disciplines' of those global markets so instrumental in its rise, and by hijacking new institutions of governance and forging new social alliances. It is a study, on the other hand, of how neo-liberal reform, as a social and political process, with its own economic and political beneficiaries and supporters, has mistaken institutions for power and failed to organise politically or to mobilise social coalitions around its agenda.

ACKNOWLEDGEMENTS

This book has been many years in the making and there are many people who have helped along the road to publication in one way or another.

First, we both would like to thank the many people in Indonesia who agreed to be interviewed or willingly spent their time with us to share information and ideas. Quite literally, so many people fall under this category that it is impossible to mention them all individually. But clearly, this book would not have been possible without their help.

We would also like to thank our colleagues – fellow academics in numerous universities and research institutes in Indonesia, Australia, Singapore, the UK, Japan and many other countries – who have been generous with their time, comments and criticism in workshops and more informally. Among these are, especially, Garry Rodan, Kevin Hewison, Kanishka Jayasuriya, Andrew Rosser, and Daniel Dhakidae.

We also thank our assistants who have helped in collecting data, arranging interviews and in editorial matters, including Arya Wisesa and Kurniawan, but especially Tauel Harper, who worked so hard and ably to put much of the manuscript into a publishable form in the latter stages. We also thank Stephanie Rogers and Zoe Botterill at RoutledgeCurzon for their help and co-operation in the book's publication.

We gratefully acknowledge the financial assistance of the Australian Research Council which funded most of the research up to the end of 1999 through its Special Centre for Research on Political and Economic Change in Asia (the Asia Research Centre), at Murdoch University. Richard Robison would like also to acknowledge support for an extended period of writing in 2001 and 2002 from the Leverhulme Trust and the Centre for the Study of Globalisation and Regionalisation, University of Warwick. Thanks go especially to Richard Higgott, the Centre's Director. Similarly, thanks are given to the Southeast Asian Research Centre, City University, Hong Kong, and to its Director, Kevin Hewison, for support for a period of writing in Hong Kong. Vedi Hadiz would like to especially thank the Asia Research Centre, Murdoch University, for its long-standing support of his work, and the Department of Sociology, National University of Singapore, for its support of much of his research since 2000.

Thanks also to Takashi Shiraishi at Kyoto University, whose various conference invitations and 'get-togethers' provided the opportunity to continuously refine some of the arguments that eventually made it to the book.

Finally, the authors would like to thank their families, who persevered through all the time spent over the years researching, writing and re-writing the book from its various drafts up to the final product.

ABBREVIATIONS AND GLOSSARY

ABRI Angkatan Bersenjata Republik Indonesia; former name for the Indonesian Armed Forces (see **TNI**).

AFL-CIO American Federation of Labour – Congress of Industrial Organisations.

AFTA ASEAN Free Trade Area.

AJI Aliansi Jurnalis Independen; Alliance of Independent Journalists.

APBN Anggaran Pendapatan dan Belanja Negara; the State Budget.

Apkindo Asosiasi Panel Kayu Indonesia; the Association of Indonesian Wood Panel Producers. A plywood marketing and export monopoly.

ASPRI Asisten Pribadi; Special Presidential Assistants.

Azas tunggal Sole principle/basis; see **Pancasila**.

Banser Barisan Serbaguna; civilian militia linked to Nahdlatul Ulama mass organisation.

Bappenas Badan Perencanaan Pembangunan Nasional; Economic Planning Board.

BBD Bank Bumi Daya; a state-owned bank.

BDN Bank Dagang Negara; a state-owned bank.

BI Bank Indonesia.

BKPM Badan Koordinasi Penanaman Modal; the Investment Co-ordinating Board.

BLBI Bantuan Likuiditas Bank Indonesia; Bank Indonesia Liquidity Assistance.

BPIS Badan Pengembangan Industri Strategist; Strategic Industries Development Board; encompassing state-owned strategic industries.

BPK Badan Pemeriksa Keuangan; the Supreme Audit Agency.

BPKP Badan Pengawasan Keuangan dan Pembangunan; the Government Financial and Development Controller.

BPPC Badan Penyangga dan Pemasaran Cengkeh; Clove Support and Marketing Body.

BPPN Badan Penyehatan Perbankan Nasional; see **IBRA**.
BPPT Board of Research and Application of Technology; Badan Pengkajian dan Penerapan Teknologi.
BRI Bank Rakyat Indonesia; a state-owned bank.
Bulog Badan Urusan Logistik; State Logistics Board.
Bupati regent; top official at subprovincial level of government.
Cendana group a name used to refer to the Soeharto family group of companies.
Cendana Palace Soeharto's private residence.
CGI Consultative Group on Indonesia; the main consortia of major creditor countries.
CIDES Centre for Information and Development Studies; a think-tank.
CPDS Centre for Policy and Development Studies; a think-tank.
DPR National Parliament.
DPRD I and II Provincial and sub-provincial parliaments.
DSP Daftar Skala Prioritas; a list that effectively set out sectors from which foreign investors were excluded.
Dwifungsi military dual function.
F-ABRI Military faction within the DPR; later F-TNI/ Polri.
FKPPI Forum Komunikasi Putra-Putri Purnawirawan ABRI; the communications forum of the sons and daughters of retired military. A military-backed 'youth' organisation.
FKSMJ Forum Komunikasi Senat Mahasiswa Jakarta; Communication Forum of Jakarta Student Senates.
FSCP Financial Sector Policy Committee. Established under the Wahid government to supervise broad policy decisions.
Forum Demokrasi Democracy Forum, a grouping of critics of the Soeharto government in the 1980s and 1990s.
FSPSI Federasi Serikat Pekerja Seluruh Indonesia; All Indonesia Workers' Unions.
GARPRI Gabungan Perserikatan Pabrik Rokok Indonesia; the cigarette manufacturers association.
Gema Madani a group of intellectuals and NGO activists led by Emil Salim in the late 1990s.
Golkar the state party during the New Order.
HANKAM Department of Defence and Security.
HIPMI Himpunan Pengusaha Muda Indonesia; the Young Entrepreneurs Association.
HPH Hak Pengusahaan Hutan; the concession right for forestry.
HTI Hutan Tanaman Industri; forestry planting concession.
IBRA The Indonesian Bank Restructuring Agency; see **BPPN**.

ICMI Ikatan Cendekiawan Muslim Indonesia; the Association of Indonesian Muslim Intellectuals.

IGGI Inter-Governmental Group on Indonesia; consortia of major creditor nations established in the early New Order and replaced in 1992.

IPTN Industri Pesawat Terbang Nasional; state-owned aircraft manufacturer.

Kabupaten subprovincial administrative unit.

KAMMI Kesatuan Aksi Mahasiswa Muslim Indonesia; an association of Islamic-based student organisations.

Keppres Keputusan Presiden; a formal system of regulations issued by the President.

KISDI Komite Indonesia untuk Solidaritas Dunia Islam; The Indonesian Committee for Islamic World Solidarity.

KKN *Korupsi, kolusi dan nepotisme*; a popular term used in the post-Soeharto period to denote corruption, collusion and nepotism.

KNPI Komite Nasional Pemuda Indonesia; National Committee for Indonesian Youth – a state-sponsored front group.

KOPKAMTIB Komando Pemulihan Keamanan dan Ketertiban; the Command for the Restoration of Stability and Order.

Kostrad Korps Strategis Angkatan Darat; an elite army unit.

Malari Malapetaka 15 Januari; refers to anti-Soeharto and anti-Chinese demonstrations which broke out on 15 January 1974.

Masyumi Majelis Syuro Muslimin Indonesia; Islamic political party of the 1950s.

MPR Majelis Permusyawaratan Rakyat; Indonesia's supra-parliament, which until recently held the power to elect the President.

Muhammadiyah Muslim mass organisation with large following among urban traders and professionals in Java and the Outer Islands.

Nahdlatul Ulama (NU) Muslim mass organisation that is predominantly based in rural Java.

OPSUS Operasi Khusus; Special Operations Command. The powerful security and intelligence unit under General Ali Moertopo.

Pancasila State Ideology based on Five Principles of Belief in One Good: humanitarianism; national unity, consultative democracy and social justice; see **Azas tunggal**.

PAN Partai Amanat Nasional; National Mandate Party.

PBB Partai Bulan Bintang; Crescent and Star Party.

PDI Partai Demokrasi Indonesia; Indonesian Democratic Party.

PDI-P Partai Demokrasi Indonesia-Perjuangan; Indonesian Democratic Party for Struggle.

Pemuda - Pancasila a state-backed youth/extra-legal organisation.
Pertamina the state company that organised and controlled oil exploitation, drilling and distribution.
Petisi 50 group an association of retired military officers, bureaucrats, and former party officials that was critical of Soeharto.
PKB Partai Kebangkitan Bangsa; the National Awakening Party.
PKI Partai Komunis Indonesia; Indonesian Communist Party.
PKP Partai Keadilan dan Persatuan; Justice and Unity Party.
PLN Perusahaan Listrik Negara; the government's electricity authority.
PNI Partai Nasional Indonesia; Indonesian Nationalist Party.
PPP Partai Persatuan Pembangunan; United Development Party.
PRD Partai Rakyat Demokratik; People's Democratic Party.
Pribumi indigenous Indonesian.
PSI Partai Sosialis Indonesia; Indonesian Socialist Party.
PWI Persatuan Wartawan Indonesia; Indonesian Journalists Association.
RAPBN Rancangan Anggaran Pendapatan dan Belanja Negara; proposed state budget.
SBSI Serikat Buruh Sejahtera Indonesia; Indonesian Prosperity Trade Union.
Sekneg State Secretariat.
Setiakawan the Solidarity Free Trade Union.
SOBSI Sentral Organisasi Buruh Seluruh Indonesia; All-Indonesia Central Workers' Organisation.
Team 39 or COLT a ten-person team to regulate foreign loans for commercial projects.
TNI Tentara Nasional Indonesia; post-Soeharto name for the Indonesian Armed Forces (see **ABRI**).
Yayasan Foundation; organisations intended as charities but often used for political funding and investment.

Part I

HISTORICAL AND THEORETICAL FRAMEWORKS

INTRODUCTION

Economic crisis and the paradoxes of transition

> This crisis consists precisely in the fact that the old is dying and the new cannot be born; in this interregnum a great variety of morbid symptoms appear.[1]

In its broadest sense, this is a study of the vast and bitter struggles that accompany the spread of market capitalism and the forging of those economic and political regimes within which it is defined. These are processes widely regarded as steps in the inevitable and inexorable triumph of liberal markets, democracy and legal-rational forms of bureaucratic authority. Yet, the very protracted nature of such conflicts and the continuing difficulties that have faced liberal reform agendas throws into doubt the assumed structural relationships between markets and liberal political and social institutions. Indeed, the spread of market capitalism has been full of seeming paradoxes as rapid economic growth and deepening integration with global markets have appeared in important instances to consolidate systems of power that are profoundly authoritarian or highly predatory. Such developments have given rise to arguments that capitalism may survive and flourish within a range of different institutional frameworks and regimes. They open the very fundamental questions of how institutions are forged and transformed.

It is within the terms of these issues and questions that we set our study of the dramatic conflicts that have defined the rise of market capitalism in Indonesia during more than three decades under the rule of Soeharto and in the period after his fall from power. Here, the consolidation of authoritarian rule and its metamorphosis towards a hegemony of politico-business oligarchies took place in the context of deepening engagement with the global economy and as progressive, albeit selective, market reforms were enacted. Even destructive economic shocks and, most recently, the collapse of the Soeharto regime itself, have not dismantled a system of power focused around the private expropriation of public authority. These seeming paradoxes raise questions about the relationship between market reform and political power, about the significance of regime change and how global markets influence institutional change. Our exposition of the Indonesian case leads us into a direct collision with mainstream neo-liberal political economy

While the question of liberal convergence and the process of transition to market capitalism has for decades occupied a central place in debates across the ideological spectrum, the Asian economic crisis of 1997/1998 concentrated the issues. The crisis appeared to confirm the fatal flaws of economic regimes constructed around the intervention of the state, whether through strategic policy agendas or the allocation of rents. Here at last, it seemed, was the decisive 'shock', long heralded by neo-classical economists and institutional political economists, that would finally discredit coalitions of vested interests and convince governments of the costs of previous policies and the need for reform (Williamson and Haggard 1994: 562–564). The crisis appeared to cut the ground from under the feet of those who had proposed that state-led developmental regimes might constitute a sustainable model of modern industrial capitalism as an alternative to liberal markets (Johnson 1982; Zysman 1994; Amsden 1989; Wade 1990; Weiss 1998). For the time being it silenced those increasingly confident claims by Asian leaders of the functional superiority of 'Asian Capitalism' over what were regarded as decaying and self-destructive Western models of individualism, free markets and liberal democracies.[2]

The crisis forced important shifts in the policies and institutional arrangements across the region, most specifically in Indonesia, Thailand and South Korea, where beleaguered governments were forced to deliver unprecedented power into the hands of the IMF as they faced looming fiscal and debt crises. As entrenched political regimes began to fracture, those who dominated them soon found they could no longer rule in the same way (Cumings 1999; Moon and Sang 2000; Robison and Rosser 1998). Yet, everywhere the reformist prescriptions of the IMF and others were resisted and confronted even at the height of economic distress. More than five years down the line we find that changes have been more ambiguous and indecisive than first thought (Jayasuriya and Rosser 2001; Hewison 2001; Robison 2001; Gomez 2002; Wingfield 2002).

Without a theory of power, neo-classical economists had few answers to the question of why governments do not quickly embrace the natural efficiency of markets even in the face of a crisis that threatens the very fabric of economic life. Clearly change was not driven by any simple and abstracted logic of efficiency. Politics mattered and neo-liberal political economists, emerging initially in the 1960s and 1970s, grappled with the problem of explaining how. For them, the central political struggle was one to protect markets from politics and from the predatory raids of rent-seeking coalitions: to impose the neo-liberal agenda as a public good that transcends vested interests. It was a struggle, at one level, to undermine the state, whose very existence was seen as dependent on the provision of rents. For others, the political problem was one of constructing a state able to insulate technocratic policy-makers from the predatory raids of distributional coalitions. It was nothing less than a political project aimed, ironically, at emptying the state of politics and creating what

Jayasuriya has termed the regulatory state, a system of technocratic, authoritarian liberalism.[3]

In this study, by contrast, we argue that change is driven neither by rational individuals operating in a world of voluntary transactions nor by the efforts of wise technocrats to empty the state of politics and to neutralise those coalitions of vested self-interest that seek to capture it. Instead, the politics of change are, we propose, inextricably embedded in what Chaudhry has termed, 'Wrenching social struggles [that] precede and shape the rules that govern markets' (Chaudhry 1994: 4). The neo-liberal agenda, in this view, cannot be understood as an abstraction driven by a collection of technopols acting above vested interests, but as an agenda backed by shifting and fluid coalitions with a concrete interest in the configuration of power and the institutions that allocate it. What is at stake is nothing less than a social order. This is why entrenched interests and alliances normally seek to preserve those arrangements within which their hegemony is embedded even at the cost of deepening economic distress and capital flight. While reforms might appear to be technical policy changes based on calculations of efficiency, in reality they often strike at the very social and economic foundations of entrenched regimes. Thus, the significance of crises lies less in economic costs and lessons learnt by policy-makers about efficiency than in the extent to which they unravel the political cohesion of entrenched coalitions and give rise to new reformist alliances.

Nowhere are these dynamics more clearly illustrated than in the case of Indonesia, the specific focus of our study. Here, we argue, the politics of institutional change and markets must be seen in the context of such 'wrenching social struggles' between shifting coalitions of state power and social interest assembled around four main socio-political agendas. Thus, the rise of Soeharto constituted the decisive triumph of the 'patrimonial administrative state' over those contending agendas of social radicalism and the reactionary populism of a declining Muslim petty bourgeoisie in the 1960s.[4] Economic life was to be decided within the framework of a highly centralised authoritarian state, and public authority was to be progressively expropriated for the private and institutional interests of its political rulers and their beneficiaries. Liberalism, as we shall discover, was never more than a slender reed, enjoying a highly ambiguous relationship with the prevailing paradigm of authoritarian rule and predatory power relations.

Thus, the central research question is to explain how those leagues of state and politico-business oligarchies that were its creators and beneficiaries have continued to preserve their ascendancy and the pervasive authority of a sometimes arbitrary and predatory state, despite periodic economic crises, deepening integration with global markets, the growth of a powerful capitalist class and even the unravelling of the political regime itself. It is, at the same time, a question of explaining why no powerful and politically cohesive coalition has been assembled behind the neo-liberal agenda.

Economic crisis, neo-liberalism and the question of convergence

It is not surprising that there was an air of triumphalism within neo-liberal circles after the Asian economic crisis. It was seen as a vindication of the functional superiority of market capitalism and an ideological victory over the proponents of state-managed models of capitalism. Far from being the functionally superior models of capitalist development previously claimed by many Asian leaders, the sorts of interventionist regimes embodied in 'Asian Capitalism' were now revealed, in neo-liberal eyes, as economic systems outmoded in an age of global markets. Such regimes had, in this view, generated those rent-seeking coalitions of predatory officials and political cronies that gave rise inevitably to inflated asset values, excessive private sector borrowing, overextended banking systems, overvalued currencies and deteriorating current accounts (Camdessus 1997: 5).

Long regarded by the World Bank and other neo-liberal economists as a model of responsible macro-economic management, Indonesia was to experience the most destructive economic collapse. A political regime seemingly invulnerable until the very last moment was to unravel more dramatically than any other.[5] An economy that had enjoyed decades of growth now faced the humiliation of economic decline and was forced to submit to sweeping programmes of policy and institutional reform imposed by the IMF. Indonesia was plunged into a destructive spiral of public debt that was to consume over 30 per cent of its routine budget outlays by 2000 and 2001, and into a crisis of private debt that paralysed its banking and financial institutions. Spreading quickly into the very heart of Indonesia's commercial world, it engulfed those major business groups and politico-business families that had enjoyed over a decade of unchecked expansion into booming and newly deregulated sectors of the economy. Highly exposed to short-term and largely unhedged loans, they were confronted by a debt crisis of massive proportions and by early 1998 most had defaulted on their loans (World Bank 1998b: 1.9, 2.3).

This was much more than a financial or economic crisis. What distinguished Indonesia from other countries in the region was the extent to which it unravelled the very institutions of state power underpinning the Soeharto regime. Unable to stem the economic collapse and deal with its deepening social effects or to maintain intact the institutional underpinning of his political and economic power, Soeharto was cornered for the first time in over three decades of rule. Attempts by elements in the military and the government to divert rising social tensions into the channels of racial and religious xenophobia did not prevent domestic dissent focusing increasingly upon Soeharto, his family and cronies. As violence erupted on the streets of Jakarta and across the country, Indonesia's President of over three decades was deserted by his own apparatchiks and forced from office. Ironically, the very architect of the regime now became the major obstacle to the survival of those coalitions of power and interests nurtured within its structures. The crisis had claimed its most important political victim. The last of the Cold War capitalist dictatorships had come to an end.

Within the IMF and amongst the inner sanctums of neo-liberal orthodoxy, there was little doubt that Asia's economies, including that of Indonesia, had fallen under the weight of their own inefficiency and dysfunction, the consequence of state intervention in the free operation of markets and the inevitable widespread cronyism that resulted (Frankel 1998; Wolf 1998a). It was the inherent structural faults of these economies, their refusal to adhere to the disciplines of global markets, that brought the hedge funds down on their heads. IMF Managing Director, Michel Camdessus, argued that, 'it would be a mistake to blame hedge funds or other market participants for the turmoil in Asia'. This was, he proposed, 'only a symptom of more serious underlying problems which are now being addressed in many countries (cited in *Asian Wall Street Journal* (hereafter *AWSJ*) 3 December 1997: 1). The crisis was regarded as a lesson for Asia's policy-makers and a 'blessing in disguise' that paved the way for better policy choices (*AWSJ* 13 November 1997: 1).

Many Western analysts were now to reassert the proposition that Asian economies were caught in an inexorable historical convergence towards Western liberal political and economic institutions (Dale 1998). Among those arguing the convergence thesis, America's Federal Reserve Chairman, Alan Greenspan was to explain the process in functional terms. The effective operation of free markets, he proposed, required rule of law that placed limits on arbitrary state action and a system of governance where transparency and public disclosure of information would enable prices and risks to be assessed on true value rather than some value based on 'moral hazard' (cited in Hamilton 1999: 47). The crisis provided a reminder that long-term growth in an era of globalisation required Asian economies and their governments to 'abandon their bad habits' and to allow market signals to set prices (*Economist* 11 March 1998). It was a convergence that would be enforced by an anonymous and leaderless herd of investors, financiers and currency dealers now operating in a global arena and able to come and go at will. Accepting the rules of these new global markets was not a choice (Friedman 1997). As Camdessus declared, 'Countries cannot compete for the blessings of global capital markets and refuse their disciplines. Hence the importance of pursuing policies that give markets confidence' (cited in Saludo and Shameen 1997).

Initially, there were good grounds for believing that a fundamental convergence was underway in Indonesia. Suddenly, the IMF was able to dictate the details of policy and institutional reform in letters of intent jointly agreed with the Indonesian government. Indonesia's corporate moguls and political families were forced to hand over billions of dollars to cover their vast banking debts while far ranging monopolies and concessions channelled into private hands by the state were cancelled. Attempts were made to break open the opaque and unaccountable organisations through which contracts and licences were channeled by the state. In the political sphere, highly centralised authoritarian rule and ideological control gave way to a vigorous and uncertain new democracy. New political interests and players, some of them from the very margins of the old regime, now jostled for power and influence within the volatile arenas of parliament, parties and elections.

Yet, the neo-liberal expectation that Asia's economies would be driven towards market capitalism in a telos of efficiency where power and conflict play no role was to be quickly disappointed. There has been no frictionless process of transformation driven by the lessons of the crisis. As we shall see, the reform agenda in Indonesia has been imposed on a reluctant but beleaguered government only after much muscle flexing by the IMF. Even so, attempts to close insolvent banks and force powerful business groups to restructure their debts and surrender assets to cover the costs of recapitalisation have met with fierce resistance. Few of the major corruptors of the old regime have been brought to court and successfully prosecuted. Attempts to dismantle the system of state controlled rents underpinning Indonesia's corporate moguls and to reform strategic gatekeeping institutions have floundered as new political rulers found that the need to control patronage and secure off-budget was central to success in the new system of parliamentary power.

In the realm of politics too, the opening of the political system and the rise of parliament is signalling the emergence of a form of democracy quite different from that contemplated by liberals. As we shall see in later chapters, President Wahid, Indonesia's first elected President under the new democratic regime, found himself confronted with the sullen, resentful military and the bureaucratic remnants of the old regime who remain entrenched in the state apparatus. He was to be drawn inexorably into the vicious world of money politics and into a conflict with a parliament where individuals and factions jostle for influence and power under the cover of broad appeals to nationalist, populist or religious sentiment. As in Indonesian parliamentary politics of the 1950s, parties appear to be evolving as machines for assembling coalitions concerned with the capture and allocation of resources and power rather than imposing distinctive policy agendas and regulatory frameworks. By the end of 1999, the new democratically elected government had already fallen victim to the common fate of many post-authoritarian reformist governments. It was unable either to retreat to tried and true methods of maintaining order or to advance the cause of progressive forces that brought them to office.

This is not, as we shall see, the first case of hijacked reform. When the populist authoritarianism of Soekarno with its truculent nationalism and strident anti-Western and anti-liberal policies was swept aside in the violent political struggles of 1965 and 1966, neo-liberals in the West welcomed a new age of rationality and free markets (Arndt 1967; Allison 1969; Pauke 1968). Soeharto not only opened the door to foreign investment but his destruction of the Indonesian Communist Party (PKI) in 1965 and 1966 had decisively resolved the most profound social conflict of post-colonial Indonesia – that between the interests of property and those of radical populism. Yet, it was not liberal markets but nationalist economic agendas that were to flourish in the 1970s and early 1980s. Although the grosser excesses of central planning had gone, state capitalism re-established itself in a more vigorous form than before reaching heights undreamt of in the Soekarno era (Rudner 1976; Robison 1986: 106–270). In the political arena, the hopes of the new regime's middle class liberal allies were swept aside as Soeharto proceeded to construct a highly repressive and exclusionary state apparatus that provided for them no democratic point of entry to power.

In the 1980s, too, a series of major economic reforms driven initially by the collapse of oil prices resulted in widespread deregulation of investment, trade and financial regimes. While liberal observers applauded these reforms, deregulation and privatisation did not lead to an open liberal market economy. Instead, as we shall see in later chapters, powerful political and economic oligarchies emerged, not to dismantle the authoritarian state but to harness it to their interests. Unconstrained by law and regulation, reform simply meant that public monopoly became private monopoly in the hands of powerful politico-business families and well-connected corporate interests. The struggles that took place in the Indonesian political system in the 1980s were not attempts to impose democratic reform on the behalf of the newly ascendant private interests. They were struggles to wrench control from entrenched state officials and to preserve it intact to serve the interests of the new oligarchic coalitions.

As we have remarked earlier, the continuing failure of governments and individuals to embrace what was considered as the self-evident natural efficiency of markets presented obvious puzzles for neo-liberals. This was especially evident in Indonesia where a close relationship between institutions such as the World Bank and a domestic economic technocracy has existed for years, and where programmes of market reform had been periodically enacted. Neo-liberals had always recognised that the economic technocrats who carried the reform agenda in Indonesia had to deal with powerful predatory alliances and the enthusiasm of state officials for rent-seeking behaviour. Nevertheless, they assumed that economic crises and external shocks would progressively undermine predatory interests and provide opportunities to convince the government of the need to deregulate the economy and release the state's grip on the market. In this war of attrition, 'good policy' could be forced on governments when 'bad times' made protection and rents less affordable (Hill 1996). Yet, when 'bad times' struck with a vengeance in 1997 and 1998, there was no rush to embrace 'good' policies and considerable reluctance to accept IMF reform agendas, despite the prevailing economic distress.

Neo-liberals, mainly within the IMF, had placed much emphasis on breaking up the very system of predatory capitalism and, in particular, the grip on business and economic life held by the Soeharto family and their cronies. As progress on reform began to stall, frustrated officials in key agencies at the helm of efforts to impose institutional and policy reforms blamed the continuing influence of old interests and ideas embedded in the bureaucracy and in business. The problem was that the regime had, in fact survived. Other neo-liberals, generally supportive of the development credentials of the Soeharto government, did not share the view that fundamental inefficiencies or dysfunction within the regime itself had caused the crisis (Hill 1999a; McLeod and Garnaut 1998). They supported the view that panic and speculation in the global financial architecture had been the triggers for crisis (Radelet and Sachs 1998; Sachs 1998a, 1998b). As for corruption and cronyism, these had long been part of the Asian scene and had not hampered growth across the region or in Indonesia in previous years.

The post-crisis disaster was attributed, in this view, not to the fact that the regime survived, but to its collapse; it was a question of state failure. A regime insulated from reformist pressures had (together with the IMF) proven unable to properly manage the crisis (McLeod and Garnaut 1998; MacIntyre 2001). At another level, the eventual collapse of a regime that had been at least able to insulate responsible macro-economic policy from predatory forces and to impose order and predictability on economic life, even upon the organisation of corruption, had now opened the door to an invasion of populist demands and a descent into policy paralysis and social disorder, ensuring that the flight of capital was prolonged (McLeod 2000b; Garnaut 1998; Hill 1999a). While it was too late to return to the old regime, the remedy lay, nevertheless, in the restoration of state authority and a return to the imagined insulation of its technocrats in their pursuit of responsible macro-economic policy. As the economist Hal Hill has argued, 'one of the big challenges of the coming years will be to find a way of separating the economic and commercial world from the political world' (Hill 2000b).

These views, we propose, embody fundamental problems. In one sense, however, they both contain important observations. The ability of critical elements of the old regime to survive the crisis, entrenched in strategic administrative and political agencies, was a fundamental obstacle to neo-liberal reform agendas. At the same time, the collapse of the institutions of centralised authoritarian rule paved the way for the more unconstrained robber baron capitalism that has undermined attempts at effective policy and institutional reform. What is missing from both these equations is the factor of power. The critical fact is that the crisis failed to sweep away the very interests and forces incubated within the Soeharto regime, which underpinned and defined it. These survived to re-establish the economic and political power relationships within new institutional arrangements. At the same time, the fall of the old regime was never going to be enough in itself to precipitate a neo-liberal triumph. If we understand the neo-liberal agenda as representing, no less than any other, a programme of political action with its roots in a specific system of social power, perhaps the central problem is that the crisis never galvanised the political triumph of neo-liberalism and its ability to impose upon the state its hegemonic interests.

The argument

It is quite true that the 1997 economic crisis was precipitated by rapid flows of speculative capital and a surge in private short-term debt across Asia that unleashed panic in a system of global financial markets. As Jeffrey Winters and others have observed, the new global financial institutions were now constructed around highly mobile private loan and portfolio capital, able to exit markets quickly and no longer bound to support conservative dictatorships by the strategic considerations of the Cold War. Where the only concern for the private equity funds and banks was the bottom line, the collapse of currencies gave rise to widespread and rapid capital flight (Winters 1997). Nevertheless, the destructive paths of corporate collapse, banking meltdown and political paralysis emerging in the wake of the currency crisis were deeply rooted within an ongoing pathology of capitalism in late New Order Indonesia.

The crisis was played out within a system of extra-economic coercion of markets that had been not only the engine of its extraordinary growth but the very cement of power relations within Indonesia's New Order. It was not the interventionism of a highly centralised system of state capitalism that was at the heart of the problem, rather, the rise of powerful politico-business families in the 1980s and the harnessing of the state to the unconstrained interests of this privileged league of oligarchies.[6] Crises in debt, banking and corporate activity were already endemic in the decade preceding the crisis. As we shall see in later chapters, over-invested and over-borrowed corporate moguls flourished in the selective market deregulation of the 1980s, propped up by government bailouts, tax breaks, cheap and discretionary credit and monopoly positions. Their ascendancy was guaranteed by the coercive power of a centralised authoritarian state and made possible by a new world of highly mobile global capital markets and a constant inflow of short-term capital.

The larger significance of the financial crisis was not that it demonstrated the inherent inefficiencies of Indonesia's economic regimes and the superiority of the liberal market model but that it swept away the system of centralised authoritarian state power as well as the financial arrangements that had held together and papered over a fragile economic order. As the regime unravelled it could no longer bail out indebted corporations or protect them from the currency collapse. Politically it could no longer contain the social tensions and contradictions that had been the very product of its rule. Pent up frustration spilled onto the streets. Financially bankrupt and unable to call on the repressive apparatus of authoritarian rule, it became impossible for the entrenched coalitions to rule in the same way. As a leading national newspaper observed, the crisis 'exposed the rotten core of the self-serving system Soeharto had built and protected with an iron fist' (*Jakarta Post* (hereafter *JP*) 31 December 1998: 4).

Yet, while the crisis might have simply kicked in a rotten door, the fall of the regime was to guarantee nothing. As Nigel Harris has noted, referring to the case of Korea in the 1980s, 'when the state establishes a system for forced accumulation, this is not simply a set of arrangements that can be changed at will. It constitutes a social order, with a weight of inertia constituted by vested interests, the immediate beneficiaries, that inhibits the creation of any other order' (Harris 1988: 47). Indeed, far from being just an insulated dictatorship or a developmental regime driven by some organisational imperative, Soeharto's New Order was embedded in a complex coalition of interests embracing the state bureaucracy, politico-business families, corporate conglomerates and commercial propertied interests descending from Jakarta to the regions and small towns of Indonesia. Soeharto resisted reform, not because he was insulated from the pressures of forces outside the state but, on the contrary, precisely because reform threatened the very basis of the social order over which he presided.

Our central proposition is that those essential relations of power and interests, and to some degree even those forces hegemonic under Soeharto, were able to survive and accommodate collisions with global markets and successive economic crises over four decades. More specifically, they have largely survived the collapse of the Soeharto regime and those highly centralised authoritarian political arrangements within which power and wealth had been incubated and hitherto allocated.[7] How this has been achieved is the principal task of this study. Specifically, we ask how those politico-business interests hegemonic under Soeharto restructured their massive corporate debt and hung onto assets in the face of demands by creditors and governments, or resisted demands that they be dragged before the courts and convicted for corruption and nepotism. We ask how they forged new alliances with political entrepreneurs drawn from the fringes of power within the old regime; the former officials and regional notables, the party apparatchiks, operators, fixers and enforcers who now surged into the newly opened arenas of parliament and elections to seek their fortunes.

On the other side of the coin is the question of why the economic crisis did not enable a deeper and more pervasive neo-liberal victory. It became clear, even to officials in the World Bank, that attempts to impose fundamental institutional changes were founding on the continued resistance of entrenched interests. Frustrated at the slow pace of reform, one Bank document observed that

> Progress on governance has been … left largely to a few key reformers who have been moving forward in their respective spheres, garnering whatever support they can muster from senior leaders. These initiatives appear ad hoc and are floundering under resistance from well-entrenched vested interests.
>
> (World Bank 2000: 43)

While such officials may have been bemoaning the fact that technocrats were not insulated from predatory raiders, insulation requires the exercise of political power. Yet, in the political sphere, within the parliament and the parties, neo-liberals were never able to mobilise a coherent alliance sufficient to enforce the reform agenda within the Ministries and the bureaucracy. No reformist political parties or even military leaders, like Chile's Pinochet, threw their weight behind neo-liberal reform policies. Indonesia produced no John Wilkes or Tom Paine, emerging to lead a popular liberal movement in the face of repressive state power. While the old order found itself fractured and no longer able to rule in the way it would like, no new order was born.

Conceiving the Indonesian problem in terms of vast struggles between coalitions of state and social power leads us into some new propositions about the way integration with global markets, programmes of deregulation, regime change and the rise of new business oligarchies and middle classes affects the reform process. In the Indonesian case, for example, a system of authoritarian state power and predatory capitalism emerged and flourished, not in the absence of a bourgeoisie or middle class but precisely as they were expanding and consolidating their economic and social power. These interests have, for the most part, also merged easily into the unconstrained capitalism of post-Soeharto Indonesia rather than assembling behind the political agenda of reform. Thus, this study confronts the highly contingent role of the bourgeoisie in the process of social and political reform, and will draw out the factors that decide its shifting relationship to liberal and predatory agendas.

At the same time, we must remember that the consolidation of this system of predatory authoritarianism took place, not in circumstances of isolation and economic decay, but as the economy surged and as Indonesia's engagement with the global economy deepened and became more intensive. The very engagement with liberal governments in the West and with an increasingly mobile system of global investment and financial regimes secured a lifeline, not only for a profoundly illiberal political regime, but also for the spectacular consolidation of a highly corrupt and unconstrained politico-business oligarchy. Hence, we argue, the ambiguous nature of the relationship between global financial and investment markets, liberal reform and good governance.

The Indonesian case also confronts questions about the consequences of liberal economic reforms in capital and financial markets and the ending of important state monopolies for the rise of market capitalism. Such reforms in Indonesia, particularly in the 1980s, were to provide the very means by which powerful private interests emerged from within the apparatus of the state itself to construct their new private corporate empires. What factors and preconditions, we ask, decide whether the expropriation of former systems of state capitalism and the ending of public monopolies will produce liberal market capitalism or just unconstrained predatory rent-seeking?

If liberal economic reforms proved to be no guarantee that the system of authoritarian and predatory power would end, then the collapse of the political regime in the late 1990s punctured expectations that the ending of authoritarian rule would finally clear the way for free markets and liberal democracy. That the essential power relations of oligarchy and the hegemonic position of many of the main players themselves have been preserved and reassembled in a remarkable metamorphosis within the political and economic wreckage of the post-Soeharto era raises questions about the relationship between institutions and social power. How important are political regimes in the larger picture of social and economic transformation? Under what circumstances can a social order be reconstructed after the fall of a political regime, and to what extent are markets and democracy defined by the specific configurations of power and interests in which they are embedded, rather than being shaped and constrained by these institutions?[8]

The book

This book is about the way in which entrenched interests and political alliances are able to reorganise in the face of successive economic crises and the fall of political regimes. Specifically it deals with the violent episodes of social conflict in which Indonesia's economic and political regimes have been forged and unravelled. These are not conflicts that pit the instrumental rationality of economics against vested interests, rent-seeking and predatory politics. Nor are they conflicts between the forces of order and those of chaos, where authoritarian rule emerged as an organisational imperative to create a cohesive and functional society. They are not contests between 'civil society' and the 'state' or between the bourgeoisie and the state. They do not reflect a contest in which a generic form of 'Indonesian capitalism' comes under assault from Western or Anglo-Saxon models. Instead, they are conflicts between fluid and complex coalitions of state and social power over questions of power, property and civil and political rights.

Chapter 1 – The metamorphosis of capitalism: theories of change

Our first task is to establish the theoretical and conceptual frameworks of the study. This will be done in Chapter 1. Here we confront those approaches that see the Indonesian case in terms of a struggle to liberate the natural efficiency of the market from the arbitrary intervention of state power and to insulate its regulatory authority from the predatory demands of distributional coalitions. We question those explanations of change that focus on the choices of individuals or state technopols, and argue how these struggles might be understood in terms of wider conflicts over power and hegemony, where often fragile coalitions of state and social power form around issues of temporary or abiding interest.

Chapter 2 – The genesis of oligarchy

This chapter attempts to explain the Soeharto state in terms of a larger metamorphosis of state and social power, as an almost Leninist authoritarian state and its ruling strata of politico-bureaucrats became progressively embedded in the creation of private wealth and social power. It examines how this slow metamorphosis was enabled as powerful officials harnessed their possession of public office and their control of strategic economic gateways to their collective private interests. It is concerned, not only with the way in which radical and reactionary populism and liberal opposition were politically neutralised in this process, but with how Chinese–Indonesian conglomerates and Western governments and investors were drafted into what was a profoundly anti-liberal project.

Chapters 3, 4 and 5 – The triumph of oligarchy 1982–1997

This section comprises Chapters 3, 4 and 5, and deals with the economic and political triumph of a politico-business oligarchy in the period from the collapse of oil prices in 1981/1982 to the onset of economic crisis in 1997. The chapters address, at one level, the question of how the economic deregulation and the deeper engagement with global financial markets that followed the collapse of oil prices became the fundamental building blocks of the new politico-business oligarchies; as public monopoly was now opened to expropriation by private interests and global financial markets provided a new engine for growth. At the same time, we examine how the new alliances of oligarchic power around Soeharto politically consolidated their emerging social position, asserting their authority within the state apparatus itself, and reinforcing state power in the face of growing populist and liberal opposition to the growing power of oligarchy. We examine the bitter conflicts over control of the various apparatuses of the state, and the intensifying disputes over civil rights, democratic reform, corruption and the concentration of wealth.

Chapters 6 and 7 – Economic crisis and the unravelling of oligarchy

In Chapters 6 and 7, we ask why an economic regime that had generated such impressive growth over more than two decades unravelled so quickly, and why a political order that had seemed invulnerable only a year earlier collapsed so dramatically in May 1998. Here, we explore the proposition that argues that the roots of the decline are, ironically, to be found in the fatal collision of politico-business families' oligarchy with global capital markets. The very factors that provided the basis for the rapid and speculative growth in investment and debt also rendered the regime extremely vulnerable to global capital markets and to episodes of panic and flight. Politically, we ask whether the regime was overthrown under the challenge of powerful progressive forces or whether it collapsed from within as its beneficiaries found themselves no longer able to rule in the old way.

Chapters 8 and 9 – The reorganisation of oligarchy

The final part of the study, Chapters 8 and 9, deals with the vast struggles to reorganise power in the period that followed the economic crisis and the fall of Soeharto. Here we ask how those alliances formerly created by Soeharto have reorganised their ascendancy in political struggles over debt, bank recapitalisation and the reform of those gate-keeping institutions so essential to the system of 'political capitalism'. We examine attempts to reorganise this ascendancy within a new political arena where elections, political parties and parliament have become the arenas of power, and in the context of alliances with new political players constituting social and economic interests formerly at the fringes of power under Soeharto.

Conclusion

At one level, this chapter deals with the question of Indonesia's future. Is this future to lie in the apparently chaotic pursuit of power and wealth in the context of a 'failed state' or will a cohesive political alliance emerge to establish a new hegemony? More generally, the chapter will reflect upon larger questions about the process by which regimes are forged and transformed and upon the relationships between markets, states, regimes and social power that emerge from the Indonesian case.

Notes

1 Antonio Gramsci used this phrase to explain the emergence of Fascism in Germany as an interregnum between old imperial Germany and modern capitalist Germany (Gramsci 1971: 276).

2 The literature on the Asian Values debate is extensive. See Mahbubani 1993; Zakaria 1994. For critical overviews, see Jayasuriya 1997; Robison 1996; Rodan 1996a.

3 This will be treated in detail in Chapter 1. The central concepts of rents, distributional coalitions and insulated technocratic policy-makers that sustain neo-liberal political economy are developed in Buchanan and Tullock 1962; Krueger 1974; Bates 1981; Olson 1982; Srinivasan 1985; Nelson 1990; Haggard and Kaufman 1992, 1995. Critical surveys of neo-liberal political economy include Grindle 1991; Evans 1995: 21–42; Schamis 2002: 2–25. A critical assessment of the neo-liberal vision of the regulatory state is that of Jayasuriya (2000).

4 As we shall see in Chapter 2, patrimonial and predatory behaviour is embedded in the state in several ways. Hutchcroft distinguishes between systems of patrimonial oligarchy where powerful business interests capture the state to distribute its resources, as in the Philippine case, and patrimonial administrative state where a powerful and cohesive state dominates a fractured and disorganised business class in the relationship (Hutchcroft 1998: 48–58).

5 The so-called Washington consensus is a term coined by John Williamson to describe those policy instruments that were generally able to enjoy a consensus in Washington. By Washington, Williamson included, 'both the political Washington of Congress and senior members of the administration and the technocratic Washington of the international financial institutions, the economic agencies of the U.S. government, the Federal Reserve Board and the think tanks'. The actual consensus included the primacy of monetary stability, fiscal austerity and some progress at least towards privatisation and deregulation (Williamson 1990: 7, 8).

6 We understand oligarchy broadly in terms of the following definition:

> Any system of government in which virtually all political power is held by a very small number of wealthy ... people who shape public policy primarily to benefit

> themselves financially through direct subsidies to their agricultural estates or business firms, lucrative government contracts, and protectionist measures aimed at damaging their economic competitors – while displaying little or no concern for the broader interests of the rest of the citizenry. 'Oligarchy' is also used as a collective term to denote all the individual members of the small corrupt ruling group in such a system. The term always has a negative or derogatory connotation in both contemporary and classical usage.
>
> (Paul M. Johnson, Dept of Political Science, Auburn University, cited on 9 August 2003. http://www.auburn.edu/~johnspm/gloss/)

A more specific explication in terms of the Indonesian case will be developed in chapters 2 and 3.

7 This point leads us into concepts of state, regime and government. These are used differently across ideological traditions. Liberal pluralists, for example, do not recognise the concept of the state, dealing only with 'government' and the institutions of the bureaucracy. Government is conceived as a neutral referee, adjudicating and articulating the demands of interest groups. For Weberian statists, the state is defined by its monopoly on coercion and is driven by certain collective institutional imperatives as well as the discreet institutional interests of its officials. It thus becomes a central player in shaping change and policy (Skocpol 1985) In this study, the state is conceived as more than just the bureaucracy or the government or the institutional expression of its officials. It is embedded also in hierarchies of social power (Jessop 1983). This conflation of institutional structures and social coalitions is also recognised by Pempel (1998), who understands regimes as a set of social coalitions, institutional arrangements and strategic policy frameworks that are mutually dependent. When one leg of the triangle unravels, the others fall. By contrast, we see systems of social hegemony as able to transcend different institutional structures. Thus we adopt a different distinction between regime and state. In our view, government is the legislative and executive branches of the state apparatus and those officials, parties and individuals who occupy its offices, while political regimes are particular institutional forms of state power and its relationship to government. There are a wide variety of regimes; liberal and social democratic, systems of parliamentarism defined by money politics and controlled by cartels of interest, authoritarian corporatism, fascism, various sorts of totalitarianism and personalist dictatorship. Hence, it may be, as in the case of Italy, that governments rise and fall with rapidity, but the essential power relations that define the state as well as the institutions of the regime remain intact. Or that, ironically, the hegemony of certain social forces may be maintained only by changing the regime as Petras has observed in the case of many democratic transitions in Latin America in the 1980s (Petras 1989: 26–32). Certainly, as Anderson has pointed out in the case of Thailand, and as we shall argue, powerful social forces and power relations that were incubated under authoritarian rule may flourish subsequently within the shell of democratic regimes (Anderson 1990).

8 See also Pereira *et al.*'s 1998 study of how markets and democracies are shaped in Latin America.

1

THEORIES OF CHANGE AND THE CASE OF INDONESIA

How can we explain why deepening integration with global markets, the assumed lessons of successive economic crises and the entrenchment of a substantial bourgeoisie and middle class has not generated a grand liberal triumph in countries like Indonesia? For neo-liberals, the problem is to explain why the inherent efficiencies and collective welfare benefits of market capitalism have not been progressively enforced by the self-interested actions of rational, utility-maximising individuals or the far-seeing calculations of state 'technopols'.[1] Set against this approach, we argue that the uncertain and volatile progress of market capitalism and democratic transition in Indonesia must be understood in the context of larger conflicts over power and its distribution. The central question is how the interests of state and private oligarchy have been consolidated in an age of rapid economic growth, the spread of global markets and the transition to democracy.

The neo-liberal thesis and agenda

Neo-liberalism has become a catch-all term popularly used to denote various strands of market fundamentalism and technocratic exceptionalism. We use the term to include an ideological movement that embodies both the neo-classical reification of markets and anti-liberal, technocratic views of politics as well as its supporting political ensemble of shifting coalitions interested in opening markets selectively, lowering taxes, restricting regulation and containing opposition from labour movements, social democratic states and environmental coalitions. Thus, neo-liberalism both constitutes a set of explanations for the seeming paradoxes of change in Indonesia and embraces some of the central political players in the conflicts that are forging the path of change.

In its most pristine form, the neo-classical view of markets is one of an abstracted and self-regulating mechanism driven by its own internal laws, forged ideally in a world of voluntary transactions between rational, utility-maximising individuals. It is a world seemingly devoid of power and conflict where the endless search for efficiency prevails as the development of private property produces a community of self-interested rational individuals who demand an end to constraints on their economic activity. Such a view became politically dominant within the corridors of power of Western governments and in the World Bank and the IMF from the early 1980s, when theorists such as Bauer, Little, Lal and Balassa challenged the former Keynesian orthodoxy and its emphasis on demand side and state managed development policies (Toye 1987: 47–94). But neo-liberalism was not simply one theoretical proposition amongst many, it became the ideological expression of a much wider and more concrete economic and political juggernaut, focused on entrenching property rights, opening markets to the free movement of private corporate interests and winding back the authority of the state over economic life.

While neo-classical economists themselves had little interest in what drove change other than the assumed costs of intervening in the natural efficiencies of markets and the various lessons of efficiency that might come from periodic crises, practical policy-makers within the neo-liberal camp have long been aware that the rise of market capitalism is not a frictionless process. Public choice theorists began to argue in the 1970s that interests vested in the allocation of rents, which were entrenched in the state and in the business world as well as in welfare lobbies, would collude to prevent the free operation of the market (Krueger 1974). Thus, the process of reform was understood as a highly political project to liberate the natural efficiency of markets from the 'irrationality' of politics and to neutralise those predatory coalitions whose raids on the state preserved and entrenched resistance to market capitalism. It is a theme that continues to define the heart of neo-liberal political economy (World Bank 1997a: 332). Whether the driving force for change was the self-interested, utility-maximising individual or state 'technopols' navigating their way through the constraints of vested interests and embedded institutional pathways, the central dispute within neo-liberal ranks, as we shall see, is whether the road to market capitalism is achieved by dismantling the state or by enhancing its capacity and insulating its officials from the contending rationality of politics.

Public choice and the Washington consensus: the state as the problem

Early neo-liberal, public choice, political economy viewed the state itself as the essential problem. Conceived as a marketplace of transactions between individual politicians, officials and lobbies, the very power of the state to intervene in the market provides the conditions for the rise of rent-seeking interests, preventing good policy in the public interest, diverting scarce resources from productive investment, and strangling economic growth (Lal 1983; Buchanan and Tullock 1962; Bates 1981; Olson 1982). The pluralist concept of the benign state committed to the common good is replaced by the concept of the predatory state in which policy and public goods are appropriated and sold by officials and politicians in return for political support.[2]

While this general characterisation of all states as necessarily predatory was a caricature of reality, public choice theory was nevertheless profoundly influential and carried two main implications for policy. First, only by restricting the power of the state to regulate capital, to raise taxes and run deficit budgets would structural opportunities for politicians and officials to profit from their positions be removed and the very rationale of distributional coalitions and rent-seeking undone. Second, it followed that dismantling state power and enhancing individual property rights would be enough to enable the spontaneous rise of market capitalism. Subsequently embodied in the so-called Washington consensus, its prescriptions for market deregulation, fiscal austerity and privatisation were imposed on governments across the world in structural adjustment packages managed by the IMF and the World Bank (Williamson 1990).

However, public choice theory was hardly helpful in explaining how change might come about. In a world of rent-seekers and free-riders, such as those assumed to prevail in Indonesia, who would dismantle the system of state power that was the very source of the rents that sustained them and provided the cement of political relationships? In this bleak environment, some neo-liberals were to look to the state itself as the only possible arena within which the reform agenda could be nurtured. Some neo-classical economists began to suggest that forms of benign authoritarian leadership could enforce free markets in the face of the vested interests of predatory officials and rent-seeking robber barons (Srinivasan 1985: 58). Lal proposed that: 'A courageous, ruthless and perhaps, undemocratic government is required to ride roughshod over these newly created special interest groups' (Lal 1983: 33).

In the prosecution of reform, neo-classical economists were forced to turn to the notion of 'enlightened technocrats or states persons who are somehow liberated from the pursuit of self-interest and thus able to see beyond short-term goals to long-term public interest' (Grindle 1991: 59–60). These change teams, or the 'handful of heroes' (Harberger 1993: 343), were able to persist in the adversity of a rent-seeking environment to pursue the objectives of normative economic analysis (Williamson 1994: 13–15). It was this view that formed the hinge of neo-liberal policy strategies in Indonesia for over three decades. While accepting the proposition that successful market reform could be achieved by dismantling the capacity of the state to intervene in the economy, market reformers were, nevertheless, to turn to 'rational' technocrats within the state apparatus itself to engineer such reforms. Continuing support for the Soeharto regime by Western governments, the World Bank and the IMF for over three decades was built substantially on their belief that the appointment of economic technocrats willing to adopt 'pragmatic' policies based on rational technical calculation that transcended the demands of vested interests was the critical factor in three decades of successful development. and would be the Trojan Horse for progressive advance towards the market economy.[3]

Despite the enthusiasm of neo-classical economists for the way in which Indonesia's technocrats managed fiscal and monetary policy, they nevertheless recognised the relatively weak political position of the technocrats in dealing with protracted resistance to deregulation and continuing corruption. It was acknowledged that Soeharto's New Order was a 'soft state' – good on macro-policy but weak at the micro level. The process of liberal reform was understood as a war of attrition in which economic crises and shocks provided the opportunities for technocrats to convince the government of the need for change conceived in the terms of the Washington consensus, focused primarily on the problems of deregulation and privatisation. It was a case of 'good policies in bad times' (Hill 1996; Soesastro 1989; Bresnan 1993; Vatikiotis 1993).

Thus, both the collapse of oil prices in 1982 and 1986 and the more dramatic crisis of 1997 were seen to be opportunities to enforce the sort of trade and financial sector reforms so long resisted by vested interests. As we shall see in later chapters, reform efforts were to flounder in the 1980s and despite the most devastating 'shock' imaginable, the bad times visited upon Indonesia by the economic crisis of 1997 clearly did not result in the movement of the technocrats to centre stage and the embrace of 'good' policies by a government that had learned its lesson. Rather than clearing the way for the spontaneous emergence of markets as expected by public choice theorists, events in Indonesia appeared to signal that the fall of the state might lead to the sort of chaos witnessed in Russia. Amongst economists writing on Indonesia the apparent dilemma was explained, in part, as the consequence of government policy errors.[4] For institutional political economists like Andrew MacIntyre, the government's inability to adopt and impose realistic responses reflected the structural nature of institutional frameworks that locked veto power over reform inside the state and insulated government from reformist pressures outside (MacIntyre 1999, 2001).

At another and intersecting level, the problem was perceived as that of a failed state, the departure of Soeharto allowing populist policies to undermine technocratic influence and precipitating a descent into instability and disorder that frightened investors, prolonged capital flight and stalled recovery (Hill 1999b). Even corruption, which had been at least centralised and predictable under Soeharto, was now more capricious and counter-productive in its impact on the economy and business in the post-Soeharto chaos (McLeod 2000a). It seemed that markets required continued authoritarian political rule so long as civil society and institutions remained weak and powerless. The paradox here, as we have argued in the previous chapter, is that the sort of authoritarianism wielded by Soeharto gave no room for the emergence of centres of countervailing power and influence in society or institutions effective in managing markets.

A problem of weak institutions

It soon became clear that efficient market economies were not emerging in countries like Indonesia despite programmes of deregulation and the unravelling of interventionist states. Some neo-liberals, under the leadership of Stiglitz in his term as Chief Economist of the World Bank, were to question the market's inherent capacity for self-regulation and equilibrium. Without the existence of strong institutions, conceived as systems of rules and rights governing the strategic interaction of individuals, it was argued, deregulation often degenerated into unconstrained rent-seeking. Programmes of privatisation became little more than a mechanism by which state monopolies simply ended up in the hands of well-connected private and business oligarchies (Stiglitz 1998a).

The idea that effective institutions were necessary mechanisms for market transitions had been long established in neo-liberal thinking (North 1981; North and Thomas 1970). Policy agendas within the World Bank and other agencies increasingly added a concern for institution-building, public sector reform and 'good governance' to their agendas for market-friendly policies (World Bank 1991b: 7, 9, 128–148). Within Indonesia, also, the World Bank became progressively concerned with the importance of a strong regulatory framework for markets in a situation where sectors now opened to private investment without effective regulation were being divided amongst powerful predators.[5]

But exactly where good institutions would come from was not clear. In the larger debate, expectations that problems of transaction costs, property rights and knowledge emerging in the new circumstances of capitalist society would spontaneously cause rational, utility-maximising individuals to establish arrangements to deal with their changing collective action dilemmas proved largely unfounded. Within Indonesia, business interests and middle classes continued to seek their advantage within the prevailing interventionist policy frameworks and networks of patronage and favour rather than in any collective political action to impose policy or institutional change. Industry associations had proven ineffective in pushing for consistent rules in regulating sectors and, in any case, were themselves simply overwhelmed and drawn into the systems of extra-economic coercion that governed economic life (MacIntyre 1991; Robison 1992).

The neo-classical interest in states and institutions was focused heavily upon how these could enable the efficient operation of markets rather than grappling with the question of where they came from. Indeed, the idea that self-interested individuals would pursue collective goals rather than their own immediate advantage in rents or free-riding was never satisfactorily reconciled with the neo-liberal idea of self-interested individuals rationally pursuing their own interests (Leys 1996: 80–106). North himself, perhaps the founding father of NIE, was to acknowledge that

> Institutions are not necessarily or even usually created to be socially efficient; rather they, or at least, the formal rules, are created to serve the interests of those with the bargaining power to create the new rules … Because it is the polity that defines and enforces property rights, it is not surprising that efficient economic markets are exceptional.
>
> (North 1995: 20)

It was at this point that Weberian ideas about the essential role of the state and its bureaucracy in providing the underpinnings of market capitalism began to influence the neo-liberal equation.

A problem of state capacity

Studies focusing on the role of the state had been making ground in the political sciences for some time as a counter to class-based interpretations (Evans *et al.* 1985). By the early 1990s, the World Bank itself was retreating from the idea that the state was unambiguously an obstacle to the rise of market capitalism. Whereas in its 1983 *World Development Report* (World Bank 1983: 47–56), it had argued that the state should be limited to repairing market failure, by 1991 it recognised that: 'without the institutions and supportive framework of the state to create and enforce the rules [to make markets work more effectively], to establish law and order, and to ensure property rights, production and investment will be deterred and investment hindered' (World Bank 1991b: 3). Its 1993 report on the rise of the Northeast Asian industrial economies admitted the importance of the state to this miracle (albeit in a desultory and grudging fashion) to the extent that it provided macro-economic stability, property rights, education and public infrastructure (World Bank 1993b). In its 1997 *World Development Report*, the World Bank extended the role of the state beyond providing macroeconomic stability to one that established the collective goods necessary for market economies: a foundation of law, basic social services and protection of the vulnerable and the environment (World Bank 1997a: 4).

But how would 'predatory' states like Indonesia realistically be transformed into systems of 'good governance' defined by transparency and accountability in economic management and by legally constituted systems of market regulation. The sort of abstracted functional affinity assumed by Weber to exist between the modern rational state and modern industrial capitalism, that would force a mutual reinforcement of the two (Weber 1964: 357, 1978: 951–952), did not eventuate in Indonesia. Neo-liberals saw the answer in terms of a voluntarist process of capacity building. Limited in the past to allocating rents, it was generally assumed that 'soft' states did not possess the administrative capacity to undertake the more rigorous tasks of regulating and monitoring markets. Such missing capacity, it was assumed, could be 'supplied' by the introduction of specific rules, mechanisms and procedures. For example, rules and mechanisms for enforcing rule of law would include an independent judiciary, the separation of powers and formation of watchdog organisations. Arm's length procedures would ensure that procurement would not be subject to special deals. A constitutional separation of the central bank from political control of parliament would ensure that its decisions would be based on 'technical' calculation rather than vested interests of politicians or social coalitions. States would be constrained and protected from capture by contracting out their activities and imposing a bureaucracy based on the principles of merit (World Bank 1997a: 28, 106).

In effect, neo-liberal policy was directed towards creating a state emptied of politics – a state with the capacity to restrain self-interested behaviour and to insulate rational decision-makers from rent-seekers and distributional coalitions. Yet, despite an avalanche of dollars provided by the World Bank and various aid agencies for programmes of good governance and capacity building, the reform agenda continued to face serious problems. While it is true that Indonesia's bureaucracy has long laboured under totally inadequate management systems and procedures and within an ambiguous legal framework, these were only part of the central problem. As we shall see later in the book, the rampant predatory behaviour in the private and public banking system did not occur for want of supervisory authority held by Bank Indonesia. The destruction of Indonesia's forests has proceeded at an obscene pace in spite of what are widely regarded as good regulatory frameworks. Clearly, institutions cannot just be 'supplied'. Neo-liberals were confronted increasingly with the fact that institutions could not be implanted without the backing of powerful and well-organised social interests.

The problem of social capital

Elements within the World Bank, frustrated with opposition and indifference to its reform agendas, began to recognise the importance of the social underpinnings of corrupt or weak governments. In its 1991 report on Governance the Bank had insisted that, 'While donors and outsiders can contribute resources and ideas to improve governance, for change to be effective, it must be rooted firmly in the societies concerned, and cannot be imposed from outside.' It exhorted citizens to be responsible, observing rather lamely that, 'Citizens need to demand good governance' and, 'Governments need to prove responsive to these demands' (World Bank 1991b: 6, 7). Following the crisis, the language of participation, civil society and social capital and community organisations became central to a new effort to create social support for market capitalism (World Bank 1998a). Stiglitz proposed that '[t]he hard part of capacity building is the development of organisational/social capital, including the institutions that enable society to function well' (Stiglitz 1998a: 12). World Bank programmes were increasingly to include support for non-government organisations and community associations, its new emphasis on social safety nets distinguishing it from the strong line on fiscal austerity taken by the IMF.

In reality, there is little doubt that the increased interest in society reflected a concern for the destabilising events in Indonesia and elsewhere in the wake of the Asian crisis and the increasing criticisms of the World Bank and the IMF. Such resentments challenged the very hegemony of the neo-liberal agenda. It became clear that many of the social interests opposed to the World Bank and IMF programmes had to be enlisted and co-opted if the neo-liberal agenda was to be successful. However, the problem was conceived as one in which social networks and institutions may simply be too weak or misguided to support markets, fitting neatly with Robert Putnam's proposition that development required social capital; the capacity of citizens to support collective action, governance and institution-building measured in terms of the density of social networks, and in the strength of values and norms (Putnam 1993). The concept of social capital enabled the social task to be approached as a technical rather than a political problem, one of creating citizens who are functional and useful to the establishment of liberal markets and, we might add, who will do what they are told (Fine 2001; Harriss 2002). Opposition could be dismissed in systemic, organic terms of dysfunction, deviance and disjuncture between social norms and the evolving division of labour.

Hence, in Indonesia, the World Bank and other aid agencies plunged into programmes of training and of harnessing NGOs to the tasks of development. As the flow of funds surged into these projects, NGOs, many of them instant creations, became a prominent feature of the political landscape. Yet, the continuing difficulties of implanting market capitalism and regulatory regimes suggested that the problem was more than simply one of the weakness of institutions or social capital but rather, about larger struggles over power.

The problems of the neo-liberal approach

The assumption that there will always be a policy, administrative or management fix for development problems can be traced back to the liberal pluralist idea of the state as an adjudicator and regulator of the demands of competing interest groups (Easton 1965). Thus, the attention of institutional political economists has been focused on the strategies and tactics of technopols negotiating their way around the veto power of various predatory interests. Building effective institutions became the key to the problem because these restrained self-interested behaviour and insulated rational decision-makers from rent-seekers and distributional pressures. Nevertheless, it was recognised that while initial reform required insulation from vested interests, reforms could not be embedded where 'the private sector beneficiaries of reform are scattered or weak' or where they were inconsistent with 'the structure of the countries' ruling coalition' (Haggard and Kaufman 1992: 27; Donor 1992: 431). But this recognition of the importance of social sector support and that policies had 'winners' and 'losers' did not change the idea of policy making as a process of choices, strategies and bargains. Once again, effective institutions could alter the structure of social power, 'empowering political elites and structuring the access of social groups' (Haggard and Moon 1990: 211). At another level, it meant simply that rational 'change teams' had to mobilise a national consensus in favour of their policies, specifically to convince private sector interests that the neo-liberal agenda served their long-term interests.[6]

Yet, such a voluntarist view of the state and the policy process neglects the overarching structures of power that define the circumstances in which policies are made. As Leftwitch has observed, such terms as 'good governance' and 'good policy' disguise an essentially political agenda as a technical matter, 'contestation about 'forms of government' is not a fool's contest; it is a contest between, different interests about power and the institutions which distribute it' (Leftwitch 1994: 377). Thus, while institutional political economists see the task of transforming states that capture and allocate resources into states that regulate markets as one of creating capacity and insulation, building more effective institutions and social capital, we propose that these tasks are not easily disentangled from the vast structures of social power and interests in which models of capitalism, developmental, liberal, predatory and social democratic, are embedded. What we are seeing are collisions of different capitalist orders.

Collisions of historical institutional pathways

Some institutional political economists have viewed the economic crises and conflicts over policies and institutions that have taken place in Asia over the past two decades as a collision between two different systems of capitalism shaped by the different historical institutional pathways in which they were forged. It is an approach based in an understanding of markets, not as abstracted mechanisms driven by their own internal laws, but defined by institutions that are the constructions of governments and politics (Zysman 1994). It follows that capitalism might exist and flourish within a range of institutional frameworks.[7] Thus, the attempts to impose neo-liberal reforms may be explained, not as an evolutionary endpoint in a universal telos of efficiency, but as a global collision of Anglo-Saxon market capitalism with East Asian developmental regimes and various forms of network capitalism based around Chinese family institutions (Beeson 1998). Attempts to transplant liberalism are prone to failure, it follows, because market institutions forged in the particular circumstances of economic change in eighteenth and nineteenth century Britain and North America do not easily translate into the logic of institutional pathways laid down in the state-led developmental models that came out of the late-industrialising East Asian experience. Indeed, it is argued that hasty and imprudent attempts to transplant liberal institutions may be destructive where they dislocate the delicate financial systems and political arrangements that once so successfully maintained growth rates in economies based on high levels of debt (Wade 1998; Wade and Venoroso 1998).

This is a seductive thesis in the case of Indonesia where entrenched interests have appeared immovable in the face of liberal pressures for structural reform. But do the inconclusive attempts to implant liberal market capitalism really signal that the authoritarian and predatory capitalism that emerged under Soeharto is an immutable set of institutional arrangements 'natural' to Indonesia? We see several problems with this approach. Soeharto's New Order was never an internally harmonious system assembled by elites making incremental rational choices within the possibilities determined by Indonesia's historical institutional pathways. In reality, the economic and political regimes that have defined Indonesian capitalism have been imposed by force in a series of violent conflicts. Just as Indonesia's predatory authoritarianism was forged in these bitter contests, it can be dismantled where global crisis and structural shifts in social power transform the political equation. The task is to explain, not how institutional frameworks have defined and shaped the actions of Indonesia's policy-makers and power-holders, but how coalitions assembled around a particular system of social power have used and discarded institutions and reorganised their power within new political and economic regimes. It is to explain how those wrenching social struggles defined by Chaudhry (1994: 4) really precede and shape the rules that govern the economy.

Collisions, social conflict and institutional change

If we understand institutions as not simply collective arrangements designed to facilitate economic efficiency, but as mechanisms for the allocation and consolidation of power, the failure of liberal market models of capitalism to take root in Indonesia and elsewhere in Asia may be explained quite differently. Because institutions are established to reinforce a specific architecture of power relations, dominant social forces will resist institutional changes where they threaten control of economic surplus and the means of economic production (Bardhan 1989). This point is important. It means that institutions that appear 'dysfunctional' or less efficient as mechanisms for growth or investment often persist for long periods because elites are prepared to sacrifice what might appear to be functional efficiency in 'rational' economic terms where their social and political ascendancy is threatened. Indeed, this is what happened to Soeharto between July 1997 and May 1998. Pressed by the IMF to undertake fundamental structural reforms in return for financial assistance, he was faced with the choice of political suicide or what appeared at the time to be probable economic collapse. Not surprisingly, he chose the latter, at least until other factors intervened.

Crises are therefore decisive, not because of the lessons they bring in terms of the costs of intervention or the benefits of reform but where they make it impossible for entrenched regimes to hold together the political and economic fabric that sustains their interests. The decisive opening to reform in Indonesia was the fall of Soeharto. But even where economic crisis delivers fatal wounds to authoritarian and predatory regimes, their collapse and bankruptcy does not guarantee a shift to specifically liberal market institutions or democratic politics. It merely opens the door to a fresh round of struggles to reshape and redefine economics and politics. To a large degree, the outcome of these struggles is determined by the extent to which the collapse of former regimes is accompanied by the unravelling of the social interests and relations of power in which they were embedded. Where the entrenched social coalitions are not able to prevent the unravelling of those political and economic regimes, they may simply reorganise themselves within new institutional arrangements. The example of Russia is enough to illustrate how military and bureaucratic elites of former authoritarian regimes are able to reinvent themselves as the political entrepreneurs, corporate moguls and criminal bosses of new market economies and democratic politics.[8]

It is this phenomenon of the reorganisation of social power following the collapse of regimes that is at the heart of this study. Here, it is important to realise that the Soeharto regime was not simply an apparatus of political repression and control. It also constituted a vast and fluid alliance of bureaucratic power and corporate wealth extending from the Presidential palace down to the regions and villages of Indonesia. Although weakened and fragmented, this social base did not evaporate with the ending of the Soeharto regime. Nor was it destroyed in the economic shocks of the 1980s. The question is how this 'pact' of politico-business interests was able to reorganise its ascendancy in the face of economic crises through the 1980s, and especially after the fall of Soeharto in 1998 when it was forced to accommodate to life without that highly centralised system of coercive and emergency rule. As had been the case in Thailand in the 1980s and in many Latin American countries following the collapse of authoritarian regimes there, established cartels and cliques were successfully to reassert their political and economic hegemony within the new political arenas of politics and parliament, and in the context of alliances with those new social forces that flooded into the world of politics. As one of Indonesia's political leaders, Amien Rais, noted:

> I believed that the process of total reform would be easy once Soeharto was out of office. I assumed he was the biggest block to reform and, after putting him aside, I believed we could push forward with reforms relatively easily. I was wrong … Now I realise the power pyramid left by Soeharto is still intact.
>
> (Amien Rais, quoted in Hartcher 1999: 56)

The second question concerns the rise of a liberal reformist coalition. As we have seen, the neo-liberals assumed that markets would emerge in a frictionless and rational aggregation of individual interests or to be driven by 'a handful of heroes' (Harberger 1993, cited in Schamis 1999: 236). On the contrary, we propose that liberal forms of markets and politics, no less than any other forms of regime, must be imposed in the arena of politics. As Polanyi argued in relation to the transformation to market capitalism in Europe, 'The road to the free market was opened and kept open by an enormous increase in continuous, centrally organised and controlled interventionism' (Polanyi 1957: 140). This not only means enforcing the policy agendas of free trade and open labour markets but also politically sweeping away those interests such as organised labour, welfare coalitions and protected sectors of business who stand to lose from market capitalism. To this extent, the significance of crises is their effect, not only in fracturing old coalitions but also in shifting power to new alliances of state and social power that are assembled around various reformist agendas.

It is true that the crisis of 1997 and 1998 propelled the IMF and the World Bank to the centre of Indonesia's political stage enabling a complex raft of structural reforms and policy shifts to be imposed on governments facing fiscal crisis and desperate for foreign money to flow back in. Even middle class reformers with little interest in market fundamentalism saw the intervention of the IMF as at least a mechanism to bring down the old Soeharto regime and its corporate clients, supporting efforts to end corruption, prosecute corruptors and reform those bureaucratic empires that allocated rents. In the end, however, no politically cohesive and hegemonic set of forces was assembled around an agenda of free markets property and individual rights, no liberal or republican party emerged to contest power with the entrenched predatory party machines that dominate democracy in the post-Soeharto era.[9]

That the liberal agenda failed to take root as expected in Indonesia was not the consequence of an early and immature capitalism where economic growth and social change was a feeble pulse and where civil society and the bourgeois interest remained weak. On the contrary, the process took place in the context of a vigorous and rapid capitalist transformation, as Indonesia's integration with global markets became more intense. This was a seemingly paradoxical outcome in the context of those relationships widely assumed to exist between capitalism, civil society, markets, legal rational forms of governance and democracy. If modern capitalism requires the transformation of predatory or patrimonial states into states that are insulated from the immediate demands of vested interests, how do we explain the pervasive role of the Indonesian state in the formation of an ascendant politico-business oligarchy? If modern rational capitalism generates a civil society or a middle class or bourgeoisie that is 'progressive' and will demand an end to the arbitrary actions of the state, how do we explain the failure of reformist politics in Indonesian society? We require ways of theoretically explaining how predatory state–society relations survived and flourished while capitalism progressed vigorously, and a powerful bourgeois interest was to embrace and dominate Indonesian society.

Why no modern rational state?

Across the ideological spectrum it has been a central assumption that modern industrial capitalism required a state that would put an end to arbitrary rule and impose the general or common interests of capitalism over the more immediate demands of individual capitalists. Thus, the modern state was necessarily conceived as being separate from and above society, possessing a substantive measure of autonomy in its actions.[10] It is true that Indonesia's technocratic elites did get their way on occasions, even overriding powerful political and business interests in the punitive tightening of credit in the late 1980s and the more recent painful attempt to force the recapitalisation of Indonesia's beleaguered banks and to restructure private debt. But their ability to impose 'pragmatic' policies never extended much beyond the sphere of macro-economic policy. The state continued to be arbitrary and instrumentally responsive to the demands of powerful private interests. Its technocratic apparatus, ironically, structurally integral to the success of the Soeharto regime, provided the conditions in which authoritarian and predatory dimensions of state authority survived. They were never permitted to stray far beyond this function.

Thus, we must consider the Indonesian state in three dimensions. One is that structural dimension where the state is forced to impose the general interests of economic growth and political order over and above vested interests. At another level, the Indonesian state must be understood in terms of what Anderson (1983) called the *state-qua-state*, defined by the abiding interests of a discrete corps of state officials and power holders in the preservation and extension of state power. It was a logic that required not only the protection of their general interests in revenue and politics but their own unencumbered possession, as a virtual political class, of the very apparatus of political and economic control that defined the state. But the question remains, why did this *state for itself* consolidate a system of state-sponsored predatory capitalism rather than the highly routinised, legal rational state that bureaucrats designed to protect their careers and status from the influences of the market in nineteenth century Germany? Why, for example, did the Indonesian *state-qua-state* not develop in the same way as that in Singapore?

The answer lies partly in the way it emerged in the chaotic parliamentary period of the 1950s where the scramble to build networks of patronage within which political support might be concentrated and public power turned into private wealth required politicians and officials to fight for control over non-budgetary state income and over the allocation of rents through key ministries and agencies. This mode of state power was extended under the authoritarian regimes of Soekarno and Soeharto that followed. Thus, even after the fall of Soeharto, there was enormous resistance in the courts and judiciary, in the public owned enterprise sector, and in the police and military, to reforms that would render greater accountability and transparency across the state apparatus.

Such a system resulted in a third dimension of state power as it was drawn more deeply into relationships with capitalists, cronies and 'fixers' that revolved around a vast system of benefices and rents. These relationships became institutionalised. This was not a world where 'rational' technocrats simply negotiated their way through the constraints of powerful interests, both within and without the state.[11] This was a vast and crudely instrumental system of state power where public authority and private interest were fused and where state capitalism gave way to the rise of politico-business oligarchies emerging from within the state itself. This was most starkly illustrated in the position of Soeharto himself, who was at once the uncontested ruler of a huge state apparatus and, at the same time, head of Indonesia's most important politico-business family, perhaps the major beneficiary of state allocated rents. It was a form of state power that survived in no small part because it proved to be an ideal shell for investors, both national and global, and for economic growth in the decades after 1965.

Why an uncivil society and illiberal bourgeoisie?

It was assumed by liberals that the development of capitalism would generate spontaneously a progressive civil society, with its autonomous institutions, to demand an end to arbitrary and authoritarian rule.[12] The collapse of the Soeharto regime was greeted with such enthusiasm partly because it was expected to liberate long pent up reformist demands and signal the triumph of an autonomous civil society dominated by a progressive middle class. This rosy view of civil society was to be disappointed. While demands for a free press and democratic elections were realised, the broader liberal agenda was soon submerged by wide ranging social violence and coercion by gangs and criminal elements, varieties of populism that were often xenophobic and fundamentalist in nature, as well as an upsurge in corruption and money politics. Civil society unchained proved neither to be uniformly middle class, nor progressive, nor civil.

That no such powerful and progressive sphere of civil society emerged after the fall of Soeharto is understandable in the context of the pervasive apparatus of repressive, corporatist and ideological controls applied so successfully by the regime to every facet of social and political life. But there are deeper problems. The very juxtaposition of a vigorous and autonomous civil society with a state that is constrained and limited is a fatally flawed premise. On the contrary, it may be argued that the emergence of a vigorous civil society requires, not the collapse of state power, but its transformation to provide the guarantees of civil rights upon which liberal society might rest. That no progressive civil society emerged in post-Soeharto Indonesia may be seen, ironically, as the consequence of authoritarian rule being replaced with an increasingly fragmented, ineffective and diffuse form of government unable to guarantee civil rights.

Perhaps more important, the very concept of civil society overlooks the vast disparities in wealth and power and the intense conflicts that shape any society (Rodan 1996a: 4; Wood 1990). But if the problem is not that of civil society versus the state but of conflicts between different coalitions of state and social power, does the emergence of capitalism itself create the very forces that will contest power and ensure the fate of some and the rise of others? The deepening resentment and frustration of Indonesia's petty property-owning classes, and their gravitation towards xenophobic and religious politics may represent the cry of a declining class swamped in a society now dominated by new systems of social power and new cultural paradigms. The central paradox, we propose, is that the rise of an increasingly significant middle class and globalised bourgeoisie in Indonesia did not generate the sort of liberal and 'progressive' drive to political and market reform assumed and expected. On the contrary, we must understand Indonesia's bourgeoisie as a profoundly anti-liberal force. One of the central political problems of neo-liberal reformers has been their inability to include Indonesia's business moguls within their camp.

How then do we explain this illiberal bourgeoisie and how permanent is this phenomenon? Marx argued that the very rise of a capitalist mode of production would be enough to transform the existing legal and political superstructures and property relations as they became a constraint on the new forces of production.[13] For Marx, this represented the triumph of the 'bourgeois interest' that ushered in an age of liberal institutions – markets, democracy and rule of law.[14] It required the ending of mercantilist restraints on the free movement of the private interest, the establishment of rule of law and equality before the law to ensure 'a public system that is not faced with any privileged exclusivity' (Marx and Engels 1956: 157, cited in Miliband 1989: 89).[15] In ongoing battles between town and country, between emerging capitalists and the beneficiaries of restrictive pre-capitalist monopolies, the bourgeois interest prevailed in the institutions of administration and law and the political system itself (Marx 1867, 1848, excerpted in Edwards *et al.* 1972: 27–31; Dobb 1947: 91–104).

Why, then has this elegant and seemingly logical process not been followed in Indonesia? As Ellen Meiskens Wood has argued, the focus of analysis should shift from 'a struggle for ascendancy between declining and aspirant classes to the dynamic of capitalist accumulation and the transformation of property relations it sets in train' (Wood 1990: 125). As capitalism is entrenched, the vast movements of populations and the rise of new industries give rise to new forces and interests and to new issues such as property rights, working conditions, social welfare and public goods around which bitter conflicts are focused. In these struggles, those who play a leading role in driving change often become its victims rather than its beneficiaries.[16] In the case of Indonesia, liberal middle classes who supported Soeharto in 1965 and who were the basis of the *reformasi* movement in 1998 found themselves subsequently pushed aside, while business interests, which played almost no active political role, found their position consolidated. More important, however, the historical circumstances of these struggles produced strange bedfellows.

Indonesia's weak and fragmented bourgeoisie were drafted into a project inherently hostile to liberal markets and political democracy primarily because the path for capitalism in post-colonial Indonesia was cleared by a repressive state that provided the investment and controlled the major gateways through which private interests might enter the economy.[17] It was, however, an accommodation that suited most of Indonesia's bourgeoisie. Sheltering under the umbrella of an authoritarian state was comfortable for a largely ethnic Chinese bourgeoisie, politically vulnerable in the face of those currents of social radicalism and reactionary populism that prevailed in Indonesia through the 1950s and 1960s. At another level, Indonesia's bourgeoisie, at least those larger corporate players among them, owed everything to the protective trade arrangements, predatory benefices and the flow of state investment and preferential bank credit. Their commercial advantage has always been constructed in directly accessing the patronage and monopolies offered by such a system of power, not in tearing it down and replacing it with open markets and transparent and general rules.[18]

But, we must ask, do the events of 1997 and 1998 show that system has come to the end of its time, signalling the resolution of that apparent contradiction between the bourgeois interest in accumulation and the bureaucrat interest in revenue, war and politics?[19] Have the system of state and patrimonial capitalism, and the arbitrariness of authority that once proved so effective in speeding up development, now become a constraint for Indonesia's capitalists as they are drawn increasingly into the world economy? Do they now require measures to ensure that the common interests of capital can shape the important policies of the state? This was the argument presented by Nigel Harris in his analysis of the traumatic transition from military rule in Korea in the 1980s.[20] Are we simply witnessing in Indonesia the end of the old arrangements and the long drawn out death throes of a decaying and outmoded form of capitalism?

Harris qualified his proposition by recognising that the state might stave off the bourgeois challenge by entering into new political alliances with populist interests, or enjoy some immunity from the pressures now exerted by a frustrated bourgeoisie because of access to windfall revenues (Harris 1988: 248). But there is little evidence in Indonesia of a growing political demand amongst the bourgeoisie for the sorts of market reforms and political accountability of the state and its policy-makers to the common interests of capital. As we shall see in Chapter 3, when those business oligarchies newly emerged from within the incubator of state protection finally moved against the constraints of state capitalism, they simply expropriated public monopoly into private hands but kept intact the larger system of authoritarian and predatory power. Where Indonesia's Chinese business interests have moved funds and investments offshore, in the early 1990s and more recently, it has not been to escape the overarching constraints of state control but in the fear and uncertainty that they would unravel.

We need, therefore, to consider two questions. Does Indonesia's bourgeoisie, specifically its Chinese corporate sector, have any option except to seek the protection of authoritarian rule or the favour of predatory parliaments so long as they do not themselves possess a broader social or political hegemony?[21] Second, will this be bypassed by globalisation as Indonesian groups extend overseas or enter partnerships with highly mobile global corporate institutions?

Will the fall of the regime mean the end of the state?

Recent US policy pronouncements have emphasised the importance of regime change as a means of transforming more fundamental and entrenched interests and practices (see Fidler and Baker 2003). But, to what extent can the dismantling of a regime, conceived as a particular set of institutional arrangements, signal the end of deeper sets of power relations and hierarchies? Can we assume that the fall of the highly centralised system of authoritarian rule that was the Soeharto regime will mean the end of that politico-business oligarchy incubated within its structures and put an end to the function of the state as an allocator of rents? Will democracy open the doors to free markets and individual rights that will pull the rug from under those capitalists embedded in the political and economic protection of an authoritarian or predatory state?

The whole question of regime change raises the question of how specific property or power relations are indeed transformed as a result of such changes or how social and political forces may reorganise themselves to capture new policy or institutional arrangements. In the case of Indonesia, as we shall see, it is important that changes in the regime in 1999 were presided over by the same interests previously encompassed by authoritarian rule. While the changes meant that they could no longer rule in the same way, the interesting point is whether the fall of the Soeharto regime meant the end of those dominant interests embedded in its apparatus. As Petras noted in relation to democratic transitions over a decade ago, to save the state (conceived as a mode of rule together with those power relations in which it is embedded) it may be necessary to sacrifice the regime (Petras 1989: 26).

While there is no doubt that with the shift from centralised rule by an authoritarian president to a system of power decided within the shifting world of electoral politics, parliament and political parties, the terms of power also shifted for different elements of the former regime. The state bureaucracy and its military were now potentially vulnerable as more power and influence passed into the hands of those controlling the parties and the parliaments. However, with most of the former politico-business alliances still intact, we must see the problem as one that involved these interests reorganising their operations within the new political arenas. There is an eerie similarity between these events and the circumstances that accompanied the transition to political democracy in Thailand more than a decade ago. There, as explained by Anderson, many of Thailand's leading business interests found parliamentary democracy an ideal framework for their activities. It opened channels to political power which bypassed the old hierarchies of the state apparatus. Not only was it the 'ideal shell' for the new political entrepreneurs and fixers within the parties, it was also ideal for the big bankers of Bangkok who took no direct part in the public politics of electoral democracy, but who could now exert 'independent political influence in a way that would be very difficult under a centralised, authoritarian military regime'.[22]

The ambiguous implications of globalisation

Finally, this study requires a more nuanced approach to the expectations that globalisation ultimately constitutes an inexorable and progressive force for political and economic transformation. Globalisation has undoubtedly given international investors, bankers and policy-makers greater leverage over trade and financial policy, public and corporate governance and even political reform (Winters 1996; Rosser 2002). Nevertheless, as we have argued, Indonesia's profoundly illiberal commercial system and its business moguls continued to receive substantial inflows of investment and finance capital through global markets until the very implosion of the Indonesian economy in 1997. Western investors and financiers were happy to invest in what they knew to be highly corrupt and poorly regulated markets. Indeed, international investors had rediscovered much of the lost era of *laissez-faire* capitalism in Asia, including Indonesia, where authoritarian governments present an environment free of the demands of organised labour, social welfare lobbies, environmental protection and progressive taxation that exist in the Western social democracies.[23]

At the same time, the neo-liberal interest in democratic reform is reinforced by the new strategic judgements of the US that its interests are now best served by democracy. This neo-conservative incarnation of US policy has been reflected in ongoing democracy promotion programmes (Smith 2000). However, the neo-liberal emphasis on private property rights and individual guarantees of protection against the arbitrary actions of the state also contains a deep suspicion that these objectives might be short-circuited by democracy. Rule of a majority intent on diverting private resources into public programmes is seen by neo-liberals as a potential threat to markets (Dorn 1993; Barro 1993: 6). The dilemma was, as stated by Gourevitch, 'Does economic growth require democracy (in order to prevent rent-seeking by those who control the state), or on the contrary, is democracy a threat to solid economic policy (because of populist raids on efficiency)?' (Gourevitch 1993: 1271). At the same time, US strategic interests clearly preferred a form of 'low intensity' democracy that was neither a door to chaos nor an opportunity for radical forces (Gills 2000).

The varied response of foreign interests to the post-Soeharto developments reflects the ambiguity of globalisation. While there was relief that an increasingly corrupt and embarrassing regime was out of the way, at the same time apprehension at unfolding disorder and a concern that the new politics would open the doors to 'populist raids on efficiency' gave rise to a clear nostalgia for the order and control of the Soeharto era. In other words, we must approach globalisation as a highly contingent process, and the interests assembled behind the neo-liberal agenda as ambiguous and selective in regard to liberal reform. In some circumstances, globalisation might be considered a force that will outflank social democracies in the West and accommodate many features of illiberal capitalism in countries like Indonesia in an ironic reversal of accepted truth.

Notes

1 As we shall see later in the chapter, technopols constituted that strata of officials and politicians acting above vested interests to introduce rational policy agendas in the larger collective interest (Williamson 1994).

2 Overviews of public choice political economy, or, as Evans refers to it; neo-utilitarian political economy, are to be found in Toye 1987: 47–95; Grindle 1991; Evans 1989, 1995: 22–28.

3 As early as 1967, economist, Heinz Arndt welcomed Soeharto's appointment of Western-trained economic technocrats to positions of authority in key economic ministries: 'if government attitudes and the climate of public opinion in Jakarta mean anything, a new era has certainly begun. There has been a willingness to eschew slogans and ideology, to face economic facts and be pragmatic, which had not been in fashion in Indonesia for many years' (Arndt 1967: 130). This has set the tone for

unswerving support for a technocratic solution to Indonesia's problems that was to pervade both the World Bank and the views of academic economists as illustrated in the pages of the influential *Bulletin of Indonesian Economic Studies*.

4 Ross Garnaut had argued that technocrats had pursued inappropriate exchange rate policies at a time of growing capital mobility and business exuberance resulting from the economic boom. In McLeod's view, Indonesia's technocrats had invited trouble by attempting simultaneously to control more than one of the macroeconomic variables of prices, money supply and nominal exchange and interest rates (McLeod and Garnaut 1998: 2–3, 36). See Jayasuriya and Rosser 1999 for a more detailed review.

5 There was no better example of this proposition than the deregulation that took place in Indonesia's financial and banking systems in the 1980s. As we shall see, although welcomed by neo-liberals because of increased efficiency in generating credit and opening up financial markets (Wardhana 1994: 85; McLeod 1994: 2, 3; Cole and Slade 1996: 353), it was to degenerate into a system of cash cows for those powerful business tycoons who dominated their ownership and flouted limits on intra-group lending (Rosser 2002: 51–84). In the booming infrastructure sector, where the government was presiding over US$15 billion of private infrastructure projects by the 1990s, the World Bank called for a clear, competitive regulatory framework for private sector participation instead of a 'deal-by-deal' approach and for transparent and competitive frameworks for bidding rather than negotiation with pre-selected bidders (World Bank 1995: vii; vii, ix, xiv, xvii).

6 This point is made by Andrew Rosser (2002: 7).

7 A range of scholars working variously within the traditions of 'statist' analysis or historical institutionalism have argued that state-led and state-managed economies like South Korea and Japan represented a highly efficient and alternative form of capitalist organisation to the Anglo-Saxon, free-market model (Wade 1990, 1992; Weiss 1998; Johnson 1982). Alice Amsden even proclaimed that the spectacular growth of Korea had stemmed from turning the neo-liberal orthodoxy on its head and 'getting the prices wrong' (Amsden 1989).

8 As Engels observed, although the democratic republic 'knows nothing any more of property distinctions', it nevertheless allowed wealth to exercise its power in a variety of ways, including direct corruption of officials, as in America, or in the form of an alliance between the state and the stock exchange, as in Britain (Engels 1968).

9 Indeed, this was a task recognised by Douglass North, who argued that reform must be carried out by organisations such as political parties and associations with an interest in perpetuating polities that will enforce and create efficient property rights, development of the rule of law and protection of civil and political freedoms (North 1995: 25).

10 For Weber, and more recently for contemporary institutional political economists, the rise of the modern rational state was a functional imperative of modern industrial capitalism requiring an end to arbitrariness in the exercise of power and a more efficient and predictable regulation of markets. The very principles of routine and regulation embodied in such a state, so different from those operating in the market, and the ascendancy within it of a corps of state managers whose authority was embedded in technical competence, were the factors that protected the self-interested world of markets from itself (see Evans 1995: 29, 30; Hutchcroft 1998: 31–44; Weber 1964, 1978). Within the Marxist camp, too, the issues of state autonomy and insulation are central in a state obliged to guarantee the common material interests of the bourgeoisie even over the immediate demands of individual capitalists; the state also manages the conflict between classes that is the natural product of capitalist society (see Draper, 1977: 237–311, 484–514). Referring to the case of Bonapartist France, Engels argued that:

> It is becoming clearer to me that the bourgeoisie doesn't have the stuff to rule directly itself, and that therefore, … a Bonapartist semi dictatorship is the normal form; it carries out the big material interests of the bourgeoisie even against the bourgeoisie, but deprives the bourgeoisie of any share in the ruling power itself. On the other hand, the dictatorship is compelled, reluctantly, to adopt these material interests of the bourgeoisie.
>
> (Letter from Engels to Marx, 13 April 1866, cited in Draper 1977: 336)

11 As political scientist, Bill Liddle, has argued, 'Classes and other interest groups, state agencies and other foreign economic and political forces create constraints and opportunities to be weighed by decision-makers faced with a problem, rather than determinants that overwhelm and deny the individual's capacity for autonomous choice' (Liddle 1992a: 796).

12 Laurence Whitehead has noted that it had always been assumed by classical liberals that 'the development of private property … was enough to produce an increasingly large, autonomous and self-reliant community of practical-minded people ('middle class' in more recent parlance) who would demand their rights and therefore underpin an open and responsive system of political organisation (these underpinnings could be called a civil society)' (Whitehead 1993: 1247).

13 Marx's quintessential statement on this issue is:

> In the social production of their life, men enter into definite relations that are indispensable and independent of their will, relations of production that correspond to a definite stage of development of their material productive forces. The sum total of these relations of production constitutes the economic structure of society, the real foundation, on which rises a legal and political superstructure … At a certain stage of their development, the material productive forces of society come into conflict with … the property relations within which they have been at work hitherto.
>
> (Marx 1904)

14 Lenin argued that 'a democratic republic is the best possible shell for capitalism, and, therefore capital, once in possession of this very best shell, establishes its power so firmly, that no change of persons or institutions or of parties in the bourgeois democratic republic can shake it' (Lenin 1963: 296, cited in Jessop 1983: 279).

15 Within modern bourgeois society, Marx argued, 'the bureaucracy and army, instead of being masters of commerce and industry, [must] be reduced to their tools and made mere organs of bourgeois business relations. It cannot be tolerated that agriculture be restricted by feudal privileges or industry by bureaucratic tutelage … It must subordinate the treasury to the needs of production' (Marx, 1849 Speech to the Cologne Jury, cited in Draper 1977: 498).

16 As Lefebvre observed in the case of the French Revolution, 'the revolution was launched by those whom it was going to sweep away, not by those who would be its beneficiaries' (cited in Hobsbawm 1990: 133).

17 Drawing on the works of Perry Anderson, and Barrington Moore (1996), Bruce Cumings (1989) has elucidated a range of historical amalgams in which capitalism developed differently as the result of specific conjunctures and collisions at particular points in world time. Liberal markets, in this view, might be seen as the result of circumstances prevailing in Britain in the eighteenth and nineteenth centuries that can't be repeated in their precise form elsewhere. In North America, Cumings proposes, migrations left the peasantry and aristocracy behind, resulting in the transplantation of Lockean liberalism in a vacuum, without the feudal or revolutionary

socialist elements of the European amalgam. By contrast, the Iberian migrations to Latin America were primarily clerical, rural and military, resulting in a caudillo system, never able to fully develop any of the dominant Southern European outcomes – liberalism, corporatism or fascism.

18 An interesting interpretation of the contingent nature of the historical role of the bourgeoisie, and of organised labour, has been drawn by Eva Bellin (2000). James Kurth earlier drew comparisons between the role of the bourgeoisie in England and in 'late industrialising societies like Germany and Japan'. In the former, he argues, the bourgeoisie had no need for state finance and protection but every interest in free trade and breaking down absolutist mercantilism. In the latter, weak bourgeoisie required the financial and protective resources of existing and powerful bureaucratic states. There was a conjuncture between the bourgeois interests and economic and political liberalism in England in contrast to Prussia where the champions of free trade were the conservative landowners (Kurth 1979: 330–335).

19 It was precisely this contradiction that, Marx argued, finally undid the bourgeois alliance with the Bonapartist regime in nineteenth century France (Marx, in Draper 1977: 430, 503).

20 Referring to the Korean experience, Nigel Harris argued that:

> What was set up to speed development becomes an inhibition to growth as capital develops, as output diversifies, as businessmen are increasingly drawn to participate in the world economy, and as the need for the psychological participation of a skilled labour force supersedes the dependence upon masses of unskilled labour: capitalism 'matures'. The old state must be reformed or overthrown, to establish the common conditions for all capital: a rule of law, accountability of public officials and expenditure, a competitive labour market and, above all, measures to ensure the common interests of capital can shape the important policies of the State. Thus, the enemy of capitalism is not feudalism but the State, whether this is the corrupt, particularist State, State capitalism, or, as is more often the case, a combination of these.
>
> (Harris 1988: 247)

21 Of course, the idea that a bourgeoisie would remain dependent upon an authoritarian state if it was unable to protect itself from various populist or radical challenges is the central theme of Marx's work on Bonapartist France. An interesting application of this idea of bourgeois hegemony as a condition for democracy is developed by Fatton (1988) in relation to Africa.

22 Anderson noted that:

> As the financial backers of many MPs, the banks can exert direct, independent political influence in a way that would be very difficult under a centralised, authoritarian military regime. Furthermore, as the representatives of a national electorate, the parliamentarians as a group veil bank power (and the power of big industrial and commercial conglomerates) with a new aura of legitimacy. This is a real and valuable asset. It can thus provisionally be concluded that most of the echelons of the bourgeoisie – from the millionaire bankers of Bangkok to the ambitious small entrepreneurs of the provincial towns – have decided that the parliamentary system is the system that suits them best; and that they now have the confidence to believe that they can maintain this system against all enemies.
>
> (Anderson 1990: 46)

23 See Robison 1996: 332. In an interesting comment on the policies of the Bush government, Zingales and McCormack (2003) draw the distinction between US policy agendas as reinforcing market reform and as simply backing the interests of the rich in low taxes, etc.

2

THE GENESIS OF OLIGARCHY

Soeharto's New Order 1965–1982

When General Soeharto swept into power amid the political turmoil and economic decay of the mid-1960s, it was a move welcomed by Western governments and by investors and economists more generally. Not only did the new regime clear away those elements of reactionary and radical populism that had stood in the way of market capitalism and the interests of private property, it was to provide the missing ingredient for an economy beset by inflation and debt and paralysed by the absence of investment capital. Agreements with Western creditor nations enabled Indonesia's huge debt to be rescheduled and restructured, and opened the door to a selective inflow of foreign investment.[1] Yet, Soeharto did not preside over a transformation to liberal market capitalism and political democracy. Ironically, the re-engagement with global capitalism was the means by which he replaced the former ramshackle and bankrupted regime with a more efficient and centralised form of authoritarian rule and extended the foundations of that vast system of state capitalism, constructed by Soekarno but never consolidated.

That this was to be no liberal regime was initially of little concern to Western powers still preoccupied with the Cold War. What mattered was that Soeharto represented an important strategic shift in the geo-politics of Southeast Asia where communist threats still loomed large in Vietnam, Thailand, Cambodia and the Philippines. Although Indonesia's small liberal middle class became increasingly apprehensive as they realised there was no real place for civilian rule in the new scheme of things, Western economists and political scientists were not immediately disturbed by a regime so profoundly anti-liberal and hostile to democratic reform. They distinguished between the sort of authoritarianism that presided over economic disintegration and political decay and that which imposed political order and economic growth on a society that was inherently self-seeking and predatory. As we have seen in the previous chapter, liberal market economists were reassured as the new government appointed Western-trained economic technocrats to supervise macro-economic policy, providing the stability needed for growth and investment.

Political scientists also initially embraced the New Order. While Soekarno's authoritarianism had been portrayed as atavistic and defined by tradition

(Benda 1964: 449; Feith 1963: 79–97; Wilner 1973: 517–541), Soeharto's brand of authoritarianism was seen as a state insulated from vested interests and an incubator for modern rational capitalism. In terms of the Huntington thesis, it provided the strong institutional cement for a disintegrating society and performed the historical role of the middle class – it was its 'advance guard', its 'spearhead into modern politics' (Huntington 1968: 222). For the leading protagonists of this approach, the New Order was the alternative to chaos and economic decay. A return to the parliamentary system offered only anomic and untrusting behaviour, instability and impotence (Emmerson 1978: 104, 105, 1976: 250; Liddle 1989: 23). Not surprisingly, this idea that authoritarian rule might be legitimised in terms of its developmental role was seized upon with alacrity by the theoreticians of the New Order (Moertopo 1973).

But why would the New Order assume the role of incubator for modern rational capitalism rather than indulging in the sort of nationalist adventurism that had characterised Soekarno or descending into predatory plundering? For political scientist, Bill Liddle, the explanation was to be found within the rational calculations of Soeharto himself who, he argued, was able to rise above the patrimonial attachments of others in the regime, escaping the suffocating cultural lenses that forced many Indonesians to resist capitalism, establishing a 'relative autonomy' from vested interests. His decisions, suggests Liddle, were a recognition that his long-term political ambitions and self-interest relied upon the neo-classical economic policies that alone could provide growth and prosperity to reinforce the legitimacy of the regime and provide a strong revenue base (1991: 403, 404, 1992a: 796–798). However, the almost complete abandonment of structural factors for the voluntarism of 'agency' and the choices of individuals brought its own dangers that were recognised by Liddle himself.

> If 'relative autonomy' rests not with the state and not with the army but with President Soeharto, an individual, is analysis reduced to idiosyncratic description? For Indonesia, does this imply that there is no telling what policies and politics the next President, also an individual, will adopt? For comparative political economy, does it imply that there is no telling, period?
>
> (Liddle 1991: 242)

Clearly, the role of Soeharto, however decisive, had to be understood in the context of broader and overarching structural frameworks. But the Indonesian case did not easily fit into the prevailing models of state and social power. Despite the prominence of its almost Leninist political institutions, its large state sector and the obsession with industry policy and planning, Indonesia was not a developmental state like South Korea, Japan, Singapore or Taiwan. As we shall see, industry policy remained wedded to the task of import-substitution, and was used to enrich state and private oligarchies rather than as an instrument for conquering global markets. On the other hand, although the Indonesian case might be seen to resemble that archetypal public choice view of the 'third world' predatory state where the 'invisible hand of the market' dominates administrative behaviour, where everything is for sale and everything has a price, Soeharto's state encompassed a highly organised system of social power not easily explained in terms of universally random and opportunist predatory practices.[2] Although the sale of political favour was a pervasive factor in defining the dynamics of corporate and commercial life, the Soeharto regime was also concerned with maintaining a specific hierarchy of social and political power. Coordinated use of state authority to exclude and repress specific interests, such as liberal reformist middle classes and labour, that might form the basis of a cohesive political challenge went far beyond the mere sale of state power and resources to the highest bidder.

One of the most interesting attempts to explain the tantalising mix of predatory and statist influences in the New Order has come from Paul Hutchcroft. In his comparative assessment of different development regimes in Southeast Asia, he draws a distinction between systems of *patrimonial oligarchy*, where a powerful business class extracts privilege from a largely incoherent bureaucracy, and those systems of *administrative patrimonialism* where power is located in the hands of a class of office-holders who are the main beneficiaries of rent extraction from a disorganised business class (Hutchcroft 1998: 52). Hutchcroft argues that bureaucracies in patrimonial oligarchic systems such as the Philippines are concerned with allocating resources, not regulating markets, and lack the capacity to impose reforms on social interests. Economic growth in these systems reinforces the power of social oligarchies resistant to reforms that would break up rent-seeking and impose general rules for business. On the other hand, he proposes that administrative patrimonial states have the capacity to impose reform, citing the example of Indonesia's successful economic reforms in the 1980s compared to the inconclusive efforts in the Philippines despite prolonged World Bank pressure and structural adjustment programmes.

Significantly, he also argues that the private sector interests nurtured in administrative patrimonial systems are more likely to tire of the demands and uncertainties of rents and to see their interests increasingly served in regimes of general rules and laws (Hutchcroft 1998: 45–64). In this Weberian-inspired approach emphasising the decisive influence of the institutional factors of relative state capacity and insulation, he compares the outlook for the Philippine case with those for Thailand and Indonesia, citing Ruth McVey's observation that the features of the bureaucratic polity in those latter countries 'have less the aspect of a developmental bog than of a container for fundamental transformation' (McVey 1992: 22, cited in Hutchcroft 1998: 48).

There is no doubt that Soeharto's New Order resembled Hutchcroft's administrative patrimonial state to the extent that power was located in the hands of a class of powerful office holders who were the main beneficiaries, at least in its initial stages, of the extraction of rents from a politically disorganised business class. Access to huge flows of revenue from oil meant that the state was not reliant on any powerful domestic sector for its tax base. Soeharto's early commitment to fiscal and monetary discipline represented the beginnings of a modern regulatory state. Yet, its potential regulatory capacity always remained subordinated to the larger needs of a system of state and social power focused around the pillars of central authoritarian rule and rents. Although a flourishing capitalist class and an associated strata of political and economic fixers and cronies had pervaded and captured the system of state capitalism in this period, there were to be few signs of growing impatience with the uncertainty and arbitrariness of patrimonial rule or demands for a shift to markets defined by general sets of rules and rights. They remained dependent upon the preservation of authoritarian rule and state allocated rents. The administrative patrimonial state in Indonesia had become the incubator, not for a class of capitalists that would outgrow and challenge it, but for that complex and fluid politico-business oligarchy of the post-Soeharto era and the system of money politics and 'savage' capitalism that now prevails.

Thus, the New Order must be explained, we argue, in the context of decisive social conflicts and a metamorphosis of state and class power within its very structures. We propose that the New Order evolved into:

1. A regulatory apparatus imposing a framework of fiscal and monetary discipline and highly organised political repression aimed at preventing the economic and social disorder that had corroded the previous regime. Within this was established:
2. A system of organising state and society relations characterised primarily by the disorganisation of civil society and the dominance of state-created corporatist institutions.
3. An extensive and complex system of patronage, personified by Soeharto himself, and the apex of which lay at Cendana Palace. This system of patronage penetrated all layers of society from Jakarta down to the provinces, *kabupaten*, towns and villages.
4. During its heyday it became a capitalist oligarchy that fused public authority and private interest, epitomised in the rise of such families as the Soehartos.

The contest for post-colonial Indonesia: statism, liberalism and reactionary and radical populism

The origins of Indonesian authoritarianism are to be found in the decaying remnants of Dutch colonial rule and its declining agrarian economy. In a world where colonialism had left neither large landowning elites nor any powerful urban bourgeoisie, the central political legacy of Dutch rule was to be a vacuum of social power set within the rambling state apparatus constructed to protect and regulate an economy based on agrarian export capitalism. It was those interests that focused around this state, its officials and their families that straddled the world of officialdom, property and middle class professionalism, that drove the nationalist agenda and formed the backbone of the dominant party of the post-colonial period, the PNI (Indonesian Nationalist Party).

With its highly nationalist and state-centred view of the world, this political class confronted powerful populist and radical undercurrents in the immediate post-colonial period. As the old colonial economy disintegrated, a reactionary, petty bourgeois populism flourished around declining and resentful rural propertied classes and small town businesses, and within the politics of Islam and anti-Chinese xenophobia. It harboured resentments against the outside world fuelled not least by the increasing marginalisation of indigenous trading and business interests (Sutter 1959: 805–808, 908–922; Anspach 1969: 180–185; Castles 1967; Robison 1986: 36–68). At the same time, various groups, including workers and peasants as well as small landowners and elements of the middle class, formed around the agenda of radical populism under the umbrella of the Indonesian Communist Party (PKI). These became embroiled in conflicts with landowners, Muslim political groups and the military over land and control of former Dutch-owned assets (Mortimer 1969; McVey 1992).

Yet, the parliamentary system that emerged in 1949 offered none of the contending forces an immediate path to power. The rapid rise and fall of fragile coalition governments meant that no singly political party was able to capture real authority. In any case, those political parties that attained office became devoted to the scramble for control of the Ministries that brought, amongst other things, the authority to allocate the concessions, import licences and contracts that were the reward of government (Rocamora 1974: 181–188; Sutter 1959: 790–98, 1311). Indeed, most of the contending interests and groups saw advantages for themselves in the retreat to authoritarian rule, welcoming Soekarno's dissolution of parliamentary politics and his moves to restore the 1945 constitution and the ultimate authority of the President (Lev 1966: 182–201; Thomas and Panglaykim 1973: 56–59; Robison 1986: 73, 99).

The transition to authoritarian rule was not an abstracted organisational imperative to impose order upon chaos in the sense argued by Huntington. Nor was it a cultural retreat from modern secular rational politics to the politics of tradition, the 'theatre state' or to the politics of revolutionary nationalism.[3] It signalled nothing less than a decisive shift in the configuration of social and economic power and the victory of these interests assembled around the state. For Schmitt (1962, 1963), it enabled the consolidation of foreign exchange regimes and other economic policies that ensured the interests of importers in Java and those gathered around the state and its nationalist programmes over those of the smaller outer island producer-exporters. It confirmed a general decline of the Islamic petty trading and manufacturing bourgeoisie already well under way as attempts to protect *pribumi* business interests through preferential allocation of trading licences under the Benteng programme in the 1950s were expropriated by party power brokers and so-called 'briefcase entrepreneurs' (Sutter 1959: 1017–35; Anspach 1969: 171–79). Islamic petty bourgeois political organisations found their increasingly xenophobic demands for protection ignored by the new Soekarno regime (Asaat 1956; Feith 1962: 481–487). When the military put down a series of regional rebellions, leading to the banning of the modernist Islamic party, Masjumi, Islamic populism suffered a major defeat.

The advent of Guided Democracy also had implications for the future of radical populism and the Indonesian Communist Party (PKI). Although it closed the door to the possibility of a PKI government through electoral victory, the communists saw the assumption of power by Soekarno as a step forward in the struggle against the alliance of feudal and imperialist forces (Aidit 1970). Thus, they supported Soekarno's National Front and embraced his ideas about Nasakom (an alliance between nationalist, religious and populist streams). While the PKI's influence on Soekarno grew, there was a price to pay. It was never able to secure an influential Cabinet post and had to tone down its class struggle rhetoric, including in the labour area, where the communist-oriented SOBSI (*Sentral Organisasi Buruh Seluruh Indonesia*) was dominant. Though a result also of growing military intervention in labour affairs after the nationalisation of foreign companies in 1957 (and the subsequent role as managers taken by military officers), the incidence of industrial strikes, for example, fell from 505 in 1956 to zero by 1962, as a partial consequence of the strategic need for moderation (Hadiz 1997: 55–57; King 1982: 115).

At the same time, the new 'Guided Democracy' set out that institutional and constitutional framework that was to mark the victory of the state and its officials over civil politics. Soekarno offered the idea of a state that transcended vested interests to embody the common good of society as a whole as an alternative to the destructive influences of 'free-fight liberalism' and to be based on self-reliance (Chalmers 1997; Soekarno 1956, 1959). It was a state set within organic or integralist ideals that were the antithesis of liberal individualism, providing the ideal legitimation of authoritarian rule with its emphasis upon the values of hierarchy, authority and discipline. It allowed political opposition to be proscribed on the grounds of its 'dysfunction' in the context of the larger organic project (Reeve 1990; Soekarno 1956). In practical terms, parliamentary government and representative democracy were replaced by Presidential rule and an appointed parliament of the representatives of 'functional' organisations, including farmers, workers, civil servants, intellectuals, veterans, the military, and so on. It became difficult for legitimate and independent organisations to exist outside these state-sanctioned and controlled groups. Most important, the problem of representation was solved by divorcing functional representation from access to decision-making (Reeve 1978: 94).

Within the economic sphere, also, 'Guided Economy' gave officials control over the policy levers of what was to be a virtual command economy. As Dutch and other foreign-owned enterprises were nationalised in 1957, most of the existing agricultural estates, trading houses and banks passed into the hands of the state. State-owned companies now sat astride the commanding heights of the Indonesian economy. While many of these were now simply milked for off-budget funding for the military and other political groups, trading houses, banks and industrial enterprises were also integrated within a plan to create a new, national and industrially based economy to replace the old colonial system. State officials determined even the priorities of trade, investment and production, deciding the allocation of import and export licences and quotas, production quotas and foreign exchange (Robison 1986: 69–101; Castles 1965).

However, it was soon evident that neither Guided Economy nor Guided Democracy was sustainable. Soekarno had attempted to contain within the ambit of these regimes a variety of contradictory and irreconcilable social interests and political forces. The political contest for control over the state remained unresolved as the military and the PKI continued to build their parallel apparatuses of political organisations and social bases. Deepening conflicts over land ownership and reform led to increasing violence and tension in rural Java as forces gathered around the competing agendas of property and radical redistribution. Meanwhile, in the cities, middle class students began to agitate against Soekarno, reflecting an increasing impatience with economic decline and corruption, and fear of increasing communist influence. In the economic sphere, the attempt to create a new industrial economy within a virtual command model insulated from Western capitalist financial and investment markets stumbled as investment quickly dried up and infrastructure decayed. By 1965, inflation stood at over 600 per cent while annual foreign debt interest payments now exceeded the value of Indonesia's exports (Arndt 1967: 130, 131).

In themselves, these contradictions did not trigger the actual circumstances of Soekarno's fall and Soeharto's accession to power. The dramatic events of September and October 1965 were the immediate result of a struggle for power within the military and amongst other elements within the military and the Presidential palace (Anderson and McVey 1971). However, they were to trigger a violent resolution of the wider social and political tensions, and most important, the murderous elimination of the PKI and its supporters by the military and their rural Islamic allies. Yet, those liberal middle classes and the conservative Islamic populists who initially welcomed Soeharto were soon to discover that the events of 1965 did not open the door to a brave new world of liberal markets and democracy or halt the declining political and ideological influence of small business and propertied interests in the rural sector. The violent elimination of its enemies and the re-accommodation with Western capitalist governments simply meant that authoritarian rule was no longer constrained by an insidious social opposition nor undermined by the sort of economic decline and decay that fed political unrest. The way was clear for the full realisation of the integralist state, authoritarian rule and state capitalism.

The New Order: the triumph of the state

How then can the rise of the New Order be understood? We have seen how economists and some political scientists viewed it as a necessary interregnum of economic pragmatism and social order at a time of chaos and adventurism, emerging as the product of an abstracted organisational imperative. Others saw the regime as nothing less than the outcome of a decisive battle between forces seeking to impose a new imperialism on the Indonesian economy and those forces assembled around various agendas of national autonomy. Thus the military and those middle classes argued by economists and modernisation theorists to represent rationality, efficiency and order were seen instead through dependency lenses as compradors, in collusion with international capital to deform and exploit Indonesia's economic progress (Levine 1969; Mortimer 1973; Arief and Sasono 1973). Yet, simply put, the state and its officials appeared to have a life of their own. They continued to frustrate the World Bank and foreign investors with continuing nationalist policies and to trample on the liberal students and middle classes who were thought to be an essential part of the New Order's support base. Even well-connected business cronies were from time to time burnt by a government willing to reign in fiscal and monetary settings at critical times. The question of state autonomy had to be confronted.

It was a question that attracted Ben Anderson who saw the New Order increasingly in terms of the 'resurrection of the state and its triumph *vis à vis* society and the nation' (Anderson 1983: 487). The New Order became, 'a *state qua state* with its own institutional memory and harbouring, self-preserving and self-aggrandising impulses which at any given moment are "expressed" through its living members, but which cannot be reduced to their passing personal ambitions' (1983: 478). It was a perspective that might be compared to Marx's categorisation of the Bismarckian state where: 'The aims of the state are transformed into the aims of the bureaus, or the aims of the bureaus into the aims of the state … The bureaucracy has the essence of the state … in its possession; it is its private property.'[4]

While the corps of politico-bureaucrats who ruled the state might be seen as the slaves of the very logic of this juggernaut, and their 'aims' limited to the preservation of its power, this was no one way street. The state and its officials operated within a range of structural constraints: the need for foreign aid and investment; the need to preserve fiscal and monetary viability; the need to maintain stable and cheap supplies of rice and other basic necessities. At a more instrumental level, the other aspect of Marx's equation is important, and we argue that it was also a process whereby the 'aims' of the bureaus began to take a different trajectory and began to subsume the 'aims' of the state. At one level, these aims were focused around the necessity to concentrate extra-budgetary funds in their own hands *vis-à-vis* their competitors and to underpin their power with networks of patronage. At another level, as the world of public authority and private interests became increasingly blurred, the 'aims' of the bureaus were extended to become the 'aims' of powerful politico-business families in constructing a social base in private capital, harnessing the power of the state to these new private interests. It was a metamorphosis that was to transform the New Order from a regime serving the interests of its own officials into a regime that produced and served a growing and complex business and political oligarchy.

The institutional framework within which this system was established necessarily enshrined the state at its centre. It was to be a highly controlled entry into global market capitalism, designed to prevent the sort of liberal outbreaks that had characterised the transition in Europe.

The political triumph of the state: subordinating, co-opting and disorganising civil society

Although initially enjoying the support of a vocal urban middle class and an active student movement, the New Order, in its early years, was essentially a military government ruling through a vast security and repressive apparatus justified in terms of the need to root out the remnants of communist and other forms of opposition that threatened the national interest. It was initially constructed around an informal cabinet of generals (ASPRI – Special Presidential Assistants). The Command for the Restoration of Stability and Order (KOPKAMTIB), initially headed by Soeharto, became the most powerful instrument of authority in the first decade of the New Order, presiding over a regime of emergency rule, screening individuals for involvement in the supposed communist plan to overthrow the government in 1965, conducting extensive political trials and imprisoning suspects. Backed by a range of draconian laws, often the remnants of the colonial administration, the emergency state was able to control the movement of individuals and their right to dissent and protest, and to impose a strict control of the press and media (Crouch 1978).

However, the tasks of running a modern state and economy meant that Soeharto quickly outgrew this narrow form of military rule and the heavy reliance on the security apparatus. Far from being a creature of the military high command, the Presidency in Indonesia became a centre of power with its own momentum. In administrative terms, the rise of the State Secretariat (Sekneg) provided the President with a discrete capacity to prepare and plan legislation and to deal with Cabinet, the Ministries and with parliament (Pangaribuan 1995). A constellation of power centres emerged to dissipate the military influence over the everyday running of government. In particular, the growing complexity of managing the fiscal and monetary aspects of economic life and dealing with Indonesia's foreign creditors and investors created another level of power and influence in the Finance Ministry, the Economic Planning Board (Bappenas), and Bank Indonesia. As we shall see, Indonesia's technocrats were, from time to time, able even to overrule military and business interests at the heart of power, imposing periodic checks on growth through sharp policies of fiscal and financial contraction.

Building on the corporatist ideas developed to legitimise Soekarno's Guided Democracy, a more sophisticated and pervasive apparatus of social control and mobilisation was designed in the early 1970s within General Ali Moertopo's Special Operations Command (OPSUS), the New Order's most important security and intelligence office. It involved the co-option of individuals and associations within a vast system of hierarchical, compulsory and state-organised front groups that would support state-defined agendas.[5] Political parties and parliament were resuscitated within a system of state orchestrated elections contested by three state approved and sponsored parties in which the state party, Golkar, was consistently to win large majorities (Reeve 1990; Ward 1974; Boileau 1983: 59–121). The other two parties were to become but artificial entities. The Indonesian Democratic Party (PDI) housed disparate elements representing the old secular nationalist and state socialist groups of the former PNI, while the United Development Party (PPP) contained the mutually antagonistic organisations of traditional and modernist Islam. As we shall see in later chapters, Golkar itself was to become one of the main pillars of the New Order, a mechanism for allocating patronage and recruiting support. Soeharto himself was to become the Chair of its powerful Board of Patrons.

Such a system provided opportunities for individuals to enter the broader state apparatus but not to represent interests, to challenge policy or change government. Ali Moertopo argued that government by direct executive authority insulated from the volatility of democratic conflict enabled the ideologies and interest groups to be subordinated to the national interest. He suggested, 'the participation of experts in the economy is evidence that development in Indonesia is no longer oriented to a particular exclusive ideology' (Moertopo 1973: 40). In practical terms, the activities of political parties were confined to the actual election period, ensuring that the masses were not diverted from the more urgent and technocratic tasks of economic development (the so-called floating mass doctrine) (1973). This eliminated the party as a permanent organisation of cadres, continually active at all levels and on behalf of a potentially coherent set of social interests. Authoritarian rule, in this Huntington-inspired view, was not only a necessary means of achieving economic development and social order, it was claimed to provide the basis for the development of civil society and democracy (Soeharto 1989, 1990; Moertopo 1973).

At the same time, the whole electoral enterprise was wrapped in the state ideology of *Pancasila*, effectively legitimising authoritarian rule as the mechanism to achieve the common will of society through consensus and the tutelage of the state. Of particular utility, Pancasila defined what was not legitimate: liberalism, *laissez-faire* capitalism, communism or any exclusive claim to represent a religious, ethnic or class interest. In 1983, legislation was introduced requiring all political parties to adopt Pancasila as their sole guiding ideology (*azas tunggal*) and to accept the larger organic unity expressed by the state. No party was entitled to claim specific representation of Islam or Islamic values. Similarly, no potential reformist party would be permitted specifically to claim representation of the working class or to have a special carriage of social justice. Later, in 1985, under the provisions of the ORMAS (*organisasi masyarakat*) legislation, the *azas tunggal* principle was extended to all non-government organisations (Lubis 1993: 166–172, 212–245). In this Orwellian world, the state and its officials became the embodiment and spirit of an organic national interest. Both political opposition and the very concepts of free competition within markets and the political arena were regarded as dysfunctional and self-seeking challenges to the collective good. It was an ideal political shell for the emergence of that oligarchy of state and social power now evolving within the very apparatus of the state.

The triumph of the state over the market: state capitalism

Soeharto's rise to power was remarkable not least for his apparent determination to rid the new regime of the persistent inflation, fiscal crisis, debt and declining levels of investment that had plagued Soekarno. He turned to a group of Western-trained economists to manage monetary and fiscal policy and to broker the resumption of economic relations with the West. Indonesia's major creditors formed the Inter-Governmental Group on Indonesia (IGGI) to restructure its debt and plan the resumption of aid and loans that were to be so critical for the development budgets of the new regime. Foreign investment began flowing into Indonesia again with the introduction of new foreign and domestic investment laws in 1967 and 1968, providing specific tax incentives and other guarantees against expropriation (Thomas and Panglaykim 1973; Robison 1986: 131–147; Chalmers and Hadiz 1997: 39–55). Yet, despite the fears of many radical critics in the 1970s that the technocrats would be the agents of a broader subordination of Indonesian economic life to the juggernaut of global corporate interests, liberal market principles were not to govern the Indonesian economy in the coming decades. Not only was the influence of the liberal technocrats confined largely to fiscal and monetary policy, even these ultimately served other and more encompassing political goals.[6]

Populist influences continued to pervade policy agendas. The regime affirmed its support for the principle that the state had a legitimate economic role in ensuring the national interest and that market forces should be tempered by social objectives. This principle was enshrined in article 33 of the 1945 Constitution, which stipulates that the economy should be run on 'family principles' (Robison and Hadiz 1993: 15; Soeharto 1989: 192–195). This was more than rhetoric intended to appeal to populist political sentiment. Flying in the face of constant World Bank criticism, the government continued to provide subsidies for a range of goods, including rice, cooking oil and fuels. Soeharto was unwilling to trust to the market mechanism the task of eliminating the shortages and speculation in rice and other basic commodities that had plagued the Soekarno period. He was well aware of the potential for resentment and social unrest among peasants and workers that could arise from volatility in this sector, and the popular support that could be gained by claiming the mandate of protector of the masses.

At the same time, economic nationalism was also to be a central driving force throughout the New Order period, aimed at both protecting domestic interests against the onslaught of powerful international investors and developing an industrial base within Indonesia.[7] Foreign investors were excluded from investment in specific sectors such as retail distribution, media and public infrastructure, or where domestic business could provide investment and expertise. However, such arrangements benefited the larger and rapidly evolving Chinese business groups. When anti-Japanese riots in 1974 exploded into a broader assault on the regime, drawing in disgruntled Islamic petty business interests as well as alienated liberal middle classes, new regulations required all foreign investors to adopt local partners, increased the range of sectors closed to foreign investors, and set out requirements for the transfer of joint venture equity to local investors. In the trade sector, a wide array of restrictions, especially in the form of non-tariff barriers, was introduced. In 1982 an 'approved traders' programme created a system of import monopolies reminiscent of the Benteng programme of the 1950s, although this quickly fell into the hands of well-connected private and state companies (Robison 1992; Robison and Hadiz 1993: 19).

But economic nationalism was also embedded in more sophisticated ideas about creating a sustainable industrial base. Import substitution strategies aimed at protecting producers in textiles, foodstuffs and other products for domestic markets had been applied throughout the 1970s and 1980s (Rudner 1991). As the initial easy period of ISI (import substitution industrialisation) began to falter, the World Bank stepped up its demands for greater market deregulation. Indonesia's economic nationalists were, however, rescued by the surge of oil prices that generated a flood of petrodollars in the 1970s.[8] Powerful economic ministers, including Soehoed and Hartato in Industry, the Investment Minister, Ginandjar Kartasasmita, and the Research Minister, Habibie, were to seize the opportunity to go beyond ISI and invest in industrial deepening, intended to develop within Indonesia a strategic economic autonomy through integrated circuits of industrial production and technology (Soehoed 1977, 1982; Habibie 1986; Kartasasmita 1985). The government underwrote a massive drive into upstream manufacture – in steel, petrochemicals, fertilisers, aluminium, cement, engineering and technology – and the development of infrastructure, public utilities and industrial estates.[9]

These populist policies of equity and stability and economic nationalist agendas were not merely ideological notions. Pervasive and protective trade and industry regimes were also the perfect mechanisms for entrenching the authority of the state and its officials over the full range of economic life. Ministries such as those of Trade, Industry, Mining and Energy, and Public Works were insulated from market pressures, able to impose various protective trade regimes and to strategically allocate contracts and licences. The agency for logistics (Bulog) now became a pivotal centre for raising extra-budgetary funds and allocating rents through its strategic control over the import, distribution and price of rice and a range of other basic commodities, including sugar, flour, wheat and cloves (Crouch 1978: 291–295; Robison 1986: 229–247). Soeharto himself headed

Bulog's predecessor, Kolognas, in 1966 and early 1967. Achmad Tirtosudiro, a 'financial' general from Kostrad, was Bulog chief from 1967 to 1973, followed by General Bustanil Arifin, a close Soeharto associate who was in charge for over two decades.

State banks were another of the main terminals through which officials were able to determine priorities in the economy and allocate rents. They accounted for 93 per cent of bank credit in 1968, and when the first of the banking reform measures were introduced in 1983, the central banks and the state banks together still provided 79 per cent of credit (Rosser 2002: 57). They played the role of agents of development, directing credit into strategic investment programmes. Those who controlled the banks were able to determine which groups and sectors were eligible for subsidised credit and what levels of subsidy would be applied. Much of the state bank credit was directed into supporting the development programmes of state corporations such as Pertamina, the government's electricity authority (PLN), and Krakatau Steel. When the government introduced credit ceilings and a variety of subsidised lending programmes after the *Malari* of 1974, its control over the allocation of credit was significantly increased (MacIntyre 1994: 250–251; Hill 1996: 99–116; Pangestu 1996: 157). Increasingly, state banks bypassed these formal strategic objectives of industry policy and credit leaked increasingly into the hands of privileged and well-connected private investors.

State-owned corporations, too, were to play a central role in nationalist economic agendas, also insulated from competition and presiding over the commanding heights of the economy. Fuelled by booming oil revenues, funding to the state enterprise sector increased from Rp.41 billion in 1973 to Rp.592 billion in 1983 (Hill 1996: 102–103). According to one estimate, the state controlled almost 60 per cent of the equity in all domestic investment and a further 9.2 per cent of the equity in foreign investment projects (Wibisono 1981: 70–71). Perhaps the most important and wealthiest of all the state corporations was the state oil company, Pertamina, which presided over the allocation of oil exploration and drilling concessions as well as the sale and distribution of fuel and oil products and construction contracts for pipelines and refineries. Under its powerful director, Ibnu Soetowo, it became a virtually unaccountable economy within an economy, providing subsidised inputs for upstream petrochemical industries, construction contracts for the private sector, and generating flexible non-budgetary finance for the state and its officials that could be used for a range of political and strategic purposes. It was to be a major financier of the military as well as funding strategic industrial projects such as Krakatau Steel and even investing in its own fleet of tankers (Robison 1986: 233–249). While few other state corporations and agencies could match the strategic importance of Pertamina and its pervasive role in allocating rents and financing the political interests of the regime, many nevertheless controlled allocation of procurement and contracting in public utilities, power generation, road building, mining and other sectors of the economy.

Although Indonesia's corps of politico-bureaucrats now stood astride a vast system of centralised industry policy and a state-owned corporate sector, such authority was not used to construct the highly planned assaults on global markets that were the results of industry policy and state economic power in Singapore, South Korea or Taiwan. Instead, Indonesia's trade and industry policies were directed primarily towards constructing an upstream manufacturing base, along the lines of Stalinist heavy industrial models, protecting domestic markets and sheltering domestic producers, increasingly the business clients of the state. However, larger strategic objectives were routinely subsumed by the search for extra-budget funding to sustain the political institutions and patronage networks of Indonesia's rulers and their own personal enrichment. Plundering society was not only or necessarily the pre-eminent focus of the New Order. Indonesia's politico-bureaucrats were also focused on a different use for the rents that were now so much a part of the cement and currency of commercial activity and corporate success. Their control of the strategic gate-keeping apparatus was now used to translate public authority into private interest, underpinning the rise of politico-business families and the construction of a social and corporate base to their power.

Forging the politico-business oligarchy

How do we explain why there have been few signs in Indonesia of an increasingly restive capitalist class for whom a system of rule-based markets has become increasingly necessary and attractive? How do we explain why there was no prolonged power struggle between those military-bureaucratic elites who presided over the allocation of rents and emerging business elites, the beneficiaries of these rents, as occurred in Thailand? There, the long transition to democracy, finally sealed by the economic crisis of 1997, represented a final political victory of elected business politicians at the head of political parties in a system where power was now fought out in an arena of elections and parliamentary government (Hewison 1993; Anderson 1990; Wingfield 2002). The answer lies in the different power relations that linked public authority and private interest in the two countries.

Not least of these were the political and social factors that precluded Indonesia's predominantly ethnic Chinese business community from playing the public political role that enabled Thailand's business interests to become politicians and to become so embedded in the processes by which power was contested in the system of parties, parliament and elections. More fundamental was the nature of the politico-bureaucratic oligarchy that emerged under the New Order. This was no simple collection of military and civil officials extracting rents to prop up their institutional bases, although this aspect was important. Beyond this, it came to constitute a broad and complex political class of officials and their families, political and business associates, clients and agents who fused political power with bureaucratic authority, public office with private interest. These were not simply the gate-keepers and toll collectors to a class of business supplicants but were embedded also in the world of business itself, in partnership with the Chinese business groups as the glue between the two. They imposed the broader interests of oligarchy over the authority of officials while preserving their control of the state.

The beginning of the metamorphosis: politico-bureaucrats as gate-keepers and the rise of Chinese business groups

When Soeharto swept into power, one of the most immediate changes in the world of business was the key role now assumed by the generals who had managed the finances of military commands in the Soekarno years. They presided over the allocation of the assets of private business clients of Soekarno now confiscated under the new regime, and assumed control of state corporations, including Pertamina and Bulog, as well as various state trading and mining companies.[10] Among the early beneficiaries of the reallocation of rents were military companies themselves. They were to receive crude handouts of confiscated assets, forestry concessions, sole agencies for the import of autos, transport and shipping companies, airlines and trade monopolies as well as banks.

Such military companies were developed usually in conjunction with Chinese business partners, many of whom had long-standing relations with the 'business' generals, in some cases stretching back to the 1950s and 1960s when they had jointly operated trading and transport companies, banks and other enterprises. Most significant among these were those that tied Soeharto himself, as commander of the Indonesian army's Diponegoro Division, with the Chinese trading entrepreneur, Liem Sioe Liong, and with Bob Hasan, who operated shipping companies in conjunction with the military (Robison 1986: 271–322).

With their forestry, trading and banking interests, a vast array of military companies and foundations was assembled in the 1960s and 1970s, the most important of which were Tri Usaha Bhakti (TUB) and Yayasan Dharma Putra (YDP), consolidating the holdings of the Ministry of Defence and the Army's Strategic Reserve, Kostrad, respectively. But military companies proved generally unable to commercially develop the assets under their control. Military-owned plantations, seized in the late 1950s, for example, had decayed through lack of maintenance and investment, and airlines and transport companies ground to a halt for similar reasons in the 1970s. They were also progressively forced out of the forestry industry after a ban was placed on raw log exports, forcing players into the more capital intensive world of plywood production, pulp and paper manufacture. Similarly, they lost control over sole agencies in the automobile industry, after the government introduced requirements to increase local content and to manufacture components and even engines in Indonesia (*Indonesian Commercial Newsletter* (*ICN*) 14 September 1981; 10 January 1983).

The military's commercial interests were consolidated into a smaller number of entities, notably the Yayasan Eka Paksi, which was effectively operated by Chinese business partners, notably Bob Hasan and the forestry tycoon, Yos Sutomo. It presided over ventures that extended from plywood manufacture and banking to property and construction, where it had received a lucrative stake in the development of the Sudirman Central Business District. But military equity in these partnerships continued to shrink as the investments became larger and extended into new and more complex sectors. Forced to progressively surrender shares, the Foundation found it increasingly difficult to generate dividends (*Tempo Online* 8–14 October 2002).

As power within the regime spread beyond its institutional origins in the military, Soeharto established a number of foundations (*yayasan*) dedicated to raising off-budget funds for purposes ranging from building mosques to providing funds for Golkar. While their operations are largely opaque, being exempted from public audit, they drew their income from a range of sources including state banks and corporations. A Presidential decree in 1978 had required state banks to transfer 8 per cent of their operating profits to two of the foundations, Yayasan Supersemar and Yayasan Dharmais (*Pikiran Rakyat Online* 8 January 1998). Chinese business, too, was forced to contribute. Yayasan Dharmais and Dakab were each given a 6.39 per cent share in Liem Sioe Liong's PT Indocement Tunggal Prakarsa, while the letters of association for the flour milling monopoly, Bogasari, stipulated that 26 per cent of its profits be allocated to various Soeharto *yayasan* (Robison 1986: 345–346; Schwarz 1990: 62–64; Shin 1989: 247–248). As we shall see, these *yayasan* were also to be the Trojan horses within which the Soeharto family entered into quite different relations with Chinese business partners.

It was soon evident that Chinese business interests were to become the main beneficiaries of the new economic regime. After all, they had dominated the domestic distribution and credit networks within Indonesia from colonial times and had increasingly developed commercial ties within a larger Chinese business diaspora that spread across the region. A vast network of Chinese wholesalers, retailers and small manufacturers and service industries dominated the countryside and small towns. In the final analysis, they were the only group with the capacity to generate the business profits that would not only provide the extra-budgetary funds needed by Indonesia's political factions, including the military but, more broadly, would underpin the economic growth that was so central to the objectives of the Soeharto regime.

The rise of the Chinese business empires in Indonesia has been studied in great detail and covered extensively by the Indonesian press.[11] We do not intend to replicate this here, merely to briefly indicate the main sites around which they were clustered and where their relationships with politico-bureaucrat patrons were constructed. Perhaps the earliest site was Bulog, where several Chinese businessmen who had been suppliers of foodstuffs to the military in the 1950s and 1960s now became the major holders of import and distribution monopolies in rice, sugar and other commodities (Crouch 1978: 291–295; *Kompas* 25 September 1969: 1; *Tempo* 28 October 1972: 44). The emergence of Liem Sioe Liong as the leading Chinese business tycoon owed much to the licence to mill flour for the whole of Western Indonesia given to his company, Bogasari, in 1970, together with a state bank loan for Rp.2,800 million, just five days after its establishment with capital of only Rp.100 million. Bogasari was to enjoy a secure and lucrative 30 per cent margin on milling and a strategic position in the domestic food industry that would enable Liem to expand into a broader dominance within the food industry through his giant Indofood conglomerate (Robison 1986: 232, 302; Schwarz 1999: 111).

Chinese business groups were to seize the major share of the 61 million hectares of forest concession allocated without any public and transparent process by the Directorate of Forests in the three decades after 1965. In an industry effectively free of attempts by the various Environment Ministers to control illegal and damaging logging practices and one that attracted remarkably low government economic rents, this was a sector that provided huge profits.[12] Concentrated in the hands of a few large groups, forestry became an integrated industry that extended into plantation development, plywood, rayon, pulp and paper manufacturing ventures. Larger Chinese conglomerates and, as we shall see, Soeharto family associates, also flourished as partners of the major, mainly Japanese, automobile companies, operating within protective trade regimes in which tariffs of between 100 per cent and 300 per cent were applied to the import of completely built up (CBU) passenger vehicles. By the mid-1990s, the industry was dominated by Willem Soerjadjaya's Astra Group (Toyota), which accounted for 54 per cent of production. Liem Sioe Liong's Indomobil Group (Suzuki, Mazda, Volvo, Nissan) was the other big player, accounting for a further 21 per cent (*Warta Ekonomi* 11 March 1996: 19).

But Chinese business interests also expanded outside those areas directly dependent upon the allocation of monopoly status. They were to be the main beneficiaries within shifting and complicated systems of protection – investment priority lists (DSP) that reserved specific sectors of the economy, either partially or totally, for domestic investors.[13] These protective arrangements were concentrated in the non-traded goods sectors and within the larger framework of the government's ISI policies. Chinese capitalists dominated the manufacture of automobile components, motorcycles, glass, televisions, tyres, batteries, cement, electronics, synthetic yarn, steel and the engineering industries that grew around the oil industry. According to the World Bank, these were sectors where it was more logical to import. Such protective regimes provided nothing less than implicit subsidies to producers who were generating negative added value (World Bank 1984: 101, 1985: 66–71).

Although unable to play a direct and public role in public politics, it was difficult for technocrat reformers and populist reactionaries alike to override the interests of these Chinese business groups. In part this was because they were so structurally essential to investment and production that policy-makers were consistently in fear of panic and capital flight (Pincus and Ramli 1998). More specifically, their growing common interests with political patrons protected them from pressures to constrain corruption and to regulate corporate governance. By the late 1970s, the relationship between state power and these private business empires was changing. No longer content to allocate rents for simple pay-offs or to generate various forms of extra-budgetary funding, politico-business families now began to establish for themselves a social base as capitalists in their own right, as investors and equity-holders in partnership with Chinese and other business groups. Potentially, here were the makings of a ruling class in Indonesia.

Metamorphosis: politico-bureaucrat rule becomes oligarchy

The quintessential oligarch was Soeharto himself. By the time he was forced from office in 1998, his extended family had muscled their way into every pore of the business world and his private fortune had been estimated variously in tens of billions of US dollars. Although this private commercial empire really became significant only in the 1980s, as the Soeharto children came more fully into the picture, and as the state monopolies in banking, finance, public utilities and infrastructure were opened to private investors, the underpinnings of its emergence were being set in the 1970s. Perhaps the key point of entry was through his private charitable foundations, the *yayasan*. These did more than attract 'contributions' for various political uses, they were also immersed directly into the world of commerce. With shares in confiscated plantations and forestry companies seized by the government in 1966 and 1967, Soeharto's *yasasan* were also major shareholders in PT Berdikari, a company established to take over the confiscated banking and trading assets of former Soekarno crony, Markham. Operating within the orbit of Bulog and not subject to state audit, Berdikari and its subsidiary companies derived their business from Bulog contracts for flour milling, rice trading and dock handling as well as covering around 40 per cent of Bulog's insurance (Robison 1986: 90–91; Shin 1989: 130–131).

In an important sense, the *yayasan* crystallised the relationship between state authority and its ascending politico-business families. They were that halfway house within which the politico-bureaucrat need for off-budget political funding merged into private commercial interests of the families. The unique amalgam of public and private power was reflected in the mix of public officials, Soeharto family members and business associates among their directors and office-holders. As well as Soeharto himself, two State Secretaries, Sudharmono and Moerdiono, and other members of the Soeharto family, including Sigit Hardjojudanto, Bambang Trihatmodjo, Tommy, Sudwikatmono and Bulog Head, Bustanil Arifin, were directors of the most important of these *yayasan*. By the 1980s we find that Yayasan Supersemar and Dharmais and Dakab held 72.6 per cent of the shares in the former Berdikari bank, Duta Ekonomi, now reconstituted as Bank Duta, and 80 per cent of the shares in the huge Nusamba conglomerate, whose investments spread across automobiles, insurance and banking, mining, toll roads, airlines and trade, the other shareholders being Bob Hasan and Sigit Hardjojudanto.[14]

However, their economic role was soon to be dwarfed as members of the Soeharto family became more directly involved in business in their own right. Soeharto's stepbrother, Sudwikatmono, had been a shareholder and director in several of Liem's companies, including, most notably, the flour miller, Bogasari. Probosutedjo, Soeharto's brother, established a large array of companies with Chinese partners in property and contracting, chemicals and glass manufacture, poultry and the automobile industry. He also held one of the two Bulog monopolies for the import of cloves. By the end of the 1970s, the elder Soeharto children began to enter the commercial world, the two eldest, Sigit Hardjojudanto and Siti Hardiyanti Rukmana (Tutut), each becoming 16 per cent shareholders in Liem's Bank Central Asia, the leading private bank in Indonesia (Robison 1986: 342–350; *Tempo* 20 March 1982; *AWSJ* 23 November 1980: 2, 3; *Prospek* 3 October 1992: 20–26; Shin 1989: 245–255)

These initial ventures laid the groundwork for the subsequent rise of more substantial private business empires made possible as the state's grip over a range of monopolies was lifted in the 1980s. As we shall see, the Soeharto children seized important trade monopolies in plastics and in the import and distribution of oil products, and began to assemble a vast collection of contracts for distribution, insurance and shipping of oil and liquefied natural gas (LNG), marking the beginning of a virtual expropriation of those contracts and licences controlled by Pertamina (Pura 1986; Jones 1986; Jones and Pura 1986). But the Soehartos were not the only political family to establish themselves in business. As early as the mid-1970s, Ibnu Soetowo, the head of Pertamina, had built an extensive private business empire that included his son Ponco and sister Endang Sulbi. In a series of partnerships that included former military officers and Pertamina officials as well as prominent Chinese business figures, the Soetowo business empire was to extend across automobile import, assembly and manufacture, property and construction, shipbuilding and repair, engineering and fabrication as well as banking (Robison 1986: 352–353).

By the 1980s, others were poised to move into the world of private business. Among them were Information Minister, Harmoko, who was to build a significant media empire, while the family of B.J. Habibie constructed a business empire around engineering contracting to the aerospace manufacturer, IPTN, and to the Batam industrial project, both of which came under the ambit of Habibie himself as Minister for Technology and Research. By 1990, the Habibie group had become the forty-ninth ranking Indonesian conglomerate, enmeshed in partnerships with various Soeharto children and with leading Chinese groups, including Salim and Sinar Mas (Aditjondro 1995b, 1995c). The replication of this pattern occurred down the chain of political families from the central government to the regions and provinces. The former Governor of North Sulawesi and Head of the Finance and Economic Branch of the Department of the Interior in the 1970s, Arnold Baramuli, who was to become an important political ally of Habibe and Soeharto in parliament, already presided over an extensive trading and manufacturing group under the umbrella of PT Poleko.[15]

It was the introduction of Presidential Decisions (*Keppres*) 14, 14a and 10 in 1979 and 1980 that brought together into a more coherent whole the complex new relationships emerging between state power and the private interests of emerging politico-business families. Under the terms of the new Keppres, the State Secretariat (Sekneg) assumed control of allocation of contracts for supply and construction related to government projects. These were formidable powers. The primary instrument for the allocation of the contracts was the so-called Team Ten, presided over by Ginandjar Kartasasmita. Between 1980 and 1986, Rp.39.5 trillion (approximately US$20 billion at the time) of projects were processed under its auspices (Pangaribuan 1988: 9–10, 205). Winters reported that Team Ten approved Rp.52 trillion in government contracts. Some idea of the significance of this figure might be gained from the fact that domestic investment approvals in the same period totalled Rp.36.2 trillion (Winters 1996: 164).

At one level, these regulations consolidated the rapid rise of the State Secretariat (Sekneg) as a formidable institution of state power in the context of Soeharto's consolidation of his power *vis-à-vis* those competing power centres in the military that had previously gravitated around Ali Moertopo and the Ministry of Defence under General Benny Moerdani. Its rise was assisted by the strategic alliance between Sekneg and Indonesia's economic technocrats, anxious to wrest control of institutions like Pertamina from the influence of the military and to check uncontrolled allocation of rents by the military, mainly to Chinese business (Pangaribuan 1995: 35–41). Responsible directly to the President, Sekneg administered the National Economic Planning Board (Bappenas), the Investment Co-ordinating Board (BKPM), Bulog and Habibie's Technology, Research and Development Board, encompassing state-owned strategic industries (BPIS). Not only was control of the massive resources from the oil boom taken from the economic ministers in Bappenas and BKPM and shifted to Team Ten, Pertamina contracts were included under its authority in 1980 and, perhaps most significant, it was given control over HANKAM and ABRI procurements in 1985 (Winters 1996: 151–164).

Sekneg's importance to Soeharto also lay in its control over Banpres (the Presidential Assistance Fund), perhaps the key mechanism through which extra-budgetary funding was channelled to the President to be spent as he pleased (Winters 1996: 159, 60). Banpres funds were drawn from a variety of sources, including the fees from the clove monopolies and forest concessions, and spent on a variety of development projects. Some idea of the extent of Banpres may be gained from the admission by State Secretary Moerdiono in 1988 that Banpres had funded over Rp.187 trillion in projects in the period of the Fourth Development Plan (Pangaribuan 1995: 32). With rampant overpricing, there was ample opportunity for excess to be divided between officials and bidders, creating a vast pool of potential money for the personal and political needs of the beneficiaries of rents.

However, the critical impact of Sekneg and Team Ten was to be in the consolidation of *pribumi* business groups through the allocation of contracts for supply and construction. Supposedly established to assist the development of the 'weak economic group', as *pribumi* business interests were known, small players were muscled aside as well-connected individuals moved in. Recriminations about Team Ten favouring certain firms split the *pribumi* business community (*Tempo* 31 March 1988: 91, 92; 2 April 1988: 88). Not least among the new players were the Soeharto children who had established themselves as the major Pertamina contractors in this period. Others also came from powerful political families, including Ponco Sutowo and Ginandjar's brother, Agus Kartasasmita. But there were also figures from well-connected *pribumi* business families, some of whom had also been beneficiaries of Sutowo's patronage in the 1970s. Among these were Fadel Muhammad, Imam Taufik, Arifin Panigoro and, most important, Aburizal Bakrie, all of whom continued to be central players through the 1990s, after Team Ten had been dissolved (Pangaribuan 1995: 51–67; Winters 1996: 123–141).

An oligarchy?

To what extent can we consider these disparate and fragmented figures and groups an oligarchy in the longer term, able to reproduce themselves across generations, to survive the departure from office of their founder and to survive even the collapse of the regime? If they survived, would they do so as oligarchs within the same set of power relationships? For the politico-business families, the factors were slightly different to those facing the major Chinese business families. While the business interests of Ibnu Soetowo had survived his dismissal as Pertamina Director in 1975, this could be attributed to the dense business and political links with other powerful political families and Chinese business groups who continued to flourish under Soeharto's rule. But would this apply if the whole Soeharto regime fell? Why wouldn't the putative oligarchy collapse just as the business clients and power brokers of the Soekarno years had been so quickly dispossessed once that regime fell? There are several factors that we will follow through the course of the study.

The first relates to the institutional strength of the state apparatus built up over the three decades of Soeharto rule and the density of interpenetration of the families and officials. How difficult would prosecution and confiscation be in the face of a civil bureaucracy, a judiciary and a state sector so embedded with the families and committed to the system of oligarchy? Second is the growing importance of property rights. Politico-business families were now owners of capital, their interests spread across dense business partnerships with Chinese corporate groups and with other politico-business families. The families had also established business partnerships with large international corporate interests, and had placed parts of their corporate empires on the public stock exchanges. To what extent, then, did the entrenchment of property rights make it more difficult to dispossess the families in the late 1990s than it was in the 1960s?

A third factor is political. The Soeharto regime had expanded its political base outside the military and the bureaucracy into other political organisations such as the state political party (Golkar), the Young Entrepreneurs Association (HIPMI), the Association of Indonesian Muslim Intellectuals (ICMI), and the communications forum of the sons and daughters of retired military (FKPPI). To what extent would they be able to preserve the structure of oligarchy through such organisations in the context of the post-Soeharto democracy? With an organisational base in the FKPPI and HIPMI, and with links to the notorious stand-over organisations, Pemuda Pancasila, and Pemuda Panca Marga, we shall see how the old politico-business families, and figures like the disgraced former General, Prabowo Subianto, were to become players in the struggle for control of Golkar in the 1980s (*Laksamana.net* 10 October 2002).

A final factor is social. The resilience of the emerging politico-business oligarchy was greater than many supposed because it did not rely solely upon the coercive power of the state to enforce its interests. An extensive set of social interests was drawn into its ambit, flourishing on the basis of economic rents and political careers; they were the benefices of the system and its institutions. In other words, can we understand the Soeharto regime and the system of oligarchy as being embedded in a social order?

The New Order as a social order

While this new politico-business oligarchy began to define itself as the core of a putative ruling class, it also fed into a broader stratum, essentially representing the old elites and/or middle classes of a bygone era. The social origins of such elites typically lay in the minor aristocratic families, old colonial bureaucracies and Western-educated privileged groups of the late colonial period. Many of these had consolidated their positions in the early post-colonial period so that members of traditional elite and usually more urbanised middle class families were frequently leading figures in the party politics of the 1950s. They constituted an extensive 'political class', which provided not only the officials of the state and senior military officers, but also some of Indonesia's leading middle class intellectuals, professionals and, sometimes, business figures. A good example is provided from the case of the usually little-discussed East Kalimantan – here the commercially-oriented Kutai aristocracy, having been in decline by the early 1960s, resurrected its fortunes considerably under the New Order within the local bureaucracy and Golkar (Magenda 1991). While it is beyond the scope and intention of this study to provide an exhaustive inventory of the complex and often contradictory broader social base of the New Order, it is useful to identify its different elements and their varying relationships with the regime.

Although outside the inner sanctums of New Order power, many old elite and middle class families were embraced by the regime at the highest levels, with many of their members, as we shall through this study, becoming Ministers under Soeharto, or serving as advisors. Others flourished as prominent members of the professions while a few entered business and enjoyed the patronage of powerful political figures. A good example is provided by the Djojohadikusumo family, who came from the lower rungs of the Javanese aristocracy. Its late scion, Sumitro Djojohadikusumo, had been a PSI (Indonesian Socialist Party) leader and Cabinet Minister in the 1950s, and a Cabinet Minister and advisor under Soeharto. His son, Prabowo Subianto, was even to marry one of Soeharto's daughters, while another son, Hasyim, became a prominent businessman who entered into several business alliances with the Soeharto children. Another prominent family, the Kusumaatmadjas, produced two long-serving Soeharto Cabinet Ministers, Muchtar and Sarwono, the latter also a senior figure in Golkar. Others were drawn from elites that were more distant from the political heart of the New Order. Frans Seda, for example, was a Dutch-educated economist-politician, a leader of the small Catholic Party in the 1960s, and hailed from a prominent family in the Eastern Indonesian island of Flores. He served as Cabinet Minister under both Soekarno and Soeharto, and developed extensive business interests, especially in textiles. Another was Emil Salim, an economic technocrat whose family had been prominent in nationalist struggles since the days of Sarekat Islam and who was a member of West Sumatra's educated elite. The point here is that members of Indonesia's colonial and early post-colonial-era elites and middle classes latched on to the New Order juggernaut nearly from the beginning – it provided valuable patronage for them and opportunities for careers within the apparatus of power.

But it was not only those who became central to the power apparatus that were significant. Other individuals representing old prominent families flourished outside in business or the professions. For such individuals, the New Order's ability to dispense patronage by way of business opportunities – through contracts, concessions, credit, etc. – was nothing less than a magnet. Nevertheless, they were often sidelined by the new group of business tycoons with closer personal links to key figures in the New Order, most crucially Soeharto himself. One of the few to really emerge from the 1950s was Soedarpo Sastrosatomo, a former PSI figure with close links to the Sultan of Yogyakarta, who built a substantial banking and shipping empire (Robison 1986: 338–339).

It was under the patronage of Ibnu Soetowo and Pertamina in the 1970s, however, that individuals from elite Jakarta and regional families were increasingly to build important business interests on any substantial level, specifically in the construction industry. Among them were Ibnu Soetowo's son, Ponco, and Siswono Judohusodo, the son of a former Deputy Governor of Jakarta, Dr Suwondo. But this was a varied group of younger clients that were associated with the Young Entrepreneurs Association (HIPMI), which also included figures from regional political and business families and former student politicians.[16] Although their patron, Ibnu Soetowo, lost control of Pertamina in 1975, the so-called Pertamina contractors remained an identifiable group often involved in joint investments and close to political figures who would become powerful controllers of rents, notably, Ginandjar Kartasasmita. As we have seen, they were to be revived under Team Ten and the provisions of Keppres 10 in the early 1980s.

Another element of the old elites drawn to the New Order – at least its outer fringes – had a historical presence in the vast rural hinterlands and towns in which small traders and religious notables constituted the traditional source of authority. Though overlapping with previous categories, it grew out of the social base of such major religious movements as the Muhammadiyah and the largely Java-centred Nahdlatul Ulama (NU). Abdurrahman Wahid, for example, whose grandfather founded the NU during colonial times, was drafted into Indonesia's supra-parliament, the MPR, in the 1980s, although he would later play the role of opposition leader and, of course, become Indonesia's fourth President. Before him, the veteran NU politician and religious leader, Idham Chalid, had worked closely with the New Order as Chairman of the MPR from 1972–1977. At the local level, these organisations' vast network simply fed into the New Order's increasingly extensive system of patronage. Schiller, for example, notes how old and entrenched families of NU notables in the town of Jepara, in Java, continued to be salient in local politics at the same time as local office was developing steadily as a means of furthering private wealth on the basis of public resources – made abundantly available by the New Order's rural development programme (Schiller 1990: 395–419).

It is significant that having made forays into national politics in the 1950s, the leaders of organised Islam representing those small town trading and rural elite interests expected to play a larger role in the New Order, especially given that they had played an important part in the destruction of the PKI in an alliance with the military. They were to be disappointed. While many such leaders were indeed given a place in the formal structures of power with the establishment of the PPP in 1973, the Muslim traditional petty bourgeoisie were effectively in decline throughout much of the New Order. As we shall see, an uneasy relationship existed between organised Islam and state power until the early 1990s when the growing presence of a new urban-based Muslim middle class of bureaucrats, professionals and intelligentsia – that had benefited from the New Order's economic successes – began to provide a natural pool from which to draw new apparatchiks during the last decade of Soeharto's rule.

But it clearly wasn't just the old NU and Muhammadiyah/Masyumi networks that latched on to the New Order at the local level. Takashi Shiraishi, for example, notes that it was the PNI local leadership that benefited from the demise of the PKI in the old communist party stronghold of 'Dukuh', near Klaten, Central Java. Here, many old local elites simply transformed themselves from PNI functionaries to become local Golkar apparatchiks (Shiraishi 1994: 87–88).

It is perhaps this dimension of New Order rule – local patterns of patronage and political mobilisation – that has remained the most understudied, though some analyses do exist. An enlightening study by Antlov in 'Sariendah' village, West Java, suggests that 'political offices' there were only 'distributed among a restricted number of families'. The most important villager, Antlov further reveals, was the headman and local Golkar leader – who came from an old wealthy family that owned hundreds of hectares of land as early as the 1930s (Antlov 1995: 145). The degree to which such a pattern is replicated in other cases constitutes an important missing element in the study of New Order Indonesia as a social order.

It is clear, however, that Jakarta-based and local elites forged extensive links to the new politico-business families of the New Order through the regime's formal corporatist political institutions, and through the various front groups that offered opportunities to ambitious civilian politicians. Indeed, such fronts were increasingly the instruments through which the middle classes and propertied and business interests accessed the patronage and protection of the state, as we shall see in Chapter 5. However, the relationship with the state was often an ambivalent one. Many were alienated as they were denied entrance into the central corridors of power or were disenchanted by the New Order's arbitrary rule and by rampant corruption. Thus, it was from these ranks that many of the regime's most articulate public critics were drawn. Others became critical NGO leaders or opposition figures.[17] But the ambivalent relationship also meant that many of these critics were vulnerable to co-option. Individuals such as Sarwono Kusumaatmadja, Marzuki Darusman and Rachmat Witoelar were to walk a fine line between reformer and *nomenklatura* as they engaged in Golkar politics in the 1980s. Perhaps the most interesting example of the type is provided by Adi Sasono, who came from a family with a long history in Masyumi politics. A leading critic of the New Order in the 1970s, his career was to progress into the ranks of the government-sponsored Muslim organisation, ICMI, and culminated as he became Minister of Co-operatives in the short-lived Habibie government.

Tensions within the regime

Notwithstanding the success of the New Order in co-opting support from outside the immediate regime, it was evident by the 1970s that important tensions were emerging. These were most evident in growing resentment of the regime by elements within the middle class elites over questions of democracy and accountability. Simmering hostility also surrounded the issue of economic policy as economic technocrats and their allies in the World Bank and amongst foreign investors and Western governments clashed with economic nationalists in the regime.

Increasing frustration amongst the urban middle classes was at the heart of student unrest and disturbances in 1971, and again in 1973 and 1974 when these concerns with corruption and arbitrary rule merged with more populist and xenophobic resentments on the part of declining *pribumi* business interests. As this potentially explosive cocktail spilled onto the streets of Jakarta in January 1974, the government responded with widespread arrests and the closure of newspapers. Subsequent student unrest in 1978 was met with more extensive measures to control political activity on the campuses and broader controls on political activity. It is true that these tensions were to remain under the surface, rising to the fore again as soon as the regime faltered in 1998. But they were not to be the forces that drove change in the 1980s or even determined the events following the economic crisis and the fall of Soeharto.

Through its apparatus of emergency rule and the overwhelming strength of its security forces, the government was able to control liberal and populist opposition. There were no avenues within which an opposition might organise to challenge or change the government. However, the regime also provided the flexibility for middle classes to survive. Although individual students and critics were imprisoned and the media controlled there was no pervasive reign of terror or violence against the middle classes, no executions or even systematic surveillance of the type that characterised regimes in Iraq, Syria or Iran. Nor was there any 'dirty war' carried out by right-wing groups as had occurred in Argentina, Chile and Brazil – at least until the final days of the regime. Many exiled students forced overseas returned with degrees from leading US universities to take up successful careers in business or the legal sector. Throughout this period, the economy kept growing and the prosperity of the middle classes increased.

At another level there was growing tension between market reformers in the World Bank and the so-called 'economic' Ministers in Finance and the Economic Planning Board, Bappenas, and the economic nationalist Ministers who had increased their grip over policy through the 1970s. The World Bank and other liberal critics became increasingly concerned with the proliferation of protected, high-cost industries in a range of sectors. They urged the removal of fuel and food subsidies and a reduction in state investment in large resource and industrial projects, and called for the abandonment of an industry policy that produced inefficient and costly industries represented in the steel and automobile sectors (World Bank 1981, 1984: 101, 1985: 66–71). As we have seen, these reports were met with hostile responses from economic nationalists who continued to argue that a vigorous and independent economy would not emerge from the free operation of global markets dominated by large international corporations, but required active state intervention to generate a broad industrial base and the capacity to produce technology.

But these continuing policy disagreements with liberal critics did not provoke any radical change in the prevailing system of state-led capitalism. The high levels of oil revenue that flooded into government coffers through the 1970s and early 1980s enabled Indonesia to drive its own policy agenda and finance its adventures in upstream industrialisation and technology industries. Indonesia became less reliant on foreign investment and loans and less susceptible to reformist pressures from external sources. Nor was there a serious conflict between the government and foreign investors even though many of them spoke out against increasing sectoral restrictions on foreign investment, requirements to take joint venture partners, over-regulation by government, corruption and bureaucratic incompetence (see for example, *JP* 5

August 1986: 7; 28 August 1986: 1, 7). For the most part, however, they found a lucrative and secure haven alongside policy restrictions and rent-seeking regimes of Indonesia. This was especially so in the oil and resources sector where they were concentrated before the Plaza Accord opened the export-manufacturing sector.

Ironically, the real fractures were not to emerge between the state and any liberal middle class, or under growing pressure from the World Bank and foreign investors, but from within the regime itself. As politico-business families formed and extended their alliances with Chinese corporate groups, the policies and gate-keeping institutions of the state were increasingly harnessed to the task of constructing and protecting their new corporate empires. In the short term, the new politico-business families necessarily became embroiled in a deepening struggle with military and other elements in the state apparatus over the control of rents as they attempted to expropriate state power to their private interests. The apparent liberal reforms of the 1980s observed by Hutchcroft were not so much the triumph of a reforming state able to push its agendas against predatory interests but a move to open lucrative state monopolies to these new private cartels and oligopolies. It was part of a victory of the new state oligarchies over the old state capitalism that was to be consolidated also at the political level. These struggles are the focus of the next three chapters.

Notes

1 New flows of aid and loans were to constitute Rp.91.1 billion or 27.2 per cent of government revenue in 1969/1970, declining to Rp.1,850 billion or 11.9 per cent in 1982/1983 as the oil boom inflated the budget (World Bank, Indonesia's Country Reports, various issues). Foreign investment, excluding that in the critical areas of oil and gas, amounted to approximately 50 per cent of total approved investment in the period to 1973 (US$2.8 billion). Most was in resources and ISI manufacture. See Robison 1986: 142–145.

2 Mobutu's Zaire had presented a caricature of the neo-utilitarian image of how state officials act; and where, as Mobutu himself noted: 'Everything is for sale ... holding any slice of public power constitutes a veritable exchange instrument, convertible into illicit acquisition of money or other goods' (cited in Evans 1995: 46).

3 In his definitive analysis of the parliamentary period and its decline, Herbert Feith (1962) saw the central and decisive struggle as that between rational modernisers intent on introducing the sort of legal-rational system of authority envisaged by Weber and 'solidarity-builders' prepared to invoke nationalist and patrimonial instruments to complete the task of nation-building.

4 Karl Marx, 'The Critique of Hegel', cited in Draper 1977: 81.

5 Schmitter's definition of authoritarian corporatism captures many of the central features of political organisation under the New Order.

> Corporatism can be defined as a system of interest representation in which the constituent units are organised into a limited number of singular, compulsory, non-competitive, hierarchically ordered and functionally differentiated categories recognised or licensed (if not created) by the state and granted a deliberate representational monopoly within their respective categories in

> exchange for observing certain controls on their selection of leaders and articulation of demands and supports.
>
> (1974: 96)

In the case of Indonesia, such groups covered its vast civil service, requiring them to vote for Golkar. Youth, women, artists, intellectuals, students, business, labour and peasants were all swept into state controlled associations.

6 For studies of the ongoing struggles in which Indonesia's technocrats sought to overcome opposition from nationalists and populists, as well as those predatory business intents growing up around the state see Sacerdoti 1980: 44–50; Rowley 1987: 70–75; Schwarz 1994: ch. 4

7 See Chalmers and Hadiz 1997: Chapter 3 for some of the basic statements of the economic nationalist agenda by its leading advocates, Ali Moertopo, Jusuf Panglaykim, Soedjono Hoemardani and A.R. Soehoed. See also Robison 1986: 147–159.

8 The oil boom consisted of two main phases. Between 1973 and 1974, the international price of oil rose from around US$3 per barrel to US$12 per barrel. Between 1979 and 1981 the price rose again, from around US$15 per barrel to over US$40 per barrel. As a result, Indonesia's gas and oil exports leapt from US$1.6 billion, or 50.1 per cent of total exports, in 1973 to US$18.4 billion, or 82.6 per cent of total exports, in 1982. At the same time, government revenues from oil and gas taxes increased from Rp.382 billion, or 39.5 per cent of total revenues, in 1973 to Rp.8.6 trillion, or over 70 per cent of total government revenues, in 1981–1982 (World Bank 1985: 207; Robison 1987: 28–29, 44–46; Winters 1996: 120–121).

9 By the early 1990s, investment in these so-called mega projects, committed or planned, was estimated at US$80 billion. According to Economics Co-ordinating Minister, Radius Prawiro, in September 1991, US$31.3 billion of this sum was private investment. Pertamina projects, mainly in refining and petrochemicals constituted US$22.9 billion of the planned state investment (Muir 1991: 19).

10 Among these were Soedjono Hoemardani, who became a central figure in the powerful security and intelligence agency, OPSUS, and a political lynchpin in a complex network of Chinese-Indonesian and Japanese business interests that surrounded it. Former Head of Military Finances, General Sofjar, assumed control of Yayasan Dharma Putra, the business group of the military's strategic command, Kostrad, while General Soeryo became head of Pekuneg, the body responsible for confiscated enterprises. Such military officers, including General Alamsjah, from 1969–73, and General Sudharmono were also to head Sekneg, a body that not only provided a critical secretariat to the President but was responsible for Banpres (*Bantuan Presiden*) a strategic source of extra-budgetary funding independent of the financial Ministries. Soeharto himself and later, General Tirtosudiro, headed Bulog in its early years, a measure of its critical importance as a source of political funding (Robison 1986: 250–269; Crouch 1978: 274, 290–295).

11 See Crouch 1978; Robison 1986; Shin 1989; Schwarz 1994, 1999.

12 Compared to the 85 per cent rents imposed in the oil and gas industry, forestry was liable only to a 17 per cent royalty (World Bank 1993a: 44–49; Ramli 1992; *Tempo* 26 October 1991: 26–32; *Far Eastern Economic Review* (hereafter *FEER*) 12 March 1992: 45).

13 Foreign investors were increasingly restricted to those sectors requiring special technologies and skills or large amounts of capital not available domestically, or were located in sectors with export earning potential. They were excluded from sectors where domestic business was considered to have the capacity to invest, such as textiles and foodstuffs. They were excluded also from sectors considered strategic, such as

media and communications and infrastructure development. In many cases, foreign investors were required to take local partners (Rice and Hill 1977; Suhartoyo 1981).

14 For details of the complexities of the *yayasan*, see Robison 1986: 345–346; Shin 1989: 130–131; Schwarz 1990: 62–64; Government of Indonesia 1985: 1–7; Yayasan Dharmais 1989–1990; Yayasan Supersemar 1989–1990.

15 These will be examined later in this chapter and in Chapter 3.

16 Among them were Guntur Soekarno, son of the former president, Suryadharma Tahir, son of General Tahir, and Tengku Sjahrul, the son of disgraced former Soekarno-era businessman Jusuf Muda Dalam. Others included former student leader, Fahimi Idriz and Sulawesi businessman Fadel Muhammad (Robison 1986: 355)

17 Despite tight controls on the press and on political organisation, several middle class figures, often with higher degrees from overseas universities, provided ongoing criticism of the regime and its policies from redoubts in the professions. Amongst these were the economist Sjahrir, and the lawyer Mulya Lubis and Adnan Buyung Nasution.

Part II

THE TRIUMPH OF OLIGARCHY 1982–1997

3

HIJACKING THE MARKETS

The seemingly untroubled evolution of the Soeharto regime was interrupted by a sudden collapse in oil prices in 1981/1982 and again in 1985/1986. With growing pressures on the budget and current account it appeared that the vast system of state ownership and public monopoly and the pervasive and protective trade and financial regimes that had constituted its backbone over the previous decades would become unsustainable.[1] Major structural adjustments were required to stimulate non-oil exports, generate new revenue sources and replace state investment as the engine of growth.

In a series of reform packages beginning in the mid-1980s, the government deregulated financial and trade regimes, relaxed foreign investment restrictions and opened a range of former state monopolies to private sector investment. Reforms were driven by the requirement that Indonesia develop international competitiveness in a range of non-oil sectors, especially manufacture. Import monopolies were dismantled where they affected inputs essential to the low wage export manufacturing industries, such as textiles, footwear and electronics, that had dramatically emerged as investors from Northeast Asia flooded into Indonesia in the wake of the Plaza Accord. Among those affected were important monopolies in plastics, tin plate and steel products held by well-connected business groups within and close to the Soeharto family (Robison 1992: 75–78). By 1995, import controls through quotas and sole import status had been replaced substantially with tariffs that were themselves brought down to levels well in line with the pace of trade reform in the region (World Bank 1995: 40–42).

Other reforms reduced the number of sectors closed to foreign investment and relaxed requirements that foreign firms progressively divest ownership to local partners. Power generation, telecommunications, ports and roads, upstream industries, including petrochemicals, long regarded as strategically sensitive, were opened to the private sector.[2] A surge of foreign and domestic private sector investment followed, coming initially into low wage export manufacture and later into large upstream industrial and resource projects and into public utilities and infrastructure projects, sectors that were to be dominated by Japanese, European and US companies.[3] Meanwhile new banking laws in 1983 and 1988 removed controls on loan and deposit rates at state banks and eliminated restrictions on the entry of new private banks (Rosser 1999: 99). Between 1988 and 1994, the number of banks increased from 111 to 240.[4] There was even an attempt to reform the administrative apparatus. Presidential Decision 4 of 1985 transferred the assessment of duty in the Indonesia ports to the Swiss firm, Société Générale de Surveillance, a move that sacrificed well entrenched and powerful elements of the Customs Department in the name of reducing export and import costs (Dick 1985: 10–11; Sjahrir 1985).

Yet the impressive flow of reforms did not lead to a flowering of liberal markets. As trade deregulation began to stall it became clear that important domestic monopolies and cartels were surviving. Even though financial reform opened the floodgates for a surge of credit into the economy, it proved impossible to impose prudential controls in the banking system. Although private interests now operated in many of the former areas of state monopoly, powerful state gate-keepers continued to determine the allocation of licences, contracts, distributorships and credit. Public monopoly had simply become private monopoly. At the same time, strategic sectors of the economy, in ISI and upstream manufacture, and in domestic trade and distribution, remained stubbornly resistant to deregulation.[5] Paradoxically, those powerful politico-business families and cronies, the expected victims of reform, were to be precisely its main beneficiaries.

Despite their periodic complaints about corruption and the slowing of the reform agenda, there was little major Western governments could do about it. Indeed, Indonesia's state budget was kept intact through this period by industrial creditor nations through the Consultative Group on Indonesia (CGI) and Japanese Overseas Development Aid (ODA), providing aid and loans fluctuating between around 15 per cent and 30 per cent of state revenues.[6]

The unexpected consequences of the reform programmes were explained by neo-classical economists within the World Bank and elsewhere as the consequence of not enough deregulation or of technical errors in sequencing reforms; deregulating trade and finance regimes before the real sector (World Bank 1996: xxvii; Hill, H. 1997; Soesastro 1989; Bhattarcharya and Pangestu 1992). However, the particular sequencing of reforms was not just a technical matter, it reflected the structural opportunities available to reformers in the context of prevailing configurations of power and interests (Robison 1997: 36). Financial reforms came first precisely because they were politically possible while an assault on the well-defended domestic manufacturing cartels or the state banks remained out of reach. As Pincus and Ramli (1998) have pointed out, not only was the sequencing of reforms beyond the control of the technocrats, decisions to open capital markets and balance budgets removed from their armoury some of the most important fiscal and monetary levers over the private sector.

In essence, the reform agenda was hijacked by those domestic politico-business alliances that had emerged in the 1970s and who now began to find the system of state capitalism within which they were nurtured a constraint upon their development. State monopolies and nationalist policies in certain areas now stood in the way of their entry into rapidly evolving sources of finance in global capital markets and those lucrative opportunities in domestic banking, public utilities, telecommunications and transportation. Applied selectively to leave intact the basic structure of the rent-seeking state and the principle of political capitalism, deregulation now suited the new politico-business families and conglomerates.[7]

In short, the attempt to liberalise the economy was made before the complex coalitions of state and social power that had underpinned the existing regimes had been dismantled or at least fractured, and where no coherent political alliances had formed to drive the reform agenda. While the World Bank remained oblivious to this simple point it was one not lost on commentators within Indonesia. Economist Anwar Nasution, for example, remained sceptical that deregulation packages would work where the monopoly and oligopoly rights of powerful state and private interests remained intact (Nasution, Anwar 1995). Laksamana Sukardi (later Minister for State Enterprises and Investment under both Wahid and Megawati) was clear that, 'So long as the ruler and businessman continue collaborating, the Central Bank and other Ministries will not be independent' (*Prospek* 6 March 1993).

How oligarchy triumphed: what they did

Far from putting an end to the system of 'political capitalism' and breaking up the emerging politico-business oligarchies, the deregulation process and deepening engagement with global financial markets proved essential to their dramatic growth and concentration through the 1980s as well as their diversification into new areas of investment. How did this happen?

- By retaining their authority over strategic gate-keeping institutions of the state, politico-business families not only preserved their authority over the allocation of rents but ensured that such power was now exercised in their private interest.
- In critical instances, the families and conglomerates successfully concentrated themselves outside those sectors most vulnerable to restructuring, within resource-based industries, domestic trading monopolies and cartels and in the domestic real goods sector and infrastructure industries.
- They now dominated the sectors of former state monopoly, moving into banking and financing and finding lucrative new opportunities in so-called mega projects in infrastructure and public utilities, and in upstream manufacture as partners in consortia that included foreign banks and engineering conglomerates.
- They forged increasingly complex and extensive business empires based on partnerships between the politico-business families, big Chinese conglomerates and foreign investors.

- They secured access to global and domestic financial markets, replacing oil revenues with largely unregulated commercial bank loans and equity in capital markets as the main source of private corporate growth.
- Most important, they began to insulate and diversify their business empires by expanding overseas, publicly floating parts of their business empires on the stock exchange and integrating some holdings with global investors.

Escaping the market: building trade and investment monopolies in the domestic sector

Despite their early victories, the advocates of free trade had not secured a complete triumph; far from it. The politico-business oligarchy was able to avoid sectors where pressures for reform were most intense. The World Bank continued to report high average rates of concentration of ownership in the manufacturing sector in areas of low export orientation, and observed that powerful interests kept the markets at bay with a range of protective arrangements. These included: cartels (cement, paper, plywood and fertiliser); price controls (cement, sugar and rice); entry and exit controls (plywood, retail marketing); exclusive licensing (clove marketing, wheat flour milling, soymeal); public sector dominance (steel, fertilisers, refined oil products) (World Bank 1993a: 91–92, 1995: 45–50).

The concentration of private conglomerates behind state-sponsored cartels became an increasing issue of public debate within the press and among Indonesia's public intellectuals (*Kompas* 5 July 1995: 2; *Warta Ekonomi* 3 July 1995: 18–22; *Gatra* 15 July 1995: 21–30; Sjahrir 1992: 9–11). Yet this pressure did little to change the situation. In particular, Bulog remained a pivotal institution in the domestic food industry with its network of sole distributorships, its pricing controls and minimum domestic content requirements. Sheltering under its umbrella, Liem Sioe Liong's flour milling monopoly, Bogasari, remained intact despite an ostensive deregulation of the flour milling industry in 1993.[8] Bogasari's lucrative 30 per cent margin on milling enabled its owners to reap windfall profits and provided Liem with a strategic position in the domestic food industry. By 1994, he controlled 75 per cent of the noodles market, 33 per cent of milk and 20 per cent of baby foods (World Bank 1995: 43–44; Schwarz 1994: 110–112).

Forestry too, remained insulated from trade and regulatory reform. With the banning of raw log exports in 1980, forestry production now offered huge opportunities for integrated forest industries in plywood, pulp, paper and rayon. As undercapitalised concession-holders, including most of the old military ventures, were forced to sell to the bigger conglomerates or to enter joint ventures or management agreements, the Soeharto children now stepped in to forge some of their largest business partnerships with the big Chinese forestry groups.[9] These major groups were assured of a high degree of control over the forestry resource base because the allocation of concessions (HPH) continued to be a closed process without public and transparent mechanisms of tender, and relatively free of environmental constraints (World Bank 1992: 44–49; Indonesian Environmental Forum 1991; *Tempo* 26 October 1991: 26–32; *Prospek* 17 August 1991: 86–96). They were able to ignore most attempts by the Environment Ministry to control illegal and damaging logging practices, unchecked fires remained a favourite source of land clearing.[10] Most important, the forestry industry attracted remarkably low government economic rents. Compared to the 85 per cent rents imposed on the oil and gas industry, forestry was liable to only 17 per cent (World Bank 1993a: 44–49; Ramli 1992; *Tempo* 26 October 1991: 26–32).

Attempts to deregulate the heavily protected automobile sector, one of the most cherished symbols of the import substitution strategy of the 1970s, were also resisted. A deregulation package was announced in mid-1991, only to be annulled a little more than a month later (*Jakarta Jakarta* 16–22 March 1991). This was not surprising given that it was an industry in which Liem Sioe Liong and the Soeharto family had gained an important foothold. Their grip on the industry was tightened after control of the giant Astra group was taken over in 1995 by a consortium including the Nusamba group, Prajogo Pangestu and government banks (*Warta Ekonomi* 11 March 1996: 19). Relying on highly protective trade regimes in which tariffs of between 100 and 300 per cent applied to imports of completely built up (CBU) passenger vehicles, these groups were in no position to compete in open markets.

When refuges were breached or abandoned, the state proved willing to rescue well-connected beneficiaries who found themselves stranded. Confronted with the removal of his monopoly on the import of cold-rolled steel products and the spiralling debt of his joint venture with Krakatau Steel in cold-rolled steel, reaching US$610 million by 1988, the government assumed Liem's 40 per cent investment in the project and indemnified him from future debt obligations (Schwarz 1994: 111–112).[11] Liem also found himself in trouble in his cement ventures. With the costs of servicing overseas debts increasingly difficult at a time of slackening domestic demand, the state stepped in to buy 35 per cent of Indocement at a cost of US$350 million. It also provided a US$120 million loan from four state banks that effectively paid off Liem's expensive US dollar loans and replaced them with cheaper Rupiah loans now insulated from concerns about adverse currency shifts (*Kompas* 14 August 1985: 1; *Sinar Harapan* 22 August 1985: 1).

But the survival of the politico-business families and conglomerates did not depend simply on barricading themselves into sectors insulated from global markets. With the ending of state monopolies in public infrastructure and utilities and in upstream manufacture, they were to move into telecommunications, power generation, construction and operation of roads and ports and air transportation and petrochemicals. These involved huge investments and took the form of mega projects involving consortia of foreign investors and bankers together with local suppliers and sole agents. By the early 1990s, investment in mega projects committed or planned was estimated at US$80 billion of which US$30 billion was to come from the private sector.[12]

Controlling the strategic gate-keeping terminals

It was critical for the politico-business families that they resisted attempts to reform the state apparatus and ensured their authority over those gate-keeping institutions that allocated licences, credit, monopolies and contracts. The first of these tasks was to prove relatively easy and the sorts of reforms imposed on the Customs Office, described earlier, were not extended into other areas. Efforts by institutions such as the World Bank and other agencies to induce reform by intensive programmes of training, notably in the Ministry of Finance and Bank Indonesia, had little impact on their performance (Hamilton-Hart 2001: 77, 78). However, capturing the strategic gate-keeping institutions from entrenched state managers and the military was more difficult.

As discussed in the previous chapter, the establishment of Sekneg's Team Ten was to be a decisive move in shifting control over public procurement and contracting into the hands of Soeharto and those officials surrounding him. This influence was to extend into other key institutions, marked in part by the movements of Ginandjar Kartasasmita, first to become Chairman of the Investment Co-ordinating Board (BKPM) and later Minister for Resources and Energy (Pangaribuan 1988: 154–159; Winters 1996: 165–170). Another new gate-keeping terminal was to emerge under Soeharto's close lieutenant, B.J. Habibie, who now wielded enormous power over the allocation of contracts and monopolies, presiding over heavily subsidised advanced technology industries in aircraft and shipbuilding as head of the Board for Technology, Research and Development (BPPT) and later as Minister of Research and Technology. The dominance of this camp reached a high point in 1995 when, under Presidential Decision 6, the day to day operation and control of public sector procurements was transferred from the Co-ordinating Minister for Economics, Finance and Development to the Economic Planning Board, Bappenas, under Ginandjar (Nasution, Anwar 1995: 5).

But the influence of Soeharto and the new politico-business oligarchies soon extended beyond these institutions across the whole state apparatus. As we shall see, no Ministry or state corporation could resist the demands of the Soeharto family as they moved more forcefully into business, and no monopoly or contract was beyond their reach. This new level of audacity reached even the smallest and seemingly most insignificant crevices. Soy meal milling, the distribution of fertiliser pellets, and even the collection of birds' nests came within the ambit of the Soeharto octopus.[13] In a virtual return to private tax farming, the Soeharto family also gained brief control of the state lottery and collection rights for automobile registration and licensing (*Forum Keadilan* (hereafter *FK*) 11 November 1993: 9–16; *Editor* 24 April 1993: 17–29). A controversial move by the Governor of Bali in 1995 to give Soeharto's grandson, Ari Sigit, a monopoly to collect beer tax ironically provoked disputes with his uncles in the hotel and tourism business (*AWSJ* 27 January 1996: 1, 4).

The real targets were, however, those key state institutions that stood astride access to the new lucrative mega projects in infrastructure and upstream manufacture, including Pertamina, PLN and the Ministries of Telecommunications. Required to deliver licences, concessions and contracts to the Soeharto children and leading Chinese conglomerates, directors and officials were dismissed and replaced where they attempted to resist.[14] Ironically, though, the ascendancy of the new private interests and their increasing expropriation of the apparatus of state capitalism required that the centralised system of state economic and political power remain intact.

Privatising state monopoly

The World Bank and the technocrats had always regarded the privatisation of the state sector as a priority, and not only for ideological reasons. As recognised by Finance Minister, Radius Prawiro, in 1995, the large number of poorly managed and inefficient state companies, classified by the government itself as unsound, constituted a clear financial burden on the state (*Tempo* 6 January 1990: 100–103; see also World Bank 1995: 51). At the same time, selling well-performing companies in local and international capital markets offered the government potential fiscal windfalls.[15] Influential business figures also made no secret of their interest in privatisation. Bambang Trihatmodjo proposed that the transfer of state assets backed with state bank credit be used as a deliberate tool for shifting the balance of power towards *pribumis*, an interesting legitimation of the Soeharto family's success in picking off several of the most commercially lucrative of the state enterprises (*Media Indonesia* 24 May 1991).

Yet by 1995 there were still 180 public corporations with a book value of assets totaling US$140 billion, producing 15 per cent of GDP (World Bank 1995: 29, 51). This was not simply the result of resistance by those officials entrenched in this sector. The bulk of state sector companies were hardly attractive commercial propositions for private purchasers, many of them unaudited and unprepared to compete in the open market. At the same time, the enthusiasm of liberal reformers for the sale of the more successful state companies to private purchasers was dampened as it became clear that where state companies were being sold they went to well-connected conglomerates and politico-business families at low prices and without open and transparent divestiture procedures (*Prospek* 18 May 1991: 1; *Warta Ekonomi* 2 August 1993: 44–45; *Editor* 13 June 1992: 75; World Bank 1995: 49; Hill, H. 1987: 23). In the prevailing circumstances, privatisation effectively meant strengthening the major politico-business families and Chinese conglomerates without necessarily satisfying the liberal reformers' objective of a shift to free markets. This was recognised by reformist ministers, including Finance Minister, Mar'ie

Muhammad, who began to talk of the state sector as a balance against concentration and conglomeration (*Warta Ekonomi* 2 August 1993: 44–45).

The state sector was to survive primarily because it remained indispensable. The World Bank recognised, for example, that the state would continue to provide the bulk of investment in public infrastructure (World Bank 1992: 114, 1993a: 103, 1995: 71). Equally important, strategic state corporations and agencies, including the state banks, were more valuable alive than dead to the politico-business oligarchy. They provided the perfect mechanism for a permanent haemorrhaging of public wealth into private hands, providing the conduit for an ongoing flow of contracts and distributorships where overpricing and over-invoicing guaranteed windfall profits. State corporations and agencies subsidised the entry of the oligarchy into sectors of the former state monopoly in public utilities, infrastructure and upstream manufacture. They provided cheap credit through the state banks, made inputs available at below-market rates and purchased outputs at inflated prices.

Once in control of key state corporations powerful interests could hand over to selected private business interests the agency functions, distributorships and contracts formerly carried out from within. This was no simple exercise in outsourcing. Decisions were usually highly opaque and involved highly corrupt relations between business and officials. Nowhere was this process illustrated more clearly than in the case of Pertamina. As we have seen in the previous chapter, Pertamina had provided the umbrella for the rise of several domestic business groups under the *aegis* of Ibnu Sotowo in the early 1970s. By the mid-1980s, however, it had become the pivot around which the Soeharto family companies (the so-called Cendana group) began their dramatic rise to corporate prominence as importers, exporters and distributors of Pertamina oil and fuel products, shipping oil and gas on behalf of Pertamina and insuring its tankers and other operations (Pura 1986). This was no ordinary outsourcing operation intended to reduce costs and increase efficiency. On the contrary, as former Pertamina Director of Exploration and Production, Priyambodo Mulyosudirjo, claimed in 1998, Pertamina had been forced to allocate contracts without tender to the Soeharto family that added millions of dollars to their cost (*JP* 10 October 1998: 1).

The extent of the haemorrhage was indeed massive. A PriceWaterhouse Coopers (PwC) audit conducted in 1998 found that Pertamina officials had received US$128 million in commissions as a result of such collaboration in the period 1996–1998 alone. Former Minister of Oil and Energy, Ida Bagus Sudjana, estimated that the fees given to the oil trading companies under the control of Cendana, Permindo and Perta were running at US$497.2 million per year just before the crisis. Losses incurred on Pertamina through these trading operations as well as shipping and insurance contracts were estimated by the same PwC audit at around US$56.8 million in 1997 and US$73 million in 1998 (*Tempo Online* 12–18 February 2002).

The same pattern was repeated elsewhere. State-owned airlines, Garuda and Merpati, found themselves forced to purchase and lease aircraft through brokerage companies owned by the Cendana group and their associates, including members of the Bakrie family, at a significant premium over market prices (*Gatra* 4 November 1995: 21–27; *Tempo* 25 January 1992: 22–26). Other Cendana companies were awarded insurance business for the airlines and for the Palapa satellites. Several of the Soeharto children now became involved in the arms procurement business, supplying aircraft and tanks to the Indonesian armed forces at highly inflated prices.[16] In controversial circumstances, Bambang Trihatmodjo's Bimantara group became the operator and manager for state-owned satellites (*Prospek* 16 January 1993: 32; *Tempo* 6 March 1993: 90–91; *AWSJ* 15 April 1994: 1, 9–10; 11 February 1995: 1, 2; Schwarz 1994: 149). Government plantations sold 90 per cent of their palm oil production to Tommy Soeharto's PT Mindo at prices reportedly below prevailing market levels (*FEER* 24 August 1989: 64, 65).

Ministries and other government agencies now allocated to pre-selected private bidders, often without public or transparent processes, the licences and contracts that provided access to those major infrastructure and upstream industrial projects that had been previously the preserves of state monopoly (World Bank 1995: 79; *Editor* 13 June 1992: 75). Apparently able to capture these licences at will, the Soeharto children moved into public utilities, including power generation, the water supply business and construction of ports and roads. Most spectacularly, Siti Hardiyanti Rukmana's Citra Lamtorogung group became the leading builder and operator of toll roads, securing over Rp.6.9 trillion worth of domestic and international toll road projects by 1996 in partnership with the state road builder, Jasa Marga (*Infobank* May 1996: 80–81; *AWSJ* 21–22 October 1994: 4; World Bank 1995: xiv–xvii).

It was their access to licences and government contracts that made the politico-business families an indispensable element for foreign investors hoping to participate in the large infrastructure projects that emerged in the 1980s and 1990s. For their part, the local partners received subcontracts for supply and construction and often paid no cash for their equity in the projects or received soft credit packages from the foreign partners. Widespread overpricing in subcontracting was to be an important means of shifting public funds into private pockets. Nowhere was this pattern clearer than in the telephone industry. When over US$2 billion of contracts were awarded in 1990 for supply, construction and local manufacture in this sector, Cendana companies were involved in each of the consortia, scooping up most of the local sole agencies and subcontracts as suppliers and manufacturers (*FEER* 8 March 1990: 54; 9 August 1990: 54; 24 January 1991: 40–1; *AWSJ* 15 January 1991: 1; Schwarz 1994: 144–5). Soft credit packages offered by one of the consortia, Sumitomo and AT&T, to their local partners were revealed to contain unspecified excess funds of US$97 million and US$90 million respectively (*FEER* 24 January 1991: 40, 41).

It was, however, in the power generation industry that perhaps the most lucrative opportunities were opened to private interests. According to another of PLN's former Director Generals, Djiteng Marsudi, PLN was forced to buy electricity in

US dollars at prices higher than those it was permitted to charge public consumers and in circumstances of an existing over-capacity (*FK* 29 December 1997: 12–21; *JP* 19 February 1998: 8). Marsudi also claimed that most contracts between power providers and PLN 'were given without tender. I was in one way or another forced to sign the contracts' (*JP* 8 June 1998: 1). The 'take or pay' contracts to supply electricity to PLN negotiated with pre-selected bidders provided an explicit state subsidy and insulated investors from currency fluctuations. It is little surprise that the companies of the Soeharto family and their associates were involved in all of the 26 licences awarded for new power generation projects.[17] Issued without public tender and with state guarantees of markets and pricing, the energy projects provided local partners with foreign suitors willing to stake their equity and with huge windfalls from overpricing.[18]

Entering the world of mega projects inescapably involved big investments. For Indonesia's business oligarchies this move was to be financed by debt, prompting a borrowing spree that began in the late 1980s, much to the concern of technocrats and liberal critics (*Infobank* February 1997: 38–43; December 1995: 160–161; Nasution, Anwar 1992; Harris *et al.* 1992). Initially, this finance was provided through state banks and pension funds, but increasingly the politico-business families forced their way into new channels of finance capital, particularly into the new, highly mobile global capital markets.[19] Commercial finance capital now replaced oil as the driving force of growth and investment.

Replacing oil funds with debt: predatory raids on state banks and agencies

As the need for loan capital soared, Indonesia's emerging private sector conglomerates and politico-business families turned to the state banks, and by the early 1990s these were providing around 63 per cent of all credit (*Tempo* 20 July 1991: 82–88). As 'agents of development', they had long been required to provide 'command credit' for strategic national investments, but through the 1980s critics were increasingly to protest the flow of state bank credit to well-connected business groups often without collateral, and on the basis of inflated and unsubstantiated cost projections. Based on inflated costings, it was alleged that much of this credit was deposited in high interest bank accounts or sent overseas to speculate on currency fluctuations (Kwik 1993).

State banks funded a range of Soeharto family projects, including Tommy Soeharto's notorious monopolies in cloves and automobiles and other major projects in petrochemicals and integrated forestry industries.[20] Special concessionary loans were made available to investors willing to establish forestry plantations (HTI), supposedly to create a long-term sustainable supply of logs and ease pressure on the diminishing natural forests. These were seen, however, as opportunities to clear existing native forests without serious intention of replanting, while channelling the subsidised loans into other areas (*Prospek* 17 August 1991: 87–97). In a time of tight money, state banks also financed well-connected borrowers to takeover other companies struggling under the pressures of debt servicing. In 1994, the textile firm, Kanindotex, owned by the Chinese businessman and convicted smuggler, Robby Tjahjadi, and with debts of US$355 million, was taken over by Bambang Trihatmodjo and Bambang Yoga Sugama, the son of former state security chief, Yoga Sugama. The state-owned banks Bumi Daya and Bapindo extended medium term concessionary credit at 8 per cent for three years then 11 per cent for the remaining five years, compared to prevailing deposit rates of 14.7 per cent (Nasution, Anwar 1995: 19).

As early as May and June 1993, Kwik Kian Gie, subsequently Co-ordinating Minister in both the Wahid and Megawati governments, claimed that the plunder of state banks had created a serious problem of bad debts, now around Rp.10 trillion at least, or 7 per cent of all debt (*Kompas* 4 May 1993; 24 June 1993: 1). The unconstrained haemorrhaging of funds was damaging the state banks. In November 1992 the government obtained a US$307 million 'Financial Sector Development Project Loan' from the World Bank of which US$300 million was used for capital injections into state banks (Nasution, Anwar 1992). According to Kwik Kian Gie, two of these banks, Bank Rakyat and Bank Bumi Daya, were technically bankrupt. While the government admitted that the banks needed an injection of US$2 billion, Kwik estimated that Bank Bumi Daya alone needed US$6 billion (interview 23 June 1992).

This simmering issue was brought clearly into the public arena in July 1993, when documents alleged to come from officials within the Bank of Indonesia were publicly circulated with details of the levels of bad debt and the names of debtors. It was claimed that the state bank for industrial development, Bapindo, had bad or doubtful debts of Rp.2,453 billion (28.7 per cent of total outstanding loans), and that the figures for Bank Bumi Daya, Bank Dagang Negara and Bank Rakyat Indonesia were higher. The list of debtors and those refusing to repay loans read like a 'whose who' of the Jakarta business elite.[21] Prajogo Pangestu, the forestry and petrochemical tycoon, was listed as having loans of around US$2 billion, of which 24 per cent was in the bad or doubtful loan category (Robison 1994: 66,67). The Jakarta press took up the story with a vengeance, drawing a picture of widespread disregard for banking regulations, collusion between officials, powerful patrons and borrowers, and the dubious nature of many of the projects receiving state bank funds. (*Tempo* 3 July 1993; *Prospek* 3 July 1993; *AWSJ* 3 July 1993; *Warta Ekonomi* 5 July 1993; *JP* 5 July 1993).

Clearly bank directors had little control over the flow of loans. In February 1994 a further scandal emerged in the state banking system when it was revealed that highly irregular loans made by Bapindo to Eddy Tansil, a business associate of Tommy Soeharto, had resulted in a US$614 million bad debt.[22] It also became clear that the loan was obtained following a series of letters to Bapindo from former Manpower Minister and State Security Chief, Sudomo, supporting the Tansil credit application.[23] Subsequent revelations from defendants in the Tansil court case that former Finance Minister Sumarlin had also lent his support suggested an embarrassing trail into the heart of power. More interesting, it appeared that Tommy Soeharto had been a partner in the project when the red clause letter of credit had been issued and remained a partner of Tansil in other ventures (*Tempo* 26 February 1994: 21–30; *Bisnis Indonesia* 27 February 1994: 7).

In the years following the Bapindo affair, the replacement of all State Bank Directors, including the influential Surasa of Bank Bumi Daya, the appointment of Standard Chartered Bank to provide management advice to Bapindo and the arrest and imprisonment of Eddy Tansil and several bank officials suggested that reform might be under way (despite the highly controversial 'escape' of Tansil in May 1996) (*FK* 17 June 1996: 12–20). Yet little changed in the state banks. In December 1994, the World Bank reported that state bank 'classified assets' (loans classified as 'substandard', 'doubtful', or 'loss'), stood at 18.6 per cent (World Bank 1995: 19). State banks remained an instrument for dispensing discretionary loans to powerful corporate interests.

State pension funds were also important sources of finance for these private business groups. In the early 1990s, the State superannuation fund, Taspen, had placed US$455 million of its funds with private firms, including some owned by Bambang Trihatmodjo, Tommy Soeharto and two Cabinet Ministers, Siswono Judohusodo and Abdul Latief (*AWSJ* 16/17 July 1993: 1, 4; *Tempo* 24 July 1993: 88–89). Although these may have been sensible commercial decisions, Taspen also made several strategic interventions to support private companies on the stock exchange at times when they faced serious liquidity problems or pressures on their stock prices.[24] In the most controversial of these cases, it was a key player in boosting equity in Prajogo Pangestu's PT Barito Pacific forestry group, to enable it to meet requirements for public listing (*AWSJ* 8 July 1993: 1, 7). The case drew criticism from amongst liberal commentators and from within parliament, regarded as a case of collusion, secrecy and misreporting in which public funds were mobilised to assist an over-exposed private investor (*JP* 16 July 1993: 1; *Detik* 14–20 July 1993: 19).

A third source of public financial support for private business groups was the government's extensive cache of non-budget funds. A report by the Supreme Audit Agency (BPK) in 2000 detailed how US$90 million of funds from the government's Reforestation Fund, originally established to develop forests and rejuvenate logged areas, had been illegally diverted through Presidential decrees to Soeharto family members and their business associates.[25] In all, the Forestry Ministry and other bodies eventually identified the misuse of over Rp.1.6 trillion (around US$160 million) (*Gamma* 1–7 March 2000: 22–23; *JP* 17 February 2000: 8).

Replacing oil funds with debt: plundering private banks

But it was the new sources of finance opened to the politico-business oligarchy by deregulation that provided the greatest potential opportunities. New banking laws in 1983 and 1988 removed controls on loan and deposit rates at state banks and eliminated restrictions on the entry of new private banks (Rosser 1999: 99). State-owned corporations could now deposit up to 50 per cent of their funds in private banks. New banks could be established with a minimum paid up capital of as little as Rp.10 billion – around US$400,000 at prevailing exchange rates (Simandjuntak 1989: 21–24). These moves not only increased the amount of credit available – they saw the number of banks increase from 111 to 240 between 1988 and 1994 (Rosser 1999: 100). Holding 14 per cent of credit outstanding in 1982, private banks were to surpass state banks in 1994 to hold over 48 per cent of credit outstanding and 53 per cent of bank funds (World Bank 1996: 17). The top twenty banks by asset value included twelve private domestic banks, four of them in the top ten (*Infobank* May 1996: 22–23).

These moves were welcomed in the neo-liberal camp. In its 1993 Report, the World Bank stated that 'as a result of ten years of financial reform, most of the policies needed to develop a robust and balanced financial structure are now in place' (*JP* 28 September 1993: 9). Yet, many private banks began to sink under the weight of bad debts long before the financial crisis of 1997, and it was the plunder of these banks by their owners that created the massive banking crisis that has continued to this day. In a situation where there was no separation of lender and borrower, the banks became simple cash counters for the big conglomerates, mechanisms for channelling loans to other companies in the same corporate empires where attempts to impose legal lending limits were generally ignored.[26] The connection between speculation and unconstrained inter-group lending was recognised, even within the industry.[27] Although introducing tighter legal lending limits in 1993 and threatening offenders with legal action, the Governor of the Central Bank found himself without the capacity to enforce these initiatives (Rosser 1999: 118–124). Few were surprised when audits of private banks concluded after the economic crisis, in 1999, revealed that inter-group lending had been in excess of 70–80 per cent of all private bank lending.

However, while the new private banks operated relatively free of regulatory controls, the state came to the rescue when some inevitably began to collapse under the weight of speculation and bad debts.[28] Rather than prosecuting these banks for transgressing banking laws, Bank Indonesia (BI) injected more funds and assembled rescue programmes together with state banks and even some of the larger private banks (*Gatra* 26 April 1997: 30). Of numerous private-banking disasters in this period, only the Soerjadjaja's Bank Summa was allowed to collapse, interestingly paving the way for a takeover of the giant automobile group, Astra, by companies associated with the Cendana group and Prajogo Pangestu (Schwarz 1994: 150–151).

Conglomerates and politico-business families also found the new Jakarta Stock Exchange a new source of cheap funding (*Eksekutif* August 1995: 12, 13). Once again, deregulation was compromised by lack of effective regulation. Inadequate rules and enforcement capacity allowed companies to go public without adequate disclosure, insider trading was rife and fake share scandals occurred frequently (Kwik 1993; *FEER* 2 April 1992: 46; *Tempo* 10 April 1993: 14–16). Often heavily exposed and highly geared, companies with privileged access to lucrative monopolies, particularly in the cement, foodstuffs or forestry and forest products industry, or with strategic positions in the infrastructure sector, found the stock exchange a valuable source of cheap equity funds as international portfolio investors scrambled to seize a share of these burgeoning industries.

Perhaps most ironic of all was the fact that Indonesia's free-wheeling predatory business interests and the new, liberalised global financial markets found each other at a critical juncture in history. Such markets proved willing to keep lending even where it was clear that projects did not meet normal criteria of commercial viability, audits and due diligence clearly taking into account the political guarantees that appeared to lay behind the projects and the fact that much of the real costs were absorbed by the public sphere – the classical moral hazard problem.[29] But it was private sector borrowing that was to prove, in the long run, the most damaging. Foreign observers expressed some concern at these growing debt levels but remained sanguine about Indonesia's capacity to handle the problems (Radelet 1995). Many of Indonesia's technocrats, though, were not so comfortable (Nasution, Anwar 1992).

What sort of oligarchy?

Within a decade of what had appeared to be a disastrous collapse in oil prices and after substantial market deregulation, economic power and wealth had been concentrated further in the hands of powerful private sector conglomerates and business networks.[30] The move beyond trade and into banking and the large resource, manufacturing and infrastructure mega projects increased dramatically the scale and complexity of their corporate empires. Politico-business families grew around officials and ministers who controlled the allocation of rents, and possessed the power to ensure that their activities would not be impeded by regulation and that the state was ready with bailouts and rescues in times of trouble.

With his authority largely uncontested, Soeharto stood at the apex of this system. Now including the six Soeharto children and various in-laws and grandchildren, there was almost no large investment in which they were not included and hardly any other major business group not drawn into some form of collaboration and partnership.[31] Outside the Soeharto juggernaut, lesser business families formed around satellite power centres in the state, including the Minister for Research and Technology, Habibie, Oil and Energy Minister, Ginandjar Kartasasmita, Information Minister, Harmoko, State Secretary, Sudharmono, and long-time Bulog boss, Bustanil Arifin. It was a pattern repeated down the power chain and in the regions and provinces as the children of former officials, generals, governors and ministers swarmed into the world of business. Often with only a tenuous hold on a few contracts or distributorships, these large numbers of predators depended on the residual influence of their parents or on other links with more powerful politico-business families to unlock the gate-keeping institutions.[32]

Despite extravagant and arrogant claims about their now-liberated business acumen and their status as a putative national bourgeoisie, these families basically constituted a political elite cemented together through their immediate or once-removed control or influence over the allocation of rents.[33] Just as elites in the absolutist states of eighteenth century Europe had assembled around state-allocated monopolies, titles, sinecures in land and state pensions, careers and commissions, these modern day rentiers dealt in the currency of discretionary credit, licences for mega projects and contracts for construction and supply to the state as their sinecures and pensions. Their significance lay not in any potential evolution as a vibrant capitalist class but in their cohesion as a political class behind predatory power relations and institutions.

Outside what we might call the gate-keepers, various business interests came from the opposite direction, seeking access to the rents. This neat distinction was not always clear. The *pribumi* contractors, produced under the patronage of Sutowo in the 1970s and later flourishing under Ginandjar in UP3DN and in the Ministry of Oil and Resources as contractors to the big resource and infrastructure projects, straddle the two dimensions. They were ostensibly the beneficiaries of economic nationalist plans to use state power to create an indigenous business class. But not anyone had entry. As detailed in the previous chapter, these contractors almost exclusively came from the families of serving or former officials and politicians or, in some cases, from *pribumi* business families with links into the Jakarta elite. Two prominent figures among them, Siswono Judohusodo and Abdul Latief, were themselves to become ministers.

At the other end of the spectrum, Chinese-owned business groups now expanded rapidly beyond their origins in trade, forestry and property. Key to the survival and growth of the largest among them was the dense web of interlocking partnerships that bound them to the Soeharto family and other powerful politico-business families as they expanded into the resource industries, public infrastructure, petrochemicals, banking, property and manufacture (*Tempo* 28 October 1989: 80–84). Perhaps the most dramatic change was the emergence of Prajogo Pangestu, to become the largest conglomerate outside the Liem group with large joint ventures with the Soeharto children in petrochemicals, pulp and paper and plantation forestry (*AWSJ* 27 August 1993: 1, 8; Schwarz 1999: 139–141). Securing alliances was only part of the agenda for the Chinese business leaders. We see a centralisation of ownership and management, diversification into new sectors and overseas, the building of horizontal and vertical integration and the pursuit of market domination in specific sectors (Sato 1993).

This large and diverse oligarchy proved to be formidable because it now straddled the formal apparatus of state power and the strategic heights of the economy. It constituted nothing less than the logic of political and economic power relations to which all else had to accommodate. A vast and complex network of business partnerships bound together the Chinese conglomerates and the large politico-business families, and included many of the *pribumi* contractors. Bakrie, for example, a supplier of steel pipes to Pertamina, now expanded his interests and involvement with the Soerharto family in plantations, mining, cattle ranching, oil distribution for Pertamina, and shares in Freeport mining (*Prospek* 8 February 1992; *FEER* 6 July 1989: 50–51; *Indonesian Business Weekly* 30 April 1993: 4–9; *AWSJ* 13 April 1994: 1, 4).

Yet this ascendancy was to confront several problems. In the long term, could the main players survive the departure of Soeharto? How would the existing politico-business families and Chinese conglomerates respectively protect their business empires when new figures controlled the state or even when centralised authoritarian rule gave way to political democracy? More immediately, the oligarchy confronted tensions within its own ranks as the split with the military and civil state apparatchiks deepened into conflicts over control of rents. Technocrats and the World Bank increased pressure to impose fiscal and monetary discipline on its activities. As growing political and economic risk in Indonesia forced an outflow of capital, the oligarchy became caught up in political disputes surrounding concentration of wealth, corruption and Chinese economic dominance.

How oligarch power was imposed and contested

As the unregulated system of markets became more volatile, the oligarchic interest was to triumph in a series of conflicts over rents, policy and the regulation of markets. These were conflicts to decide who controlled rents and who the beneficiaries should be. They were defined by bickering amongst the Soeharto children and by the deeper resentments from those consigned to the margins of the rentier system. Increasing centralisation of power and wealth, growing debt and the re-emergence of nationalist industry policies threw the oligarchy into conflict with both middle class liberal reformers and populists. It was a conflict the oligarchy would win, at least until 1997, carried by their grip on political power and by the continuing inflow of foreign investment and loans.

Conflicts over rents

As we have seen, Soeharto pursued a ruthless programme of centralising and concentrating the control of rents under his authority. His creation of Team Ten and UP3DN[34] within Sekneg had been only the beginning of a systematic exercise that brought the Soeharto family and other politico-business interests into increasing conflict with other centres of state power, notably the military, and with state managers. Nor did Soeharto hesitate to cut across the authority of his closest lieutenants when it suited his political needs. For example, the move by Minister for Oil and Energy, Ginandjar Kartasasmita, to organise two consortia of *pribumi* contractors to undertake a US$1.8 billion design and construction project for the oil refinery, Exor 4, and the US$1.7 billion project to construct a catalytic cracking plant in Cilacap in central Java were rudely disturbed in 1990 when Soeharto ordered Ginandjar to make Liem Sioe Liong and Prajogo Pangestu the principal domestic partners for both projects. This was a move widely regarded as a payoff for their role in the bailout of Soeharto's Bank Duta after its huge foreign exchange losses (Schwarz 1999: 128–129).[35]

The military, too, saw their independent sources of discretionary funds and patronage progressively diminished. Resentments and tensions flowed under the surface. Nowhere were these better illustrated than in the public dispute over Habibie's 1994 purchase of thirty-nine former East German warships. In a rare victory, the Finance Minister and the ABRI commander, Feisal Tandjung, combined to resist the Technology Minister's authority. At US$12.7 million, the cost of the dilapidated ships themselves was not the main issue. Rather, it was the US$1.1 billion Habibie planned to spend on repairs and refurbishment in Indonesia, involving procurement and contracts favouring the shipyard, PAL, a part of Habibie's own empire in BPPT and the Board of Strategic Industries (BPIS). The military resented Habibie's appropriation of spending in the defence sector. For his part, Finance Minister Mari'e Muhammad was entirely unconvinced that the proposed figure of US$1.1 billion was genuine. In the context of public and often bitter exchanges the allocation was restricted to only US$319 million, which included cuts to the cost of repairs in PT PAL from US$64 million to just US$9.5 million (*Tempo* 11 June 1994: 21–27).

The politics of rents was not only about who controlled their allocation, it also revolved around the question of who got what. It was no surprise that squabbles over the spoils soon broke out amongst the Soeharto children themselves. In his bizarre attempt to establish private control of the beer tax in Bali, the hapless Ari Sigit managed to galvanise into action a raft of interests, including brewers and hoteliers, among them his uncles, forcing the cancellation of the licence (*AWSJ* 27 January 1996: 1, 4). More serious disputes were to break out, as we shall see, when Tommy was awarded a licence to produce a national car that included exemptions from import and luxury taxes giving him a decisive commercial advantage over other family members in the automobile industry required to pay normal tariffs and taxes.

As monopolies and cartels were put together with breathtaking speed and daring by the Soeharto family and their associates, they had little compunction squeezing other interests. In one small but intriguing episode, Bambang Trihatmodjo's attempt to corner the citrus monopoly in Kalimantan brought him into collision with traders entrenched in the industry who were, surprisingly, able to mount effective opposition, eventually forcing him to withdraw (*Prospek* 20 June 1992: 70–80). Resistance to monopolies was not so successful elsewhere. Bob Hasan was able to impose an iron discipline within the plywood industry through his export cartel, Apkindo. With state-sanctioned authority to require the membership of all Indonesia's plywood producers, he had the power to set prices and quotas, and to allocate markets to members in a strategy to seize control of world prices and markets from Japan and Korea. Plywood producers became increasingly agitated at the constraints of Apkindo as prices plummeted in 1995 and Japan was able successfully to secure alternative sources of supply

(Pura 1995b; *Tempo* 4 December 1993: 83–84). It was not until Soeharto fell, however, that Apkindo was deregulated.

It is the notorious clove monopoly of Tommy Soeharto that provides perhaps the most interesting insight into the way centralisation of rents was established and how the state used its coercive power to stamp out opposition from other business interests hurt in the process. Claiming to act in the interests of clove producers against the big cigarette companies, Tommy received monopoly rights from the Ministry of Trade to purchase and sell cloves through a Board controlled by Tommy (BPPC). Bank Indonesia provided a low interest liquidity loan of US$345 million (*Prospek* 7 March 1992: 70–81; 10 July 1993: 67; *Tempo* 12 January 1992: 79–86; 10 July 1993: 74; 25 September 1993: 88–89; Schwarz 1994: 153–157). The outcome looked uncomplicated. Encouraged by the promise of higher prices, growers had flooded the BPPC warehouses.[36]

As BPPC began to encounter difficulties, it was plunged into a deepening spiral of dispute with both the clove producers and the cigarette manufacturers who were the ultimate losers in this adventure. Tommy Soeharto found the cigarette manufacturers association (GARPRI) surprisingly resistant and critical. Initially, they refused to purchase BPPC cloves, thus reducing the cash flow to a cartel with huge debt-servicing obligations (Schwarz 1994: 155). At this stage, the government was once again mobilised on the behalf of BPPC to break this resistance. The Minister of Trade, Arifin Siregar, reminded cigarette manufacturers that they required certification that their cloves came from BPPC. Substantial fines and confiscation were threatened for attempts to break the cartel. In the end, the manufacturers capitulated.[37] In the end, however, it was at the supply side that BPPC came to its sticky end, unable to cope with the sheer scale of cloves that now flooded into its warehouses.[38]

Nevertheless, the struggle over the distribution of rents was to create serious tensions in the oligarchy. When the two consortia of *pribumi* contractors bidding for the Exor and Cilacap oil refinery projects were shouldered aside for Liem and Prajogo in 1991, as described above, the scramble for contracts and bank credit opened tensions between *pribumi* and Chinese business interests. With the government drawing up plans for organising up to US$70 billion in loans for mega projects, *pribumi* businesses became concerned they were going to be overlooked for the big Chinese groups. It was claimed that the government had channelled Rp.90 trillion in credit to the Chinese conglomerates, largely through state banks, and that the Chinese had received the bulk of overseas loans.[39] A delegation of angry *pribumi* businessmen visited Soeharto to discuss access to projects and loans. They were reportedly supplied with guarantees about access to state bank credit (*Tempo* 20 July 1991: 82–88).

Harnessing economic nationalism to the interest of oligarchy

The interests of oligarchy were also to be pursued within the framework of economic nationalism. In what has become popularly known as the contest between the 'engineers' and the 'economists', leadership of the 'engineers' was to be assumed by B.J. Habibie, a German-trained engineer who was to draw heavily on the historical experiences of Japan and Germany in developing an industry policy intended to leapfrog Indonesia from the stage of reliance on natural resources and cheap labour directly to that of an advanced economy. By investing in long-term technological projects such as aircraft and ship manufacture, Habibie and his researchers in the think-tank, CIDES, argued that a new competitive advantage could be created through the by-products of technology transfer and a highly trained workforce (Habibie 1983; *Tempo* 10 October 1992: 21–33; *Tiara* 23 May 1993).

Neo-liberal critics, within the Bank and among the economics community generally, ridiculed Habibie's policies. Not only were his theoretical treatises greeted with derision, it was pointed out that his industrial ventures experienced heavy losses, that sales were confined largely to captive markets among Indonesian state corporations and in politically organised swap deals with other countries such as the planes-for-cars deal with Malaysia and the planes-for-rice deal with Thailand (Hill, H. 1984; *FEER* 11 June 1987: 110–116; *Prospek* 1 May 1993: 16–25; *AWSJ* 11–12 August 1995: 1, 5).

Nevertheless, with the support of Soeharto, who was clearly attracted by the vision of Indonesia as an advanced economy of real substance, the nationalist project went ahead. Habibie became Head of BPPT (the Board of Research and Application of Technology), and then Minister for Research and Technology. In 1989, he was appointed to manage a new strategic industries board (BPIS), which included ten state enterprises to be isolated from the privatisation process, among them the state-owned aircraft factory (IPTN), a shipbuilding enterprise (PAL), Krakatau Steel and various engineering, transport and munitions firms with a total asset value in 1992 of US$4.6 billion. From these bases, Habibie wielded considerable authority over state procurement and tendering processes, and was reportedly influential in determining key appointments in PLN and Telkom (*Prospek* 31 October 1992: 16–27; *Warta Ekonomi* 8 October 1989).

The idea that somehow deregulation and industry policy could be strategically reconciled was to gain ascendancy amongst policy-makers. Ginandjar Kartasasmita, now Head of the Economic Planning Board, Bappenas, argued that industry policy was a precursor for a strong market economy, 'We can no longer rely so heavily on what is usually called comparative advantage in such areas as natural resources, low labour costs and soft foreign loans' (*Economic and Business Review Indonesia* 19 June 1993; *JP* 23 March 1993; 7 June 1993). At the same time, Ginandjar also argued that liberalisation by itself led to a concentration of wealth and economic power, harnessing populist sentiment to the nationalist policy cause (*JP* 29 May 1995: 1; 10 July 1995: 4). In a similar vein, Co-ordinating Minister for Industry, Hartarto, claimed that industry policy had reduced imports in petrochemicals and generated a trade surplus in industrial products (*Tempo* 17 April 1993: 90–91).[40]

It was no surprise when several figures from the state-sponsored Muslim political organisation, ICMI, chaired by Habibie, were appointed to the new Cabinet of 1993, coinciding with the departure of the last of the 'economists', Sumarlin, Wardhana and Mooy (*FEER* 1 April 1993: 72–75). But little changed. The new economic ministers, Saleh Afiff (Co-ordinating Minister for Economy and Finance), Mar'ie Muhammad (Minister of Finance), and Soedradjat Djiwandono (Governor of the Bank of Indonesia), did not abandon established conservative fiscal and monetary policies, as some had feared. The government, on the other hand, kept injecting funds into the highly protected industrial projects, including an estimated US$1.6 billion into the aircraft manufacturer, IPTN, of which Rp.400 billion (US$160 million at the time) came in controversial fashion from the Special Reforestation Fund (*Kompas* 5 July 1994: 2). Plans to build a US$2 billion jet aircraft in the United States were also initially supported by Soeharto, although without promises of money. Suggested sources of funding included an offering of public shares in the project and further selling of the assets of state companies (*AWSJ* 11–12 August 1995: 1).

Economic nationalists in Indonesia often referred to the examples of the (then) successful state-led industrialisation in Korea and Japan to justify industry policy and protection. As Ginandjar Kartasasmita noted, 'why was it that within total protection, they [Japan and Korea] were able to develop efficiently and effectively to penetrate the world market?' (*Kompas* 24 August 1985: 1). But the Indonesian case was different. Indonesia's economic nationalist agendas increasingly became devices to supply politico-business families and corporate conglomerates with monopolies and subsidies rather than attempts to build long-term strategic positions for the economy. No criteria for success in global markets determined the flow of protection and state credit in an agenda that was captured and driven by predatory commercial interests.

This complex interplay was to be illustrated in the development of Indonesia's automobile industry. Liberal reformers, including Finance Ministers, Sumarlin and Mar'ie Muhamad, together with the World Bank and private liberal critics had long pressed for deregulation as the answer to high prices and overcrowding in the industry (*Kompas* 20 November 1990; World Bank 1981; Sjahrir 1988). Within the industry itself, the beneficiaries of existing protective arrangements fiercely defended their positions. Citing prospective floods of foreign vehicles and loss of a potential domestic industry with its flow-on into technology, the automobile industry association (Gaikindo), as well as leading assemblers and components manufacturers, invoked the Japanese experience which, they argued, had involved a thirty-year incubation period within strategic protection to achieve an internationally competitive local base of industry and technology (*Jakarta Jakarta* 16–22 March 1991; 15 June 1991; 15 September 1991).

But leading advocates of nationalist policy, Industry Minister Hartarto and Research and Technology Minister, Habibie, had long been concerned with the continuing low levels of domestic components in cars assembled in Indonesia and with the problems of technology transfer despite continued protection. They favoured an industry policy that would give market advantages to those companies willing to develop high levels of domestic content (*Suara Pembangunan* 25 May 1991; *Kompas* 8 November 1991; *Jakarta Jakarta* 16–22 March 1991; *FEER* 10 December 1992: 39). In the meantime, the automobile industry remained insulated from the reforms that had swept across the trade and financial sectors. When an initiative did arrive, it was not a decision for deregulation but one to build a national car.

Presidential Instruction No. 2, 1996 granted exemptions from import duties and luxury sales taxes for automobile manufacturers who produced cars with Indonesian brand names, that were developed with domestic technology, engineering and design. Two Ministerial Decrees allocated this facility to PT Timor Putra, a company 70 per cent owned by Soeharto's youngest son, Hutomo Mandala Putra (Tommy), and 30 per cent owned by Kia of Korea. Timor was required to use 60 per cent domestic components by the end of three years. In return, no other company was to be granted the same package of exemptions for this period (*AWSJ* 20 February 1996; 18 March 1996: 1; *JP* 29 February 1996: 1). Industry Minister, Tunky Aribowo, and other officials defended the policy in terms of the need to acquire technology transfer and a domestic capability in engineering production as well as reducing the US$3.6 billion used each year to import automobile components. Kia's involvement with Tommy's Humpuss group was likened to the way Austin and Renault were linked with Nissan and Hino in Japan in the 1950s as the basis for the development of a competitive Japanese industry (*Tiras* 14 March 1996: 28; *Asia Times* 1 March 1996: 8; 18 March 1996: 16).

More difficult to explain, outside the logic of crude political patronage, was the appointment of Tommy Soeharto as the sole beneficiary of the policy. Questions about the absence of any transparent tendering process or the capacity and experience of Tommy were simply ignored.[41] The decision of the government in 1996 to grant Timor exemptions from import duties and luxury goods taxes provoked complaints from both the US and Japanese governments. The Japanese Ambassador argued that the policy transgressed standards of transparency fairness and equity expected by foreign investors, while the Japanese government was reportedly considering taking the case to the World Trade Organisation (*JP* 16 March 1996; 23 March 1996). US companies, General Motors, Ford and Chrysler, planning a comeback into the Indonesian market, immediately indicated they would also reassess or cancel their planned investments (*AWSJ* 4 April 1996: 9; 2 April 1996: 11).

It set the scene also for bitter protest from Indonesia's other car producers who still had to deal with 300 per cent taxes on CBU vehicles (Robison 1997: 55–57; *JP* 7 March 1996: 1; 18 March 1996: 4). Facing declining sales, a slump in profits and, in the case of Astra, a fall in share prices, several announced the cancellation of planned investments while others indicated plans to produce their own national car.[42] Claiming that only themselves and Astra had the capacity to achieve local content requirements, the directors of Indomobil clearly indicated their doubts about the seriousness of Timor's intentions to fulfil its obligations (*JP* 16 March 1996). One of the ironies of the situation was that it pitted members of the Soeharto family against each other. The President's step-brother, Probosutejo, was involved in a joint venture with General Motors, intended to produce Opel and Chevrolet, while Bambang Trihatmodjo's Bimantara group was allied with Hyundai. Vowing to produce a cheap car to compete with Timor, Bambang criticised the new regulations as 'inadequate and monopolistic', vowing to press for the same privileges as Timor (*JP* 7 March 1996: 1; *Asia Times* 3 April 1996: 5).

In the end, Soeharto imposed the Timor decision in the face of vigorous resistance from technocrats and the World Bank, at the same time derailing even the mainstream economic nationalist agenda aimed at creating a viable automobile industry within Indonesia before it was forced to open itself to competition from the Malaysian and Thai automobile industries under the AFTA provisions.[43] The decision to appoint Tommy's Timor consortia as the national producer was greeted with considerable scepticism within the press, in parliament and within the automobile industry. It was generally thought that Timor possessed neither the capacity nor the intention of meeting the domestic content provisions, and there was similar scepticism that the government would be able effectively to monitor and enforce the provisions of the regulation. It was a decision also that injected a level of bitterness and antagonism within the oligarchy itself, an ominous development that threatened to escalate as the regime moved towards its end.

This complex struggle between neo-liberal reformers, economic nationalists and the oligarchy also pervaded the petrochemical industry. It was an industry dominated by the Soeharto family and their associates and one where foreign investors and banks had rushed into arrangements with well-connected licence-holders in a sector that enjoyed protection from foreign imports and received subsidised inputs from state corporations, including electricity, cement, steel and Pertamina feedstock (*AWSJ* 6–7 January 1995: 1, 4; *JP* 16 February 1996: 1, 4; Nasution, Anwar 1995: 16). Tensions were to focus around the giant US$2.4 billion Chandra Asri olefins project, owned jointly by Prajogo Pangestu, Bamabang Trihatmodjo's Bimantara group and Henry Pribadi, a business associate of Liem Sioe Liong.[44]

Backed initially by a US$550 promissory note from Bank Bumi Daya, Chandra Asri relied heavily on its expected position as supplier of ethylene to the domestic chemical producers, PENI and Tri Polyta (in which the principles of Chandra Asri were also major shareholders), and upon guarantees of cheap Pertamina feedstock.[45] Chandra Asri confronted difficulties when both Tri Polyta and PENI proved reluctant to conclude an agreement so long as they had access to international markets in which prices looked set to collapse as a world-wide glut of olefins loomed. This confirmed the views of a range of observers who had long argued that Chandra Asri was not viable without protection, a perspective justified when Chandra Asri director, Peter Gontha, requested a 40 per cent protection for olefin producers for eight years (Gontha 1992).

Defenders of Chandra Asri within the government, including Minister for Investment, Sanyoto Sastrowardoyo, and Soeharto himself, argued that it would reduce imports and build local capacities in technology and engineering, contributing to the creation of a self-supporting national industrial structure (*AWSJ* 6–7 January 1995: 4; Hobohn 1995: 29). Chandra Asri was viewed as virtually a state corporation, part of a broader national industry strategy. Others feared that high production costs resulting from protection would cause potential harm to downstream producers and exporters. In particular, a tariff on ethylene was expected to directly affect the synthetic fibre industry, essential to textile exports. A number of Japanese and other North Asian investors indicated that they would seek other sites for investment if tariffs raised production costs (*Tempo* 30 October 1992; *AWSJ* 6–7 January 1995: 4). Another threat to Chandra Asri came as Indonesia's technocrats and the World Bank became increasingly alarmed at the growing levels of debt.[46] The government was therefore faced with pressures to reign back the domestic money supply, curtail overseas loans for mega projects and impose systems for prudential control in the banking sector.

The formation of a ten-person team (Team 39 or COLT) in late 1991 to regulate foreign loans for commercial projects resulted in the postponement of four projects worth US$9.8 billion, including Chandra Asri (*FEER* 19 September 1991: 80–81; 24 October 1991: 76–77; *Kompas* 14 October 1991). Chandra Asri sidestepped these controls when its Indonesian owners reorganised investment through Hong Kong subsidiaries, thereby making ownership technically fully foreign. In the end, regulations were changed to allow domestic companies with investments over US$50 million to register as foreign-owned and escape the restrictions on overseas borrowings. State bank loans to Chandra Asri were unfrozen and a new finance package was put together that included a US$550 credit facility and a US$213 million loan from Bank Bumi Daya and Bank Dagang Negara (Schwarz 1999: 152, 153; *Tempo* 24 August 1991: 91; 20 June 1992: 94; *FEER* 12 March 1992: 45; 7 June 1993: 52; *Warta Ekonomi* 29 April 1991: 30–39; *AWSJ* 12–13 August 1991: 4, 9).

Soeharto met protests by the technocrats with a decision in 1994 to abolish the COLT team headed by Finance Minister, Mar'ie Muhammad, and replace it with a team headed by the Co-ordinating Minister for Trade and Industry, Hartato, and including Investment Minister Sanyoto. Its function was reduced subsequently to one of regulating state-related foreign borrowings (World Bank 1997c: 14, 15; *AWSJ* 16–17 December 1994: 1; Robison 1997: 14, 15). Chandra Asri was also to prevail on the question of tariffs. As late as September 1994 Industry Minister, Tunky Aribowo, announced there would be no tariff. However, the threat of bankruptcy to a US$3.26 billion project belonging to the President's son as well as a major setback to plans for a domestic petrochemical industry made defiance difficult. In the end, the President imposed a 20 per cent surcharge on imports of propylene and ethylene in February and March 1996, measures designed to induce both PENI and Tri Polyta to sign supply agreements with Chandra Asri, thereby ensuring its survival (*JP* 16 February 1996: 1, 4; *AWSJ* 4 March 1996: 1).

As the activities of the politico-business families and the conglomerates became increasingly audacious, concerns about the erosion of financial discipline, the inertia of government and the negative impact of conglomeration on income distribution and competition were on the increase. Some of these attacks came from officials and former ministers and officials.[47] Yet the oligarchy was able to resist pressures to control debt and began to reorganise its capital base out of the reach of reformers.

Sweeping aside pressures to control debt

The abolition of the COLT Team raised the larger issue of debt. While the conglomerates' dependence on loans was increasing, continuing pressure on the current account and a rapidly expanding money supply in an overheating economy forced Indonesia's financial managers to try and stem the flow of credit from overseas.[48] In the end, they were never able effectively to impose controls on private borrowing. Within the neo-liberal camp this was assumed to be a technical problem. Policy-makers had deregulated the financial sector before the real sector. However, the sequencing of deregulation was no matter of technical choices. Financial reform was simply easiest because it was a politically available option.

The technocrats had opened the capital account in 1970, partly as a move to pre-empt corruption in trade and partly to attract Chinese investment back into Indonesia by providing them with a right to exit. Introduction of a balanced budget law had been introduced to prevent the huge deficits of the Soekarno era. In the process, the technocrats bargained away their main instruments of control. As Pincus and Ramli have argued, having effectively 'surrendered control over capital movements, interest rates, credit creation and (to a large extent) fiscal policy, the monetary authorities were left with interest rates on Bank Indonesia securities (SBIs) and the exchange rate as the main levers of macroeconomic adjustment' (1998: 731). Soeharto, in any case, retained substantial non-budget sources of expenditure. As we have indicated earlier, shadowy agencies such as Sepdalopbang and Banpres, as well as the ubiquitous *yayasan* were repositories for funds from a mixture of public and private sources.

In the face of persistent inflationary pressures as the money supply surged, and given an unwillingness to repeat large devaluations of the currency to protect the rupiah, the government could only resort to harsh liquidity shocks in 1987 and 1991 to contain the pace of credit creation and relieve domestic pressures on the rupiah. In the so-called Sumarlin shocks, the government drastically reduced the liquidity of state banks and in 1991, Rp.8 trillion was withdrawn by state companies from eight state banks, sending interest rates to levels above 30 per cent. This restricted access to credit at a time when the big conglomerates and politico-business families carried heavy debt exposure, often in the form of short-term loans, exacerbated in some cases by the strengthening of the Yen. In several high profile cases, large companies went to the wall, notably Bank Summa, Bentoel and Mantrust. A large number of companies simply took their money offshore (*Tempo* 17 August 1991: 86).

While state managers attempted to impose monetary discipline in a vain attempt to contain foreign loans, the state banks, as noted earlier, continued to provide funds for selected borrowers, often at concessionary rates. Both the private and public sectors ignored requirements about disclosure and reporting. The government refused to take action even as technically insolvent state banks continued to dispense discretionary loans in contradiction of other government policies intended to tighten liquidity. In short, the regulatory power of the state was in the hands of those with no interest in regulation.

Insurance policies: securing the oligarchic interest

Conglomerates proved able to reorganise their capital in the face of a range of political and economic risks, and to take advantage of new opportunities for investment overseas. Through the 1990s many of Indonesia's largest groups established holding companies that were then partly floated on the Jakarta Stock Exchange (JSX). Going public made them less a focus for public resentment and less vulnerable to the sort of asset confiscation that had taken place at the end of the Soekarno era. This had to be balanced against pressures for disclosure, admittedly meagre in Jakarta but more stringent overseas. But public listing had other advantages. It facilitated sales within the group. For example, Liem's publicly listed PT Indocement purchased part of the flour miller Bogasari and 51 per cent of PT Indofood, an internal rearrangement of Liem's empire that netted US$500 million in cash (*AWSJ* 22 June 1992: 1, 4). Further internal reshuffling was to take place in 1994/1995. Conglomerates were now using the JSX and the state banks to raise cash, not solely for expansion but to buy other companies and expand overseas.[49]

The issue of capital flight intensified and came to a head in 1991 when Christianto Wibisono, the head of a business intelligence agency, PDBI, published claims that US$76 billion had been placed in Asian Currency Unit funds in Singapore (*Tempo* 17 August 1991: 86; 28 September 1991: 26). Although the extent of these funds was disputed (Sjahrir, cited in *Kompas* 7 August 1991: 1, 11), it was clear that large sums of money were leaving the country. For a range of commentators, the flight of US$800 million in 1991 was intended to disorganise the Indonesian economy and force a devaluation, allowing a return of the money at a premium (Sjahrir, cited in *Tempo* 9 March 1991: 86–90; Kwik cited in *Warta Ekonomi* 17 August 1991: 22). Whether or not this was the intention in this instance, it soon became clear that the movement of capital offshore was structural in nature and that conglomerates had begun diversifying their holdings as insurance against the coming transition of power and to take advantage of economic opportunities outside Indonesia.

In 1991, the Dharmala conglomerate was already generating 35 per cent of its earnings overseas and, by 1994, Anthony Salim reported that the Liem group also received 35 per cent of its revenues from overseas operations based around his First Pacific group in Hong Kong and his United Industrial group in Singapore (*Tempo* 28 September 1991: 86; *AWSJ* 25–26 February 1994: 1, 8). In 1993 large Indonesian conglomerates, including the Salim group, Lippo and Sinar Mas, had made investments in China estimated at US$797 million, and Liem had plans to invest US$1 billion in his former home province of Fujian (*JP* 14 April 1993: 1; *Asiaweek* 4 May 1994: 62–64). As various conglomerates concentrated their holdings into holding companies, these were floated, not only on domestic bourses but overseas (*Asia Inc* December 1994: 48). In 1997 Liem sold shares in the highly profitable Indofood to QAF in Singapore, moving funds offshore at a time of impending crisis, and transferring control out of Indonesia. Clearly, this was both capital flight, as argued by Econit's Rizal Ramli, and a case of raising cash by internal sales, as argued by Kwik Kian Gie (*Gatra* 2 August 1997: 31; *FK* 11 August 1997: 19).

Although Indonesia's conglomerates had made themselves less vulnerable to economic and political volatility in Indonesia, putting much of their wealth out of the reach of legislators and reformers in Indonesia, their movement overseas and concentration into giant holding companies focused attention firmly on the issue of conglomeration and on the commitment to Indonesia of Chinese business. Within parliament itself and in state-sponsored organisations such as ISEI (the Association of Indonesian Economists), conglomerates were criticised for their reliance on protection and monopoly and failure to contribute to the broader economic growth in Indonesia, as well as their moves back into China and Hong Kong (*Detik* 1 December 1993: 1; 7 December 1993: 20). The spectre of anti-Chinese sentiment was being raised.

But it was the triumph of oligarchy in the political sphere that was to pave the way for their hegemony and keep at bay the tensions and resentments that resulted. Ignoring demands for reform, trampling over opponents within the state and critics without, they consolidated a system of power that made them seemingly invulnerable. It is to the dynamics of this political victory that we now turn.

Notes

1 Whereas oil and gas taxes had constituted Rp.8628 billion, or 70.6 per cent of total domestic revenues in 1981/1982, this was expected to decline to Rp.6338 billion or 57.1 per cent in 1985/1986 (World Bank 1993a: 185). Net oil exports peaked at US$6,016 billion, or 81 per cent of the total in 1981/1982, declining to US$1,426 billion in 1986/1987 (World Bank 1994: 205).

2 The new sectors opened in 1994 included sea-ports, electricity, telecoms, shipping, civil aviation, drinking water, public railways, nuclear power and media. Information Minister, Harmoko, overrode the opening of media on the grounds it contravened existing law. It should be noted that Harmoko himself was a major media owner who

would have been affected by the deregulation. Ten further sectors opened in 1995 included cooking oil, motor vehicles and advertising (World Bank 1995: 40–41).

3 After having declined to around US$2.5 billion in 1985, foreign investment approvals soared to almost US$40 billion in 1995 (World Bank 1996: 12–14; *Suara Karya* 25 September 1986). Total private fixed investment increased from 31 per cent of the total in 1979/1980 to 73 per cent in 1993/1994. Private foreign investment now created 12 per cent of total value added in the late 1980s, while the figure for domestic private investment was 57 per cent (Pangestu 1995; Hill, H. 1990).

4 Holding 14 per cent of credit outstanding in 1982, private banks were to overtake state banks by 1994 to hold over 48 per cent of credit outstanding and 53 per cent of bank funds (Econit 1996: 4–7; World Bank 1996: 17).

5 While strategic import monopolies in cold-rolled steel, plastics and tin plate in the hands of Liem Sioe Liong, the Soeharto family and Bob Hasan were three high profile casualties of reform, the importer/producer nexus was not fundamentally unravelled because the bulk of imports was destined for domestic use and did not come under the aegis of the May 1986 regulations. Industry Minister, Hartato, claimed that upstream producers would not be hurt because the domestic market absorbed 80 per cent of their products (*Tempo* 17 May 1986).

6 For example, the main creditor consortia (CGI) pledged US$5.26 billion in 1996, of which Japan provided an aid package of US$2.6 billion, almost all in the form of official development loans (ODA). These are a double-edged sword. Since 1987–1988, debt servicing has exceeded aid receipts in the state budget. For 1994/1995, debt servicing of Rp.17.6 trillion compared with receipts of Rp. 10.7 trillion.

7 See for example the comments of prominent *pribumi* businessman, Sukamdani Gitorsardjono (*Kompas* 26 December 1993: 2).

8 The June 1993 package that ended Bogasari's monopoly allowed three new competitors into the market, including Siti Hardiyanti's Citra Latorogung group. Bulog, however, remained the sole importer and controlled prices. The new entrant had no control over the import of wheat and was required to export 65 per cent of production in any case. In a situation where a new mill cost US$40–50 million, Bogasari retained the decided advantage.

9 Among the biggest Chinese operators in the forestry industry were Prajogo Pangestu (Barito), Eka Tjipta Widjaja (Indah Kiat and Tjiwi Kimia), Sukanto Tanoto (Inti Indorayon Utama and Riau Pulp and Paper) and Bob Hasan (Kertas Kraft Aceh and Santi Murni Plywood). Siti Hardiyanti Rukmana was to join with Prajogo in PT Tanjung Enim Lestari, a US$1 billion pulp and rayon complex with associated forestry and plantation interests. Sigit Hardjojudanto and Bob Hasan, together with the government, formed PT Kertas Kraft Aceh, while Hutomo Mandala Putera joined with Eka Tjipta Widjaja in PT Inah Kiat and took over the government paper mill, PT Kertas Gowa. Bambang Trihatmodjo's Bimantara group joined with Nusamba and the military *yayasan*, Kartika Eksa Paksi, in a US$600 million paper and pulp venture (*AWSJ* 25–26 May 1990: 1, 4; 17 May 1994: 4; *Tempo* 26 October 1991: 22–32; *Warta Ekonomi* 29 April 1991: 30–39; *Editor* 23 December 1989: 14–23).

10 See statements by the members of Parliamentary Commission IV and Minister Surjohadikusumo on illegal logging and fires (*JP* 18 September 1993: 1, 22; 22 September 1993: 8). An attempt to impose a Rp.11 billion fine on Prajogo Pangestu's Barito group in 1992 was simply ignored (*FEER* 12 March 1992: 45).

11 It should be noted that Liem had been appointed sole importer of cold-rolled steel, selling it at a domestic price of US$550 per tonne when the international price was US$380 per tonne.

12 According to figures released by the Co-ordinating Minister for Economics, Finance and Industry, Radius Prawiro, in September 1991, US$31.3 billion of the US$80

billion was private investment. Of the total investment, petrochemicals accounted for US$10.4 billion; pulp and paper, US$ 9.9 billion; mining, fertilisers, industrial estates, property development and other manufacture, US$11 billion; and infrastructure and transport, US$21.4 billion. Pertamina projects, mainly in refining and petrochemicals, constituted US$22.9 billion of the state investment (Muir 1991: 19). Reporting a detailed survey of mega projects, *Warta Ekonomi* estimated the total value of the largest thirty-three at US$35 billion (29 April 1991: 22–39).

13 In 1988, Bulog awarded a monopoly to mill soymeal to PT Sarpindo, a company owned by Hutomo Mandala Putera and Bob Hasan. Sustained by a price differential of 23 per cent above imported parity, Sarpindo was to earn US$12 per tonne in a milling fee from Bulog as well as US$72 per tonne from the soybean oil by-product it was able to retain (World Bank 1994: 94; Schwarz 1994: 132). A controversial monopoly on fertiliser pellets was allocated to PT Ariyo Seto Wijoyo, owned by Soeharto's grandson, Ari Sigit (Econit 1996).

14 Among these were Soeparno of Garuda who had been required to purchase MD11 planes through a Bimantara company, allegedly an expensive and unnecessary process (*JP* 17–18 January 1992; *FEER* 21 September 1993: 71–72). An attempt by the head of the state airline, Merparti, to refuse to lease IPTN's C235 from IPTN on the terms offered led to his dismissal (*FK* 20 November 1995: 36–40; *Gatra* 4 November 1995: 21–28). The Head of TVRI, Ishadi, was also to be a victim of the struggle following his criticism of the awarding of an educational TV channel to Siti Hardiyanti Rukmana, that used TVRI time and facilities as well as generating advertising revenue (*Tempo* 25 August 1990: 75–78; *Matra* 7 August 1992). Conflicts over the awarding of contracts to supply telephone systems and to construct power generation plants resulted in the dismissal of Cacuk Sudarijanto of Telkom and Ermansjah of PLN (*JP* 22 April 1992; 2 May 1992; 10 October 1992; 12 October 1992; *FEER* 5 April 1990; *Prospek* 27 June 1992: 70–73).

15 Over US$760 million was raised from the sale of equity in Telkom and Timah in domestic and US capital markets and used to prepay high interest debt to the Asian Development Bank and the World Bank (World Bank 1996: 10).

16 For example, companies owned by Soeahrto's eldest daughter, Siti Hardiyanti Rukmana (Tutut), PT Bheering Diant Pramata and PT Surya Kepanjeng, supplied British Hawk fighter aircraft and Scorpion tanks to the armed forces at highly inflated prices, extracting millions of dollars from state coffers (*Jakarta Post.Com* 15 July 2002).

17 Bambang Trihatmodjo and his brother-in-law, rising star Hashim Djojohadikusumo, were to secure the contracts for the huge Paiton power plants. Other main local players were to be Liem Sioe Liong, Sudwikatmono, Hutomo Mandala Putera, Sigit Hardjojudanto, the Bakrie group and Poo Djie Gwan, a long-time supplier to PLN (*Prospek* 27 June 1992: 70–73; *Editor* 1 February 1992; 8 February 1992; 27 June 1992: 68–69; *Warta Ekonomi* 22 June 1992: 24; *AWSJ* 9 February 1994: 1; 14 February 1994: 1; 13 September 1994: 1; 20 October 1994: 3; 3 November 1994: 1, 2; 24 April 1995: 4, 5).

18 For example, Hashim Djojohadikusumo and his sister-in-law, Soeharto's daughter Siti Hutami Endang Adiningsih (Mamiek) were given a 15 per cent equity in the power generation project, Paiton I, ostensively set against future earnings (*AWSJ* 21 December 1999: 1). A state audit of the Paiton Energy Corporation in 2000 revealed that the Indonesian partner, Hashim Djojohadikusumo, never paid cash for his stake in the consortium with American and Japanese companies, Mission Electric, General Electric and Mitsui. The audit claimed that the multi-billion dollar contract was awarded to Paiton with no open bidding process and project costs had been inflated by over US$600 million. It was alleged that over US$22 million had been paid in bribes. Paiton also purchased coal from a company owned by Hashim at around

US$15 above the average market price, charging PLN 8.5 cents per kWh compared with 6.4 cents per Kwh charged by other producers (*JP* 7 January 2000: 1; *AWSJ* 5 January 2000: 1, 2).

19 For example, the US$1 billion Tanjung Enim Lestari pulp and rayon venture of Prajogo and Tutut, originally recipient of a US$450 million state bank loan, was reportedly to proceed with a US$700–750 overseas loan and with a 20–30 per cent shareholding by Marubeni. Public floats of forestry industry companies on the Jakarta Stock Exchange proved highly successful. Among the listed companies were the forestry giant, Barito, which raised US$295 million in a controversial 1992 float, the paper maker, Pabrik Kertas Tjiwi Kimia, the pulp and paper venture, Indah Kiat, which raised US$348 million, and Inti Indorayon Utama, which raised Rp.253 billion in 1991 (*FEER* 11 April 1991: 56–57; *AWSJ* 25–26 May 1990: 1, 4; *Warta Ekonomi* 29 April 1991: 30–39; *AWSJ* 11 April 1994: 1, 7).

20 Among the more notorious cases, a Bank Bumi Daya loan of US$550 million was provided to Prajogo Pangestu for the Chandra Asri petrochemical project, and a concessionary Bank Indonesia loan of US$345 million was provided to Tommy Soeharto to finance the controversial cloves monopoly. State banks gave US$450 million for the Enim Lestari pulp and rayon project jointly owned by Prajogo and Siti Hardijanti Rukmana, and a consortia was to provide a US$600 million guarantee for Tommy Soeharto's national car project in 1995 (*Tempo* 9 March 1991: 86–90; 11 May 1991; *Warta Ekonomi* 29 April 1991: 30–39).

21 The document, titled simply 'Information on State Banks', listed both banks and debtors. The Soeharto children figured prominently. In the uproar that followed, Prajogo Pangestu denied having any bad debts at all, alleging conspiracy (*JP* 23 June 1993: 8).

22 It was revealed that Bapindo had loaned US$430 million to Eddy Tansil of the Golden Key group, resulting in a total debt to the state of US$614 million with interest. A 'red clause' facility to draw a letter of credit for US$241 million had been illegally issued for the import of equipment for a proposed petrochemical plant whose asset value had been grossly overvalued. It appeared that no attempt had been made to import the equipment and that part of the credit had been paid to Tommy Soeharto (*Tempo* 16 February 1994; 19 February 1994; 5 March 1994; 12 March 1994; 2 April 1994; *FEER* 3 March 1994; 23 June 1994; *AWSJ* 9 May 1994; 11 May 1994).

23 So-called *surat sakti* (magic letters) or *katabelece*, from powerful and influential figures to officials, were common means of influencing decisions about allocation of rents.

24 It intervened to buy shares in Bank Danamon when it faced a serious liquidity problem, and in the Soeharto family's Bank Duta when its share price slumped in the wake of its 1991 foreign exchange scandal (*AWSJ* 10 January 1991: 10, 14).

25 While the most notorious case involved the diversion of Rp.400 billion to fund continuing production of Habibie's ill-fated aircraft venture, IPTN, another Rp.250 billion was also handed to Bob Hasan to assist his debt-laden pulp and paper project, PT Kiari Lestari, while Soeharto's grandson, Ari Sigit, received Rp.80 billion for a urea fertiliser pellet plant (*JP* 12 June 1998: 10).

26 Bank lending to other companies in the group was officially limited to 50 per cent of total equity capital and only 20 per cent of total equity capital could be lent to any single borrower.

27 See, for example, the comments of Robby Djohan, head of Bank Niaga (*Matra* July 1993: 14–24).

28 Bank Yama (belonging to Soeharto's daughter, Siti Hardiyanti Rukmana – Tutut) along with Bank Continental and Bank Pacific (owned by members of the Sutowo family) were rescued with help from Bank Indonesia and Bank Negara Indonesia in 1995, and creditors were paid out of state funds. Sixty per cent of Bank Industri,

belonging to Soeharto's brother-in-law, Hashim Djojohadikusumo, was taken over by the state bank, Bank Rakyat Indonesia, when it got into difficulties in 1990 (*Infobank* February 1997: 21; Rosser 1999: 122). Private business interests rescued the Soeharto-associated Bank Duta in 1992 when it encountered heavy losses from illegal foreign exchange speculation. No formal charges against the bank were ever laid (Schwarz 1994: 112, 128, 141).

29 The government planned to inject US$4 billion into projects in 1991, raised from overseas banking consortia and domestic banks. At a time of tight monetary policy, Finance Minister Sumarlin argued that, 'The money to be used to finance these projects is not government money. If it were to come from the government in the shape of Bank Indonesia Liquidity Credits, then the government should be called inconsistent' (cited in *JP* 29 July 1991: 4). Public foreign debt, however, rose dramatically, reaching US$100 billion in 1996.

30 Private investment was initially hit hard by declining demand and overcapacity in a range of ISI industries, including textiles, cement and automobiles, and by increased debt-servicing costs as the Yen soared in value after the Plaza Accord of 1986 (*Bisnis Indonesia* 17 July 1986; 19 July 1986; *Suara Karya* 15 May 1986; 12 July 1986). Nevertheless, it was estimated that by 1990, the top 200 conglomerates had a turnover of around US$50.6 billion, or 35 per cent of the GDP. The top five conglomerates accounted for over US$16 billion of the total and the top fifty-five were estimated to control around 70–75 per cent of all private domestic capital in Indonesia (PT Data Consult Inc. 1991, cited in World Bank 1993a: 91, 92).

31 The Soeharto children were Sigit Hardjojudanto, Bambang Trihatmodjo, Siti Hardiyanti Rukmana (Tutut), Siti Hedijanti (Titiek), Siti Hutami Endang Adiningsih (Mimiek) and Hutomo Putera Mandela (Tommy). Also involved in the family business were son-in-law, Indra Kowara, from an old business family and married to Tutut, and Hashim Djojohadikusumo, son of Sumitro Djojohadikusumo, whose brother, General Prabowo, was married to Titiek. Most recently, Sigit's son, Ari, entered the business world. There is a vast literature on the business holdings and activities of the Soeharto family. Among these are Robison 1986; Shin 1989; Jones 1986; Pura 1986; Schwarz 1999; *Swa* August 1985: 12–55; *FEER* 22 May 1986: 40–42; 23 August 1990: 56–59; 25 August 1994: 47–49; *Prospek* 6 March 1993: 18–24. Perhaps the most detailed accounts are to be found in the successive investigations of George Aditjondro (1995b, 2000).

32 Some of the more interesting examples are: Bambang Yoga Soegama, the son of former Intelligence Chief, Yoga Soegama, who became involved in several joint ventures with Tommy Soeharto; Biaki Sudomo, the son of Sudomo, the former admiral and Chief of Security; and Amris Hassan, son of former Education Minister, Fuad Hassan. Outside Jakarta, a similar process was taking place. Dwi Setyo Wahyudi, son of Central Jakarta Mayor, Ismail, established a substantial business group. Such examples are almost endless (Aditjondro 1995b; *Prospek* 6 March 1993: 12–24).

33 The Soeharto children had often suggested that their business success was made more difficult by having a president as father. Tantyo Sudharmono gave an extraordinary genetic explanation for the success of children of political leaders and officials in business. 'If the father is a winner, should the son or daughter be a loser? … Well, the offspring of a racing horse runs faster than that of an ordinary horse, doesn't it? Should we repress the racing horse's offspring? We shouldn't. Or, because the ordinary horse cannot run fast, should the racing horse be required to slow its pace?' (*Prospek* 6 March 1993).

34 Legislation to give domestic business favoured treatment in government contracts for supply and construction, such as Keppres 10, 14 and 14a, were introduced under the ambit of a broader programme known as UP3DN (the promotion and use of domes-

tically produced goods). Ginandjar Kartasasmita was appointed Minister of UP3DN to take charge of the implementation of the programme. See Pangaribuan 1995: 51–73; Winters 1996: 95–141).

35 With Ginandjar's backing, BBG, a consortium of former Team Ten contractors including Bakrie, Fadel Muhammad of the Bukaka group and Iman Taufik of the Gunanusa group, was formed to undertake the US$1.8 billion Exor 4 project. After talks with Fluor and Mitsubishi and the issue of a letter of intent from Pertamina, it seemed that BBG was indeed to be the main contractor. A second consortium (CNT) comprised other former Team Ten contractors, Ponco Sutowo of the Nugra Sanatana group and Wiwoho Basuki of the Tripatra group as well as Ginandjar's brother, Agus Kartasasmita, together with the foreign partners, Mitsui and Toyo.

36 Tommy proposed to purchase cloves at Rp.7,000–8,000 per kilo and sell at Rp.13,000 per kilo, doubling the farmers' income and, in the process, generating an expected profit of US$100 million per year for his consortia (BPPC).

37 So bizarre was the BPPC saga that it was followed in detail in the Jakarta press and by analysts. See, for example, *Editor* 7 July 1990; *Kompas* 30 April 1991; *Suara Pembaruan* 25 May 1991; Schwarz 1994: 153–157; Robison 1997: 44, 45).

38 It was soon unable to absorb the burgeoning stocks of cloves and prices fell to levels lower than they were before. Tommy Soeharto helpfully advised growers to burn half their crop (*Prospek* 12 August 1991; *Warta Ekonomi* 12 August 1991; *Matra* May 1992). As the disaster intensified, it was announced that interest payments on the BI loan could not be met by BPPC. The government was again forced to come to the rescue. In 1993, new loans were provided from state banks, BRI and BBD, at subsidised rates of credit. In 1992 the government arranged for the buying monopoly, the costly and troublesome part of the arrangement, to be handed over to the Federation of Co-operatives while BPPC retained the selling monopoly (*Prospek* 7 March 1992: 70–81; 10 July 1993: 67; *Tempo* 12 January 1992: 79–86; 10 July 1993: 74; 25 September 1993: 88–89; Schwarz 1994: 153–157).

39 It was claimed that offshore loans to Chinese business had totalled US$6.7 billion in 1988 compared to US$888 million for *pribumi*. By 1991, the figures had improved to US$7 billion for Chinese business and US$4 billion for *pribumi* (*Tempo* 20 July 1991: 82–88).

40 For extensive expositions of the sort of economic nationalist ideas gaining currency and political backing in this period, see Hartato 1985; Ginandjar 1985; Habibie 1986, excerpted in Chalmers and Hadiz 1997.

41 When asked why it was he who obtained the monopoly, Tommy suggested reporters should ask his father (*Tiras* 14 March 1996: 23–26).

42 Astra announced the shelving of Rp.1.2 trillion in investment plans, while Liem Sioe Liong's Indomobil postponed a Rp.600 million investment programme (*Asiaweek* 22 March 1996: 50; *JP* 6 March 1996: 8). Indomobil stated its intention of producing a car priced at around Rp.20 million within four or five years based on the exemptions already available under the May 1993 package.

43 Further crowding of the small Indonesia sedan market of 350,000 vehicles by such a costly programme of protection for a small, new and vulnerable producer was also seen as a questionable move, particularly when over 50 per cent local content had already been achieved by Astra's Kijang in the production of multi-purpose vehicles (*Kompas* 1 March 1996: 1; *JP* 1 March 1996: 4; 18 March 1996: 4).

44 The vast and complex takeover of this industry by the Soeharto children and other business conglomerates close to Soeharto are treated in detail in Schwarz 1999; Robison 1997: 60; *Warta Ekonomi* 29 April 1991: 30–39; *AWSJ* 17–18 September 1990: 1, 6, 7; *FEER* 2 May 1991: 40–41; 7 January 1993: 52; 11 August 1994: 70; *JP* 23 June 1995: 1; *Tempo* 24 August 1991: 89–92.

45 Chandra Asri received an implicit subsidy of US$416 million buying LNG (liquefied natural gas) and LPG (liquefied petroleum gas) from Pertamina at subsidised rates (Nasution, Anwar 1995: 16).

46 The investment in mega projects potentially involved US$70 million in domestic and foreign loans. It was estimated that there were some thirty-seven mega projects of Rp.1 trillion each (around US$500 million at the time). These included larger projects such as Chandra Asri and Enim Lestari, the pulp and paper venture, which each involved over US$2 billion (*Warta Ekonomi* 29 April 1991).

47 See the comments of former Oil Minister, Subroto, in *Kompas* 6 May 1995: 4.

48 Overseas borrowing in 1990 stood at US$4.6 billion with debt repayment of US$6.6 billion, an outflow of US$2 billion. Total foreign debt stood at US$65.7 billion or 69 per cent of GDP with a debt service ratio of 35.2 per cent (Ramli 1991: 5). According to the Governor of the Bank Indonesia, Adrianus Mooy, money in circulation had risen 64 per cent in two years and was heading to danger levels (cited in Sjahrir 1991: 88).

49 See Kwik Kian Gie's comments in *JP* 15 April 1993: 1. The economist, Anwar Nasution, also noted that capital exports came from the ease with which large companies got bank credit (*Tempo* 28 September 1991).

4

CAPTURING THE POLITICAL REGIME

Within Indonesia, the small but vocal liberal intelligentsia greeted the state's partial withdrawal from economic control as signalling not only the emergence of a new free market era, but also the imminent loosening of the authoritarian reins over politics. In other words, they saw a relatively uncomplicated relationship between economic liberalisation and political democratisation. Hopes for liberal, democratic reforms in the political sphere among this intelligentsia were bolstered by the popularity of Huntingtonian notions of 'waves of democratisation' in academic discourse, and later, the fall of the Soviet Empire.

It is argued here, however, that economic liberalisation in Indonesia was not accompanied by political democratisation. Instead it took place together with the appropriation and re-fashioning of the existing authoritarian corporatist framework in the interests of a newly ascendant coalition of politico-bureaucrats and their families as well as large business conglomerates, *pribumi* and Chinese alike. The immediate purpose of this coalition was to harness the state, its military and civilian bureaucracies, to the task of constructing and safeguarding private wealth. In this task it was faced with two major challenges. First it had to bind the military and civilian bureaucracies to its will, substituting loyalty to powerful families for loyalties to state institutions, replacing careerists with dependable figures, not only to ensure that state monopolies were delivered into the right hands but also to guarantee the protection of the new oligarchs. At another level the new oligarchy had to keep civil society paralysed and disorganised, a task that was complicated by the growth of a middle class intelligentsia that felt increasingly constrained by the state and by the spread of an urban working class, as manufacturing exports became more important. Social change associated with economic development and industrialisation had unleashed new social forces that gave rise to new contradictions in the existing order.

Thus, the new oligarchic coalitions that emerged in the economic sphere were in fact constructing a new political shell for themselves in the 1980s by appropriating state power without democratising it. Important aspects of this development included the gradual marginalisation of the military from the political process, the moulding of Golkar into the political vehicle of oligarchs, and the establishment of new political laws which governed parties and mass organisations, and enshrined Pancasila as the sole state ideology (*azas tunggal*).

Consequently, in contrast to the expectations of the liberal intelligentsia, the rise of private capital in Indonesia did not necessarily mean open markets or the loosening of the grip of political authoritarianism. Unlike in Thailand, where political parties and parliament had achieved an ascendancy over the military and the centralised state bureaucracy by the early 1990s, in Indonesia an authoritarian corporatist state remained firmly in control while parties still served as the instruments of state power holders.

Moreover, as discussed in earlier chapters, economic deregulation in Indonesia did not result in the sort of economic liberalisation envisaged by the World Bank or neo-liberal economists. Instead it resulted in the selective shifting of monopolies from the public to the private sphere, frequently involving state subsidies and protection for politically well-connected conglomerates (Robison and Hadiz 1993: 25). The opening of financial and capital markets as part of the economic liberalisation process opened new opportunities for state-protected private businesses, including those of the Soeharto family. No regulatory framework existed to constrain the unfettered activities of the new 'robber barons'. Indeed, it is argued here that the oligarchs themselves were busy developing a political format that would further protect their economic ascendance and insulate them from threats, whether emanating internally from within the state, or externally, from wider civil society.

Public debates over political 'openness' and democratisation, which raged in Indonesia in the late 1980s and early 1990s, appeared on the surface to be a contest between a new reformist middle class and an outmoded authoritarian regime. In reality, they also reflected a deeper conflict for ascendancy over strategic institutions of power within elites. In this contest a coalition of oligarchic interests led by Soeharto confronted military and bureaucratic interests led by former intelligence tsar General Benny Moerdani. Significantly, a range of opposition groups, moderate and more radical, were either drawn into these struggles, or attempted to create political space for themselves by capitalising on a political environment characterised by partial elite fragmentation.

Within such struggles, nationalist and populist rhetoric was mobilised by both sides. Sections of the burgeoning middle class were among the more salient of forces eventually drafted into the process of contestation. The culmination was the establishment in 1990 of ICMI, the Association of Indonesian Muslim Intellectuals, intended to provide Soeharto with a new power base to deal with the military and to redress his previously strained relations with organised Islam. ICMI primarily became a vehicle to develop support among the urban Muslim middle class, which welcomed the overture and the avenue it presented to bureaucratic power (Ramage 1997: 7; also see Hefner 1993).

Elite conflicts enabled a new vibrancy within the political opposition, although not necessarily a new effectiveness. Sustained industrialisation prior to the economic debacle of 1997 had produced an increasingly complex civil society that the rigid security-oriented framework of Soeharto's New Order was ill equipped to accommodate. In spite of stringent state controls, for instance, industrial action increased dramatically in the 1990s as the urban working class grew in size. Workers, as well as journalists, lawyers and students, demanded that they be allowed to establish independent associations outside the monolithic corporatist framework organised by the government in the 1970s and codified by the 1985 laws on political parties and mass organisations.

An important development occurred in 1996, when Soeharto's violent, forced removal of Megawati Soekarnoputri as head of the Indonesian Democratic Party galvanised and momentarily unified sections of this opposition. However, in general, a state strategy of selective co-optation and repression left opposition movements fragmented and largely immobilised. Thus, while liberal critics in particular were successful in setting the agenda of public debate on political change, they remained unable to shake the foundations of an authoritarian corporatism increasingly harnessed to the interests of oligarchy.

Appropriating state power

Soeharto versus the military: the contest over state institutions

The dislocation of salient military and bureaucratic interests by the ascendance of new oligarchic interests resulted by the 1980s in increasingly open rifts between Soeharto and sections of the military establishment. Their political struggles focused around the concentration of power in the President and the theme of political succession – which was widely discussed publicly due to Soeharto's advanced age. They represented a polarisation between two elements of state power: those who continued to be based predominantly within the institutions of the state apparatus and those who increasingly built upon social and economic alliances that were assuming an oligarchic character. Clearly exasperated by developing intra-elite challenges, Soeharto threatened in startlingly candid terms that he would *gebuk* (clobber) anybody who tried to remove him unconstitutionally, a warning generally understood to be directed at the military clique under Moerdani (*Gatra* 30 May 1989).

The problem for the military as an institution was that the President had not only developed alternative sources of power, but also imposed his authority on the military high command. This was a process with some history. As we have seen, in the 1970s Moertopo's OPSUS had given him an alternative source of advice and new institutions through which power could be extended. A critical turning point came in 1973 and early 1974 as a power struggle intensified between Moertopo and General Soemitro, Deputy Commander of the Armed Forces and Commander of KOPKAMTIB. Soemitro's failure and subsequent removal from office set back the military agenda.

The tension between those who represented the institutional interests of the military and defended its autonomy within the regime and those who saw the military as merely its 'bodyguards' was to be a continuing and central theme in Indonesia's political life. Those defenders of the institutional interests of the military, now led by Moerdani,[1] were to become increasingly resentful. In particular, they recognised that the influence of Sekneg and of Research and Technology Minister B.J. Habibie challenged not only the political influence of the military high command, including its control of Golkar, but impinged on its authority over the procurement of military equipment, and removed many of its lucrative economic privileges (see Pangaribuan 1995: 67). Both Soeharto and Sudharmono, a military lawyer by training, were of course, strictly speaking, part of the 'military family' themselves. However, a clear distinction was developing by this time between those who had a background in the military apparatus but whose base of power lay outside, and those who continued to depend on the formal institutions of the military to maintain their authority and influence.

By the 1980s and 1990s the privileged position of the military as an institution within the state apparatus had become increasingly ambiguous. According to Schwarz, military personnel held ten out of forty-one Cabinet positions in 1993, compared to thirteen out of thirty-two positions in 1982. While this only represented a slight decline, the 1993 military Cabinet members, according to Schwarz, 'had considerably less clout' than their 1982 counterparts. It worried the military leadership that Soeharto was by this time apparently 'no longer listening' to them, and seemed intent on making the military 'weak and subservient'. One retired general was distressed that only Soeharto had the power 'to get anything done in Indonesia' (Schwarz 1994: 284).

The military's political decline was most clearly reflected in the protracted conflict between the military and the State Secretariat, Sekneg. A capable administrator, Sudharmono had been steadily rising as a key figure in the New Order, using his position as State Secretary to act as a gate-keeper, governing the access of others to the President. However, a real shift in power was signalled by Presidential Decisions (Keppres) 10 and 14a, which effectively cut the military out of that strategic intersection where contracts for supply and construction were decided. Under Sudharmono, Sekneg used its position in the economic sphere to develop a support base among a *pribumi* bourgeoisie by awarding government contracts on a preferential basis (see Pangaribuan 1995).

At the same time, the military's lucrative forestry and transport monopolies had largely evaporated as better capitalised conglomerates and new family enterprises moved in. With the introduction of legislation in 1980 banning raw log exports, all holders of forestry concessions were required to establish processing mills. Holder of numerous such concessions, the military had of course never entertained the idea of investing in or managing them in any serious fashion, but had simply let them out to subcontractors to strip. In the early 1980s, therefore, these forestry interests were progressively being absorbed into the larger corporate empires such as the Barito group (*Tempo* 26 October 1991: 27).

In 1988, Soeharto was initiating further moves to undercut military autonomy and influence. Military Commander Moerdani was made Minister for Defence, which was an effective downgrading of his power. The military-dominated security command, KOPKAMTIB, was replaced by a new organisation, Bakorstanas, which formally reported directly to Cabinet and to the President. The new body, although still headed by the armed forces commander, was supposed to be more accountable and subject to legal constraints, a change in status which represented a downgrading of the autonomy and power of the military (*Tempo* 17 September 1988; 19 November 1988; *FEER* 22 September 1988).

Not surprisingly, 1988 was also the year of the most dramatic direct confrontation between the military and the President. This came when the military failed in a very public attempt to block the appointment of State Secretary Sudharmono as Vice-President after Soeharto's 1988 re-election. Interestingly, it was suggested that Sudharmono had a history of leftist sympathies and associations, or in the official parlance at the time, that he was not *bersih lingkungan* (of a clean background) (Cribb and Brown 1995: 147). That the President simply overrode military objections brought home the realisation that its power and influence was less than was assumed, even though Soeharto clearly still relied upon the military as the coercive base of the New Order. This began a long struggle on the part of military leaders to regain its lost power.

Sudharmono's elevation to the Vice-Presidency was to prove a double-edged sword. It effectively separated him from his power base – Sekneg – and gave him a largely ceremonial role in politics. Nevertheless, the military had to go into full attacking mode to seize back much of the ground taken by Sekneg and the President. They realised that the only way to achieve this was to exert more control over the Presidency, or, at least, to reduce Soeharto's authority over institutional military affairs. Political reform involving the imposition of some form of accountability and transparency upon the state apparatus, especially the Presidency, from the military's point of view, was a logical means of redressing this imbalance of intra-regime power in its favour. Thus, senior military leaders were to make frequent public comments on the need for political 'openness' and encouraged debate about political succession.

General Soemitro, who retired after 'Malari', took the lead in these debates. In his view, now that communism had been destroyed and Pancasila was established as the sole basis for politics, there was no need for the strict control and management of the DPR and the parties. Furthermore, he argued that political parties should be mass based with initiatives coming from the bottom rather than being imposed from above, and thus, Golkar's subordination to the state apparatus should be ended. Political leadership, he also proposed, should be cultivated within the parties and experience in the DPR and MPR should be a prerequisite for candidacy to higher office (Soemitro 1989, 1990). Aiming directly at Soeharto, Soemitro argued as well for the reduction of some Presidential powers, including the power to nominate the membership of the DPR and Golkar, because authoritarian controls were no longer necessary to guarantee continuation of the existing policy regime or to protect the social order.

Serving military leaders also went on the attack. One channel of protest was the military faction in the DPR (F-ABRI). In a forum usually noted for its supine nature, F-ABRI members, notably Police Colonel Roekmini Soedjono and Major General Samsudin, were to take the lead in using the DPR as the vehicle for public discussion of a range of issues. These included government controls on press freedom, the accountability and transparency of the bureaucracy and its operations, and the subordination of parliament to the bureaucracy (*Kompas* 28 June 1989; *Pelita* 28 June 1989; *Merdeka* 28 June 1989; *Tempo* 8 July 1989). The DPR's Second Commission, under the Chairmanship of Samsudin, was used in particular as a forum for public airing of these issues (*Tempo* 8 July 1989; 15 July 1989).

Social issues were also especially targeted by the F-ABRI, which became much more active than the other parliamentary factions in receiving disputants to the DPR, particularly those protesting compulsory land acquisitions in often highly publicised cases. General Samsudin himself was to shake the DPR with the claim that 'the land in Irian Jaya has already been divided up by elements from Jakarta' (*Tempo* 15 July 1989). Public knowledge of such matters was particularly sensitive given the identity of those 'Jakarta elements' that were expropriating Indonesia's land. These allegations therefore constituted a direct assault upon those leading political and business figures that could be seen increasingly to constitute a putative ruling class.

Thus, another dimension of the military–Soeharto conflict related to the question of social justice, in particular the economic dominance of the Chinese business conglomerates and the position of the President's family in the world of business. Given that Soeharto was extremely vulnerable on this issue, such a focus on social justice was an obvious strategy to undermine his public legitimacy. However, the attack by the military on the new bureaucrat-capitalist alliance was not simply a cynical device in the struggle between competing bureaucratic institutions. While the military clearly saw the emerging bourgeoisie, embodying the integration of conglomerate wealth and bureaucrat authority, as an amalgam of power likely to marginalise its own institutional position, this hostility also derived from a genuine tradition of petty bourgeois populism. As General Wahono, the chairman of Golkar from 1988 to 1993, was to lament, big businesses only 'think of their own success and problems', and fail to use their position to help smaller, struggling businesses (*Tempo* 20 July 1991: 80).

Outside the DPR, Moerdani and other senior officers propagated a 'reformist' position in public seminars and military colleges. This position tended to focus upon three major themes. First was the need to move from authoritarianism to a more open political system that would nurture initiative and new ideas. Second was the need to abandon feudal attitudes in favour of a more modern, rational approach. And third, there was the need to eliminate gross distortions in wealth and to protect the weaker sections of society (*FEER* 18 January 1990; 3 August 1989; *Kompas* 28 June 1989; 13 November 1990; 12

January 1991; *Suara Pembaruan* 25 May 1991; *JP* 4 March 1992). Significantly, Soeharto himself employed much of the same rhetoric to secure his own position *vis-à-vis* his opponents. In a famous speech commemorating Independence Day in 1989, Soeharto proclaimed that Indonesia had reached a stage at which it was important to allow the free airing of differences in opinion and to encourage creativity (Soeharto 1989).

However, the military found itself in a dilemma, for it had an organic view of society that fundamentally contradicted the notion of democracy. Not surprisingly, therefore, its vision of political reform, arguably like Soeharto's, was framed by an insistence that it would not involve any disturbance of the existing social and political hierarchies. Senior military officers, including then-Military Commander (and future Vice-President) Try Sutrisno and then-Army Commander Edi Sudrajat endorsed political openness but with the proviso that it did not interfere with economic development, bring about a fundamental change of values (liberal, 'Western'), or disturb national discipline and order. That Sudradjat did not have liberal reform in mind was illustrated by his claim that the concepts of majority and minority did not concern Indonesia because it adhered to the family principle (*Kompas* 19 December 1990). Sutrisno was anxious that in any political reform everyone continued to 'know his place, function and responsibility' (*Kompas* 10 December 1990). A range of military spokesmen expressed concern that vigilance be exercised to ensure that 'openness' was not exploited by 'small groups' (*Kompas* 7 August 1987; 9 November 1990; 17 November 1990; 21 November 1990; 9 December 1990; *Kedaulatan Rakyat* 3 August 1987). Wahono stressed that openness should not lead to 'disintegration' and instead must 'bolster national unity' (*Kompas* 26 December 1990).

The military dilemma, essentially, was to reconcile greater political openness and the reduction of Presidential power with the maintenance of the existing social order and the prevailing political ideologies and state policies; to change the regime and preserve the state. In the system of Pancasila democracy, some centre of power must fulfil the role of arbiter and enforcer of the 'national interest'. Just as the populist social reformism of the military posed no real blueprint for greater social justice other than moral exhortation or arbitrary expropriations, it presented no real solution to the political dilemma.

In the final analysis the military's reformism was always severely constrained. On the one hand, in order to reduce the power of an authoritarian Presidency that it saw as being increasingly a component of a hegemonic alliance of corporate conglomerates and powerful bureaucrat families, it had to open the political system and enter into alliances with social forces outside the state. On the other hand, these strategies would have, by their very nature, undermined the political dominance of the state and its strata of bureaucrats.

Military leaders would also have been quite aware of the dangers of letting the 'openness' debate get out of hand. Throughout this time, the military's own role in social and political affairs was the target of intense criticism from political opposition circles. For example, labour activists were to routinely campaign, domestically and internationally, against military intervention in labour disputes. Such interventions were lucrative for local commands in industrialised areas because it allowed them to make deals with factory owners (YLBHI 1994) in which the military would act as a guarantor of industrial peace. Thus, there was a clear danger that the openness debate would spill over uncontrollably to the issue of removing the military's cherished 'dual function', under which it had maintained the power and ability to forge political and business alliances.

Moerdani and his allies thought that they had won a victory over Soeharto in early 1993 when they successfully promoted the appointment of General Try Sutrisno as Vice-President, thereby blocking any Soeharto intention of selecting someone less amenable – in this case, long-time Soeharto aide, B.J. Habibie. However, former Presidential adjutant Sutrisno might have been Soeharto's first choice in any case. More importantly, Soeharto was powerful enough to quickly strike back by omitting Moerdani altogether from the new Cabinet formed that year, as well as prominent allies such as General Harsudiono Hartas, the influential head of military social and political affairs (Schwarz 1994: 286; MacDougall 1993: 2).

The flurry of military resistance to Soeharto signalled the last stand of the officials of the corporatist, authoritarian state against the encroaching ascendancy of oligarchy. In the decade from the mid-1980s, Soeharto was to gain the upper hand over the military. With the eclipse of Moerdani and his allies, Soeharto increasingly decided on strategic appointments in the military high command. As a result of these interventions, Presidential adjutants such as Kentot Harseno, Sjafrie Sjamsuddin, Hendroprijono, as well as son-in-law Prabowo Subianto, were automatically on the fast track to promotions and vital command positions. Both Harseno and Sjamsoeddin served as military commanders in Jakarta, and Prabowo eventually ascended to the command of elite Kopassus and Kostrad troops, while Hendroprijono headed military intelligence before serving as Minister of Transmigration. Many of these military high-flyers, together with Generals Hartono (army chief of staff and later Minister of Home Affairs) and Feisal Tandjung (armed forces commander and later Co-ordinating Minister for Politics and Security), came to be regarded as close allies and protectors of the oligarchy. Their function as 'bodyguards' to the oligarchy was perhaps best expressed in General Wiranto's pledge, made right after Soeharto's resignation speech on 21 May 1998, that the military would continue to protect him and guarantee his well being. Wiranto, Military Commander and Minister of Defence, was yet another former adjutant whose swift rise through the ranks owed much to Soeharto's patronage ('Who's Who', *Jakarta Post.Com* 7 August 2003).

Golkar as the party of oligarchs

A critical problem confronting the generation of military and civil bureaucrats who successfully established themselves as capitalists was the forging of a new relationship with the state. The abiding interest of generals and bureaucrats and their associates in maintaining the institutions of state power as well as the political and ideological regime that guaranteed autonomy for the state apparatus would not necessarily secure the emerging oligarchy over the long term. In other words, the leading elements of the bureaucrat capitalist families and the major client bourgeoisie were faced with the problem of making the state accountable to their class interests as their patrons retired from office. The fact that Soeharto would eventually leave the political stage, one way or another, must have been a worrisome prospect for political and economic allies who had constructed business empires under his patronage and the system of predatory state authority. Clearly, the departure of Soeharto would threaten their continued access to state favour.

One means of securing their longer-term interests was to transform Golkar into a political party of the oligarchs, independent of the military and civilian bureaucratic elites. By stacking Golkar with political and business clients of Soeharto, the election of any new President could be decided by the new oligarchy rather than the military. Not surprisingly, this endeavour to transform the party encountered stiff resistance from the officials whose authority a new Golkar would circumvent.

As discussed earlier, Golkar itself was originally the creation of the military. Control over Golkar remained in the hands of Soeharto and the military through the Supervisory Board, the *Dewan Pembina*, which exercised authority over selection of candidates and officials and the policy set by the Central Executive Board (DPP) (Reeve 1990: 151–173). An important change of direction was to take place in 1983, however, with the appointment of Sudharmono as Chairman, signalling the decline of military control over Golkar.

It soon became clear that Sudharmono's objective was to establish Golkar as an alternative mechanism to the military for regenerating the leadership of the New Order. This would not only make Golkar more autonomous from the military, but would also allow Soeharto a base of power that was relatively independent of it.

In achieving his objective, Sudharmono brought into the party several prominent civilian politicians and businessmen. The most important of these were Sarwono Kusumaatmadja and Rachmat Witoelar, members of the so-called Bandung group, former student activists in 1966. Although careful to acknowledge the prevailing political and ideological realities, it was clear that this group held a somewhat different vision of what Golkar should be. For the group's members, the two major issues were Golkar's independence from outside control – including the military – and its internal democracy. In essence they were proposing that Golkar should become a new and independent source of power based on the direct mobilisation of popular support – although perhaps in the corporatist rather than the liberal model (*Tempo* 2 July 1988; 16 July 1988; 23 July 1988; *Kompas* 2 August 1988; 16 October 1988).

Witoelar, for example, believed that as the largest party, Golkar was the best institutional platform on which 'a political structure responsive to the needs of the people' could be built. He saw the Bandung group's role as facilitating the process of *regenerasi* within Golkar – passing the torch to the younger generation – while helping the party achieve independence from 'outside forces' (*FEER* 11 August 1988: 22). Witoelar was to become Golkar General Secretary in 1988, succeeding his colleague Sarwono, who occupied that position from 1983.

It is important to note that while such 'young blood' was brought in by Sudharmono, their elevation did not seem to be opposed initially by the Moerdani camp (*FEER* 1 December 1983: 41). As we shall see, in later stages of Indonesian politics, they would ally themselves with the Moerdani and Sudrajat camp, especially once Soeharto had turned to more overtly Muslim-oriented politicians in a further bid to extricate the institutions of the state from military control.

A first step in Sudharmono's plan was to reorganise membership basis from organisations to individuals. The implications of this were significant. Power thus theoretically shifted from organisations such as the military and KORPRI to the leadership of Golkar. While intellectual circles in Jakarta spoke enthusiastically of Golkar becoming an instrument for democratisation (*FEER* 20 October 1983; 1 December 1983), it was, however, more a case of whether or not Golkar was to become the party of an emerging ruling class, separate from state institutions from which it had emerged.

Significantly, Golkar flourished under Sudharmono's management. Its financial base was substantially strengthened. In part this was achieved through the contributions of various Soeharto-controlled organisations, most notably Yayasan Dakab. Under the patronage of Soeharto, Yayasan Dakab provided Rp.50 billion in the period 1988–1993. Also, private business, primarily the major Chinese conglomerates, was expected to provide Rp.30 billion for the same period. Another Rp.20 billion was to be raised from other business sources (*Prospek* 7 September 1991). In 1991, it was estimated that Golkar was in control of Rp.700 billion originating from business donations and accumulated interest (*Prospek* 20 July 1991: 80). Thus, Soeharto's demonstrated capacity to raise large funds outside the formal state budget provided him an important source of autonomy from the military.

The military was clearly restive at the prospect of an independent party with access to mass politics, which might have been the result of Sudharmono's initiatives. From its point of view, Golkar became another arena within which it attempted to contain the ascendancy of the new oligarchs. A confrontation between the military and Sudharmono was therefore inevitable and this occurred in 1988, at the five-yearly Golkar Congress – the Munas IV. Opponents of Sudharmono's changes moved to reimpose upon Golkar tighter military control and greater conformity to the organic political vision. Figures as diverse as Albert Hasibuan, head of Golkar's intellectual section, Jusuf Wanandi of the Centre for Strategic and International Studies, and Minister of Home Affairs Soepardjo Rustam thus argued that Golkar's primary function was to unify and draw in various social groups in society rather than represent interests. Before the Congress they propagated an anti-liberal vision of a party whose function was to integrate, institutionalise and secure political stability (*Kompas* 19 October 1987; 2 August 1988; 16 October 1988).

Significantly, this did not contradict the similarly organicist vision of Soeharto and Sudharmono; the contest over Golkar was not about the organicist vision but about who would preside over the party. Sudharmono may have sought to develop Golkar as an alternative mechanism for renewal of the political leadership, but neither he nor Soeharto had ever contemplated the prospect of Golkar as a party in the liberal vein. Thus, its board was not elected by the party but comprised the President as Chairman and various Cabinet members as Deputy Chairs and other key personnel. As political observer and former Golkar parliamentarian Rachman Tolleng pointed out at the time, Golkar was not the ruling party but the party of the rulers; it did not determine the membership of the government, but on the contrary, its membership was determined by the government (*Kompas* 16 October 1988; 20 October 1988).

Nevertheless, the military's challenge on Sudharmono did serve to contain the influence of the civilian and intellectual rising stars of Golkar – such as Marzuki Darusman, Sarwono and Witoelar – who largely owed their newfound power to his patronage. There were also initiatives to stop 'newcomers' and 'fence-jumpers' from gaining executive positions over long-term Golkar *apparatchiks* (*Kompas* 2 August 1988; 16 October 1988; 19 October 1988; *Tempo* 29 October 1988). While Golkar's reorganisation under Sudharmono had encouraged politically ambitious individuals previously unrelated to the party to join, according to a ruling of Golkar's fourth National Congress (Munas IV), only those who had been with the Golkar 'struggle' for at least ten years could take up national leadership positions. Thus the ruling hindered the progress of such 'newcomers' as Abdurrachman Wahid and Nurcholish Madjid, prominent Muslim intellectuals and leaders who were then flirting with Golkar by serving as MPR members for the party. Also affected were former PPP politician Ridwan Saidi, a past leader of the HMI, and up-and-coming businessman Fahmi Idris (*Tempo* 22 October 1988: 35). Interestingly, the rule was exercised somewhat arbitrarily, as former NU youth leader, Slamet Effendy Yusuf, managed to get appointed to Golkar's youth department, in spite of having stood for election to the DPR under the banner of the PPP as recently as 1982 (*Tempo* 5 November 1988: 23).

The most visible outcome of Munas IV was thus the re-establishment of military authority over the party. Sudharmono was replaced as chairman by General Wahono, a low-profile military officer generally regarded as cautious, conservative and unlikely to take dangerous initiatives. Although Rachmat Witoelar was named General Secretary, Sarwono and Akbar Tandjung were elevated out of Golkar to the Cabinet and none of the prominent younger reformers, notably Marzuki Darusman or Theo Sambuaga – a 1974 student leader – were moved on to the central leadership. Significantly, military nominees were to secure around 70 per cent of regional Golkar seats, a major setback for civilian members (*Tempo* 29 October 1988; 17 December 1988; *FEER* 10 November 1988; *Kompas* 16 October 1988).

Power over Golkar for the time being was therefore to shift back into the hands of the military and civilian bureaucrats that had challenged Sudharmono. Golkar was also to remain a party designed to extend the state apparatus into the social sphere and to prevent the operation of popular politics. In other words, it was confirmed for the time being as an instrument of the military and civilian bureaucracy-proper, not of newly emerging alliances of corporate conglomerates and bureaucratic families whose social base essentially lay outside.

Nevertheless, the battle over Golkar was not over and the tide was to turn yet again against the military. After all, Soeharto remained the ultimate power in Golkar due to his position as head of the all-powerful *Dewan Pembina*, which had extensive veto powers over the party executive. Thus in 1993 the Golkar Congress elected to the chairmanship Soeharto stalwart and Information Minister Harmoko. He was the first Golkar chairman without a military background. In a clear affront to the military, Minister of Research and Technology Habibie was also charged with the task of selecting Golkar's new executive board. As we have seen previously, the military hierarchy had been increasingly resentful of Habibie's interventions into the procurement of military hardware, which it saw as a mere pretext to 'drum up business for his strategic industries'.[2] It was resentful too of his obvious close personal relationship with Soeharto, which it probably saw as being at the expense of military influence (Schwarz 1994: 95). Though Ary Mardjono, a military man, was named General Secretary to replace Witoelar, he was widely regarded more as a Sekneg official or a confidante of Habibie rather than a representative of the military establishment (*Tempo* 9 October 1993).

Again losing their grip over Golkar, senior military officers were uncharacteristically forthcoming in voicing their displeasure over recent developments. One Major-General, Sembiring Meliala, argued that without Soeharto, Harmoko and Habibie were 'nothing', and insisted that the military would eventually choose Soeharto's successor. In widely publicised comments he also suggested that no one could attain the Presidency without military support, and even that the people wanted a President from the military (Schwarz 1994: 355). Given that the military still controlled the bulk of regional Golkar organisations, he argued that the military would simply takeover again once Soeharto went (*Detik* 27 October–2 November 1993).

The Golkar experience of the 1980s and 1990s was instructive. It clearly showed contestation over an extended period of time between the interests of those whose power firmly resided in the institutions of the state bureaucracy, including the military, and those who were attempting to appropriate these institutions for interests that lay outside. The latter essentially comprised newly emerging oligarchic alliances of politico-bureaucratic families and business conglomerates that had been incubated by state power but whose longer-term survival now lay in their ability to appropriate its institutions. It is significant that positions in Golkar from 1993 were to be increasingly filled by members of these oligarchic families – most prominently, the Soehartos. Thus, Bambang Trihadtmodjo and Siti Hardiyanti Rukmana – otherwise known as 'Tutut' – were included in the Executive Board, as were other children and relatives of bureaucrats and business people (*Editor* 4 November 1993), including Ponco Sutowo (son of Ibnu) and Gunarijah Mochdi (sister of Ginandjar Kartasasmita), who were named deputy treasurers. Furthermore, among the department chairs under Harmoko were Bambang Soegama, son of former intelligence chief Yoga Soegama, Bambang Oetojo, son of former army chief Bambang Oetojo, and Tantyo Sudharmono, son of the former Golkar chairman himself. As MacDougall aptly puts it, each of these *anak pejabat* (children of officials) is the 'offspring of some figure who has been close to Soeharto, and so they are eminently politically reliable in a crunch' (MacDougall 1993: 5). For such people, control over Golkar also meant guaranteeing continued access to state patronage and the power to elect a President once Soeharto was gone.

ICMI: co-opting the new Muslim middle class

Another organisation that was drawn into the contest over state power was ICMI, the Association of Indonesian Muslim Intellectuals. Although the political creation of Soeharto and his close aide, B.J. Habibie, the original initiative for its formation is often attributed to grassroots activists.[3] However it actually came into being, ICMI was to become an institutional channel within the regime that enabled Muslim political activists to advance their political and bureaucratic careers. For Soeharto, ICMI presented an opportunity to conjure up a corps of new mandarins that could countervail the influence of the military bureaucracy in particular, while offsetting calls for democratic reform by providing a place for members of the Muslim middle class within the state. This new Muslim middle class was of course itself the product of sustained economic development during Soeharto's rule.

Significantly, ICMI also served to redress Soeharto's previously strained relations with organised Islam, the basis of which was the suppression of Islamic-based social and political organisations and activists in the 1970s and 1980s. The most important Muslim political figure to reject ICMI was NU leader Abdurrahman Wahid, who argued that it was set up to 'cater to the bureaucratic and intellectual aspirations of Muslim intellectuals within and outside the bureaucracy' and to ensure that Soeharto received their support. He also rejected it for its 'exclusivist' nature, which he argued would alienate non-Muslims and nominal Muslims (Wahid 1990: 4). However, the establishment of ICMI won over to Soeharto's side a number of political activists who were previously amongst the harshest critics of his regime, including NGO figures such as Dawam Rahardjo and Adi Sasono, Muhammadiyah leader Amien Rais, and Islamic scholar Imaduddin Abdulrahim. Indeed, such figures as Sasono were to praise Soeharto for what he had done for the *ummat* (Ramage 1997: 105) after the establishment of ICMI – though significantly, they continued to be highly critical of the military as an institution and of its role in political life. Former PPP politician Sri Bintang Pamungkas, for example, argued that the military should not have automatic representation in parliament and that, in fact, this was an obstacle to democratisation (quoted in Samego *et al.* 1998: 208).

It was no surprise that the military was inclined to see the establishment of ICMI as yet another threat to its already eroding political position and an undisguised tool of Soeharto. Ramage quotes one senior retired military officer as suggesting that ICMI was 'pure political manipulation by Soeharto', who was unsure that the military would support his candidacy for President in 1993 (Ramage 1997: 42). Indeed, in a break with tradition, the military took its time in expressing support for a renewed Soeharto Presidency that year.

Naturally, the fact that Habibie headed ICMI would not have endeared it to the military in the first place. Moreover, sections of the military were clearly wary that ICMI provided a platform, and more importantly, access to bureaucratic power, for intellectuals and activists whose calls for democratisation were construed as partly a euphemism for de-militarisation. In 1993, a number of ICMI-related bureaucrats, Habibie allies such as Wardiman Djojonegoro and Haryanto Dhanutirto, were appointed to the Cabinet, though none of the political activists within ICMI had yet scaled these same heights. ICMI was also well represented in the Golkar Executive Board under Harmoko with Habibie associates such as businessman Fadel Muhammad and communications scientist Marwah Daud Ibrahim winning top positions (MacDougall 1993: 5).

Furthermore, military figures were worried too that ICMI would provide a base for those who advocated an Islamic state and were opposed to Pancasila. Though a major component of ICMI consisted of entrenched technocrats and bureaucrats, a more vocal element urged the empowerment of Muslims economically and politically, though few would or could speak openly in terms of an Islamic state. By contrast, defence of the Pancasila state at all cost had been well ingrained in the political ideology of the military, not least because it provided the basis for the organisation of state and society relations, under which it could play a legitimate role.

Significantly, the establishment of ICMI itself was to create cleavages within the military apparatus. The military was soon popularly perceived to be divided between anti-ICMI 'nationalists' and pro-ICMI 'greens' (the preferred colour of Muslim parties in the past), who were simultaneously considered to be Soeharto loyalists. Indeed, the latter included close allies of Soeharto and his family, such as one-time Army chief of staff and later Minister of the Home Affairs General Hartono, and Co-ordinating Minister for Defence and Security General Feisal Tandjung. The fast-rising Soeharto son-in-law, Prabowo Subianto, was later also to forge alliances with ICMI figures, first as head of the elite Special Forces and then of the Strategic Reserves. With Hartono he was to support a think-tank, the CPDS (Centre for Policy and Development Studies), which was mostly inhabited by ICMI intellectuals. The CPDS is widely regarded to have included such individuals as A.M. Fatwa (a former political prisoner), Amir Santoso (University of Indonesia), Din Sjamsuddin (a director general of the Department of Manpower), Yusril Mahendra (University of Indonesia), Lukman Harun (Muhammadiyah), and Fadli Zon (a fast-rising young Muslim activist closely allied to Prabowo). Though the loyalty of some of this group to Habibie and the mainstream ICMI leadership was questionable at best,[4] it was the site of a new alliance between military officers strongly connected to oligarchic interests and Muslim intellectuals and activists, some of whom were once counted among the Soeharto's foes.

Worries over ICMI among sections of the military leadership must have reached new heights in 1993, when Soeharto increasingly identified himself with ICMI by agreeing to become its chief patron (*FK* 16 September 1993: 15–16). Yet in conjuring up ICMI as a device for balancing military power, Soeharto potentially facilitated important contradictions: he raised the spectre both of Islamic politics and of a populist challenge to the organic state. As Soeharto would find out, ICMI figures such as Amien Rais and Sri Bintang Pamungkas were unpredictable and difficult to control.

But what did the former anti-Soeharto firebrands who helped formed ICMI receive in return from the President? As mentioned earlier, ICMI provided them with both a legitimate voice and, for the first time in the New Order, real access to political and bureaucratic power and protection. For Muslim political activists once persecuted and gaoled, ICMI provided a base to propagate populist ideas without fear of the security apparatus, a privileged position never known to them before. To many, the establishment of ICMI signalled a new era in which Islam finally had 'arrived' as a social and political force. In the euphoria, some may have even seen it as a possible tool to strike back at those deemed responsible in the past for Muslim suppression, including within military circles. 'Enemies' would have included the Catholic Moerdani, as well as the Catholic and Chinese intellectuals and business figures that gained influence in the 1970s under the tutelage of Ali Moertopo. In an interview with the writer V.S. Naipaul, Imaduddin, for example, argued that the Moertopo group used to threaten Islam and was afraid of 'Islamic development'. With the rise of ICMI, however, he boasted that due to 'God's will', he now had 'friends in the cabinet' (Naipaul 1998).

Individuals such as Imaddudin – who claims to have been instrumental in the establishment of ICMI – were quite aware that the organisation would be used by Soeharto to help legitimise his rule and assuage critics. The choice to approach Soeharto through Habibie was a calculated one, however, as 'nothing can be done in this country without the approval of the first man' (Naipaul 1998).

Nevertheless, in many ways ICMI was an artificial entity and embodied important contradictions. Significantly, ICMI chairman Habibie was a proponent of a high-tech vision of economic development, which contradicted the populist vision supported by many of ICMI's political activists, premised on a redistribution of wealth favouring the majority of the indigenous and Muslim poor (Schwarz 1994: 182). Sasono and Rais, for example, were known to support a Malaysian NEP-style programme that would raise the status of the indigenous petty bourgeoisie. This difference was mostly ignored, nevertheless, as the marriage of convenience with Habibie (and by implication, Soeharto) was far too attractive to abandon. Thus an early confrontation between 'activists' and 'bureaucrats' was averted when Dawam Rahardjo accused Habibie of running a 'one man show' in 1991, by virtue of Rahardjo's acceptance of a position in the ICMI Board of Experts (*Tempo* 23 February 1991).

Nevertheless, this populist vision represented not only a contradiction with Habibie's favoured policies but also logically meant opposition to the rapacity of bureaucratic and capitalist families. It certainly meant attacking the dominant position of the Chinese conglomerates, but by extension, also the avarice of the Soeharto family itself. Amien Rais was to incur Soeharto's wrath by drawing public attention to the Busang goldmine scandal, which implicated Soeharto family members, as well as the lucrative operations of the well-connected American mining giant, Freeport, in Irian Jaya.[5]

Tensions also grew as some members eventually began to question ICMI's effectiveness as an avenue to power (Ramage 1997: 106) – because no one from the activist component of ICMI, as mentioned earlier, had secured a Cabinet position by 1993 in spite of Habibie's protection. Indeed, Soeharto could have deliberately stalled the rise of the firebrand elements within ICMI, while fuelling their hopes by elevating the less threatening technocratic component.

In spite of these contradictions, the point to be made is that ICMI was an important instrument in the protracted struggle being waged between Soeharto and entrenched military and bureaucratic interests for dominance over the state. There was also significantly an intersection between the economic and political populism of ICMI and the interests of a section of the oligarchic elites coalescing around Soeharto. *Pribumi* businessmen such as Fahmi Idris and Fadel Muhammad, who had gained much from Sekneg patronage in the 1980s, were to play an important role in both ICMI and Golkar, while expounding on the need for a more equal distribution of the economic development cake. Additionally, ICMI's importance has been that it provided a launching pad for the political ambitions of Habibie – Soeharto's eventual successor – by ironically casting the champion of a highly elitist economic development programme in the role of protector of populist and Muslim interests. ICMI was thus never to act as a genuine democratising force, but just another conduit within the organic state.

Notes

1 Ironically, a protégé of Moertopo within the elite circles of military intelligence.
2 Habibie was also chairman of BPIS, which presided over the development of hi-tech state industries.
3 For an account of ICMI's establishment, see Hefner 1993.

4 Hartono, for example, was widely regarded as a rival to Habibie for the Vice-President in 1998. See *Gatra* 16 August 1997: 33–38.

5 Rais claimed that Indonesia was being exploited by foreign operations in the mining sector, which garnered huge profits. Ironically, the Busang gold discovery was later revealed to have been an elaborate hoax on the part of a Canadian mining firm.

5

DISORGANISING CIVIL SOCIETY

It is significant that however bitter the struggle was between Soeharto's oligarchy and representatives of such institutions as the military, they shared a common vision of authoritarian rule through a form of organic corporatism. No less than the military, Soeharto was determined that state power would not be challenged by popular politics. Yet the consolidation of the oligarchy and the process of rapid economic growth both produced challenges from outside the state.

A prominent source of opposition to Soeharto had emerged in the 1980s from the periphery of the regime itself, in the form of a conservative reaction represented by the so-called Petisi 50 group, an association of retired military officers, bureaucrats, and pre-New Order party officials shunted aside by Soeharto. They were invariably early supporters of Soeharto and the New Order's suppression of communism, but had grown increasingly restive of Soeharto's apparent unaccountability to anyone else. While the Petisi 50 group were active in publishing pamphlets and critical statements (e.g. Petisi 50), and developed some links with disenchanted grassroots Muslim activists that provided a potential social base, its significance was more symbolic than real. Indeed, the quarrel of the Petisi 50 was basically with the individuals in power, rather than the system of power itself. Thus, detractors labelled its members as the *barisan sakit hati* (Disappointed Front).

Nevertheless, the links forged between the Petisi 50 – as well as 'Fosko', a discussion group formed by retired military officers – and Muslim activist radicals, however tenuous, were sufficient to elicit fears of a potentially dangerous source of challenge. Thus Soeharto moved quickly to clamp down on this alliance. Prominent Muslim leaders such as A.M. Fatwa and H.M. Sanusi (a former Cabinet Minister) as well as former General Dharsono were imprisoned in the mid-1980s for alleged complicity in a range of subversive activities (Bourchier 1987). These included the planting of a bomb at the Liem Sioe Liong-owned BCA office in Jakarta, and complicity in the Tanjung Priok riots, which saw possibly hundreds of demonstrators from the working class port shot dead by security forces. Petisi 50 members were also variously banned from leaving the country and hindered from conducting business, as well as subject to a state-enforced news black-out.

In contrast to such repression, during the period between 1988 and 1994, a degree of freedom in public debate emerged and an associated relaxation of controls upon the press allowed reformist ideas a wider arena of expression.[1] This situation was partly attributable to elite conflicts described earlier. But new social and economic contradictions arising from Indonesia's rapid industrialisation were also to create the conditions for challenges to state authority and for the ascendancy of the oligarchs. Such contradictions notably involved contests between urban workers and manufacturing capital, customary land-holders and investors seeking land for industrial or commercial purposes, and elements of the burgeoning middle class and state officials diverging over such issues as rule of law, transparency and accountability, and freedom of expression. Thus, an environment more conducive to the revitalisation of political opposition movements was created.

The broad oppositional movements to emerge included middle class groupings of intellectuals and NGO and student activists, and increasingly, informal or semi-formal associations of dispossessed peasants and members of Indonesia's new urban, industrial working class. The values explicitly or implicitly espoused were diverse and showed a multiplicity of 'indigenous' as well as 'foreign' influences. They included versions of liberalism, social democracy, radicalism, socialism and various forms of nationalism and populism, as well as the traditions of social and political activism of Indonesian Christian and Muslim communities rooted in the anti-colonial struggle.

Yet a range of factors hindered the effectiveness of opposition, not least of which was this diversity itself, which prohibited ideological coherence. Thus, for example, alliances between the rising middle class and organised labour or the bourgeoisie never emerged. The most inhibiting factor was surely the institutional framework of the New Order, which offered no avenue for political activism and organisation outside of the instruments of state control. This framework, which was particularly geared to ensure the continued disorganisation of civil society, was codified by a series of legislations on politics in 1985.

Harnessing authoritarian corporatism: Pancasila and the oligarchic project

It is clear that throughout the 1980s and most of the 1990s, political opposition in Indonesia operated in a different institutional framework to that of Thailand, which was then experiencing a process of political liberalisation. The concepts of representative and opposition politics continued to be rejected as legitimate propositions within the official state ideology in Indonesia, despite frequently stated assertions of a new phase of 'political openness' among wide sections of the political elite. Thus, the institutions of political life, including parties, parliaments and elections, remained the instruments of state control and thus, did not offer the opportunity for either representation of interests or capture of government. Whereas in Thailand, liberalism in the system of political representation was kept in check by ruling elites through periodic military coups, in Indonesia liberal democracy was regarded as inherently hostile to an established form of corporatism, known as *Pancasila* democracy. Consequently, there remained no effective institutional avenues outside the existing corporatist apparatus for political participation and contestation, although attempts to establish independent associations continued to be made.

The rules that governed Indonesia's brand of authoritarian corporatism, though already effectively in place since 1973, were further entrenched by a set of controversial laws on political parties and mass organisations that were passed in 1985. These laws essentially formalised state recognition for only a limited number of political parties: Golkar, the PPP and the PDI, the three contestants of the 1977 and 1982 elections. They also confirmed the corporatist monopoly of government-created organisations in the 'representation' of groups in the social sphere – in reality aimed to pre-empt the development of effective independent associations. Significantly, the government was empowered according to these laws with wide-ranging rights to ban parties and organisations deemed a threat to Pancasila, which was enshrined as the ideology to which all had to profess adherence. This particularly enraged Muslim groups who were affronted that their organisations had to replace their Islamic base with a mere secular creed, Pancasila (HMI 1984). Thus, in a letter to Soeharto, the highly respected former leader of the Masyumi, Sjafruddin Prawiranegara, accused him of contravening the original intentions of Indonesia's 'founding fathers' by disallowing associations to be formed on the basis of religion (Prawiranegara 1983).

The importance of Pancasila to the oligarchic project, however, was that it effectively nullified the very notion of legitimate political opposition on the pretence that the state ideology emphasised co-operation and harmony between different social groups, under the aegis of a benevolent state. The official interpretation of Pancasila thus cast state–society relations in a similar light to relations in a family, with the state taking up the traditional, wise father-figure role. But this also allowed it wide-ranging powers to harshly punish any disobedient member of the 'family'.

It was no surprise that sections of organised Islam were among the fiercest critics of these laws, especially prior to the establishment of ICMI. As discussed earlier, the weakening of Islam as a social and political force clearly suited the logic of Soeharto's New Order, which was premised on state control and the demobilisation of mass-based groups and associations. Indeed, by the early and mid-1980s, relations between Soeharto and political Islam were so strained that overt, bloody confrontations involving security personnel and various Islamic-oriented groups became quite common. Apart from the infamous Tanjung Priok case of 1984, the brutal suppression of local Muslim insurgents in Lampung by a security force led by one-time Presidential adjutant Hendroprijono is a prominent example ('Who's Who', *Jakarta Post.Com* 7 August 2003).

But opposition to the corporatist format as enshrined by the 1985 laws was not confined to Muslim groups. Legal and human rights organisations were at the forefront of this opposition, as were non-government organisations that fought for their exclusion from the stipulations of the law on mass organisations (Setiawan 1996: 40). Taking advantage of political space partly created by brewing elite conflicts, liberal as well as more radical opposition groups vigorously pursued the issue of the concentration of political and economic power in the hands of large business conglomerates and oligarchic families, well into the last years of Soeharto's rule.

While overt civil society-based political opposition regained vibrancy during these latter years, it remained severely constrained by an inability to develop effective organisational vehicles from which to mount a challenge to state power. The draconian laws of 1985 succeeded in negating both the very notion of opposition to state authority, as well as the legitimacy of associations outside of the direct reach of state control.

Consequently, as late as the mid-1990s, opposition figures were landing in gaol or experiencing state intimidation for challenging the way that politics and society were organised. Prominent among these was former parliamentarian Sri Bintang Pamungkas, incarcerated for ostensibly defaming Soeharto, but who had also challenged the political arrangements of the New Order by setting up an unrecognised political party, the PUDI (Indonesian United Democratic Party). Labour leader Muchtar Pakpahan, who formed the independent union SBSI (Indonesian Prosperity Trade Union) – which the government refused to recognise for years – was also gaoled, ostensibly for being behind rioting in Medan in 1994 and in Jakarta in 1996. Student and labour activists who had formed the People's Democratic Party (PRD) were among those receiving stiff prison sentences as well. They were 'guilty' of both forming an unrecognised political party and a fledgling labour organisation (the PPBI – Indonesian Centre for Working Class Struggle). In addition, journalists active in the Alliance of Independent Journalists, AJI, were also harassed, intimidated and imprisoned. AJI itself was set up to challenge the state-sanctioned Indonesian Journalists Association, PWI, in the wake of the closing by the government of three news publications, including the respected *Tempo*, in June 1994 (Hill, D. 1994: 72).

In any case, the overturning of the laws that helped to insulate oligarchic power from popular pressure would have to wait until the fall of Soeharto in 1998. For the time being, the New Order continued to be successful in setting the agenda and prescribing the institutions that defined state–society relations at the expense of opposition movements. In other words, in spite of the emergence of salient new social forces in the context of rapid economic growth, and the rise of new demands, civil society remained effectively disorganised.

New social forces

Middle class reformism

Much of the debate about democratisation in Indonesia has focused on the potential role of the middle class as a democratising agent. The middle class, however, is widely varied and includes among its ranks such diverse groups as government bureaucrats, urban professionals, managers of national and multinational firms and banks, teachers, students, journalists, artists and rural *ulama*. While sections of this middle class, especially the urban intelligentsia, have expressed interests associated with economic and political liberalism, others have displayed a political conservatism rooted in the privileged position enjoyed while Indonesia's economy was growing, as well as the lack of institutional channels through which they could politically stray. Other groups – such as NGOs and student activists – made forays into more radical politics, involving themselves in environmental, labour, and peasant movements. Thus, the middle class has certainly produced yuppies, but also secular liberal reformers, supporters of authoritarian corporatism, rural populists and other kinds of radicals (see Robison 1996). In any case, the growth of the middle class, especially its urban component, tied to rapid economic growth, underpinned many of the hopes within Indonesia for the emergence of a democratising force as it liberalised economically.

Usually it was the representatives of the urban intelligentsia, consisting of academics, journalists, lawyers and business practitioners that expressed the hope that the liberal reforms underway in the economic sphere would be replicated in the political life of Indonesia. This link between the market economy and political democracy was a logical expectation in terms of the liberal view of society, because both were natural expressions of the sovereignty of the individual over the state. These expectations were mirrored in confident and self-congratulatory articles and features on middle class lifestyles and attitudes frequently printed in such publications as *Kompas*, *Eksekutif*, *Matra* and others.

Liberal reformers essentially thought that economic deregulation and the protection of individual civil rights would break up the agglomerations of monopoly power in both the economic and political spheres. This was a central theme running through the arguments of economic liberals, including Kwik Kian Gie, Christianto Wibisono and Sjahrir, all prominent public commentators and activists. In their view, deregulation would, by definition, undermine the system of *dirigisme* that enabled officials and political power-holders to build networks of patronage by constraining markets and allocating rents. By reducing the state to a manager of rules and laws of the marketplace, the sorts of political and economic cartels that emerged from state-sponsored monopolies would be eradicated.

Such institutions as the World Bank also, indirectly, contributed to the debate. The World Bank had become interested in questions of governance and the weakness of regulatory legal and administrative frameworks in the early 1990s (World Bank 1993a: 91, 92, 1995: 49, 50). Complaining of the 'present lack of transparent, predictable and enforceable rules for business', it cited malpractices in state procurement, the allocation of licenses and tendering processes for infrastructure contracts (World Bank 1993a: 135–164, 1995: xv–xvii). Assuming that institutional capacity was the core of the problem, the World Bank supported a range of programmes aimed at improving economic legislation, increasing skills and improving procedural regimes in procurement and tendering as well as in the stock exchange, government departments, banks and the legal apparatus (Trubek *et al.* 1993; World Bank 1993a: 135–164).

Lack of capacity was, however, not at the root of the problem. Liberal critics within Indonesia recognised that it was one primarily of power and politics. Thus, economists such as Sjahrir and Kwik Kian Gie characterised the banking system as one out of control and manipulated by big business (*JP* 15 April 1993: 12; *Tempo* 9 March 1991: 44). They recognised that the regime's lack of interest in the sort of institutional reforms proposed by the World Bank did not suit the interests of the politico-business oligarchy, which wanted a state with the power to control and allocate rents. In particular, control of large sources of off-budget funds was necessary, not only in the building of personal politico-business empires but in order to assemble the networks of patronage and support which constituted the heart of the regime. Even after the fall of Soeharto, despite attempts to eliminate the *yayasan* and to reform such institutions as Pertamina and Bulog, Ministries and Agencies reported the existence of non-budget funds of around Rp.7.7 trillion (US$860 million) (Government of Indonesia 2000: 8).

Significantly, the power struggles between the military and the oligarchy opened the political arena for unprecedented public debates and criticism of the government in the early 1990s. But as early as 1985, former Finance Minister, consultant and businessman, Professor Sumitro Djojohadikusumo, had launched a virulent public attack on costs imposed on business by practices of corruption, collusion and protection (*Kompas* 14 August 1985: 2; 23–24 August 1985). He entered the arena once again in 1993, alleging that up to 30 per cent of development funds were leaked through over-invoicing and as a consequence of a general culture of corruption (*Editor* 9 December 1993: 60–61).[2] Corporate and financial scandal, bad debts, collusion, budgetary leakages, illegal levies and the economic costs of corruption, all became ammunition for the reformist critique (*Editor* 9 December 1993: 60–61; *Kompas* 13 May 1996; *FK* 25 March 1996: 95–108).

It is interesting that many Indonesian economic liberals were to propose a functional link between capitalism and democracy. They argued that only political democracy could provide mechanisms that mediated increasingly complex and autonomous social forces outside the state. The economist Sjahrir, for example, drew on the example of the enthusiastic response of the Thai stock market to the fall of Suchinda and his replacement by the democratic Anand to suggest that democracy is necessary to the social stability of industrialising economies. Sjahrir was also to propose that as governments rely more heavily on taxes from society they are subjected to increasing demands from new taxpayers with the consequence that rights of representation and citizenship became more difficult to deny (*Warta Ekonomi* 22 June 1992: 8)

The Chinese-Indonesian intellectual Christianto Wibisono argued that only a democratic political system would be able to release the productive forces necessary to compete internationally.[3] Former Lippo Bank executive Laksamana

Sukardi who, like Kwik Kian Gie, became a key figure in the PDI, was critical of the increasing concentration of economic power on the basis of huge state bank credits and the transfer of public monopolies to large private business groups without the consent of the people. Expressing concern about the gross mismanagement of the economy that had led to mounting foreign debt, he also argued that controlling these depredations required democratic controls (*Kompas* 11 July 1993).[4]

It was significant that the upper ranks of Indonesia's urban middle class were increasingly well placed to oppose government policy. This was owing to the fact that many of them had become increasingly independent of the state in their employment and activities as consultants, bankers, businessmen and lawyers working with private firms and think-tanks.[5] Kwik Kian Gie and Laksamana Sukardi, for example, helped to develop the PDI's capacity for systematic policy critique. While these liberal reformist critics did not condemn conglomeration *per se*, they drew a distinction between firms whose position was achieved in competition in open markets and those they claimed were founded in political manipulation, monopolies, cartels, trusts and distorted markets as well as collusion in banking and on the stock exchange. Thus, the PDI proposed anti-trust legislation as a central and distinguishing part of its policy agenda.

Crucially, the views of such liberal reformers were widely disseminated in public given that the press has frequently been in the hands of sympathisers. Indeed, this allowed liberal reformers to attain a degree of intellectual ascendancy over public debates on economic and political development. Moreover, to the government's displeasure, non-governmental organisations increasingly monopolised debate on reform in a variety of areas, from human rights to environmental policy, under a leadership drawn from the urban middle class, and made use of their broad international connections. Nevertheless, in spite of their intellectual ascendancy, middle class reformers remained spectacularly unsuccessful in achieving concrete political results.

There were a number of reasons for this failing. First, liberal, social democratic and more radical reformers were not always the dominant component in civilian and opposition politics. Various corporatist reformers, whose strategies were primarily aimed at achieving greater influence over policy making and autonomy of institutions within the existing corporatist structures, rather than their replacement with representative institutions and ideologies, were much more involved in the actual political manoeuvrings. The ICMI intellectuals and the reformers brought into Golkar by Sudharmono in the 1980s are cases in point. Second, liberal reformism relied almost entirely upon an urban intelligentsia that was unable to draft the bourgeoisie into the project of liberal opposition.[6] Whereas the bourgeoisie in Thailand was attracted to what Anderson terms 'electoralism' as a mechanism for outflanking the central bureaucratic apparatus, in Indonesia it threw itself behind the status quo, or was more attracted to a reform of corporatism. Moreover, the middle class itself did not constitute a coherent force for liberal reform in Indonesia, for the most part basking in the material privileges accorded to it with sustained economic development. Finally, social democratic and radical reformist projects that might have emerged from alliances with the growing working class never took off because of the latter's organisational weaknesses. Thus such alliances did not attract the interest of most middle class activists, only a limited number that worked in the labour rights area. It is significant that the urban industrial working class, though growing in numbers, still suffered from the legacy of 1965, which included the violent crushing of leftist labour organisations, followed by the development of stringent controls over organising (Hadiz 1997).

In the end, the regime was able easily to contain the political effects of growing discontent. The prospects for reform, therefore, appeared to lay not in any liberal political victory through the agency of the middle class but either within the incrementally progressive corrosion of oligarchy by engagement with global markets or in the political collapse of the Soeharto regime from within.[7]

Ultimately, in any case, the view of the middle class as inherently liberal and reformist is difficult to sustain. With a primary concern for security and stability, for careers, property and improved living standards, there is a natural middle class fear of economic decline and political chaos that forces them into the arms of regimes which offer economic growth and political order. Hence the enthusiastic support of the middle classes for the overthrow of Soekarno in 1965 and 1966. While the leadership of such reformist political institutions as the Legal Aid Institute (YLBHI) and the various student movements were drawn from the middle classes, so too were the leaderships of conservative organisations and think-tanks integral to the regime. These included the CSIS, as well as the CPDS and CIDES (Centre for Information and Development Studies) – both with ICMI links. Moreover, state technocrats, both *dirigiste* and free market, were clearly middle class, as was the Golkar civilian leadership devoted to the organic statist ideals of Pancasila. It is instructive that extra-judicial killings of alleged criminals in the early 1980s, known as the *penembakan misterius* (Petrus) or mysterious killings (Cribb and Brown 1995: 129–130) had also attracted the widespread support of a middle class fearful of criminal violence. In the end, whether the middle class serves authoritarian or democratic regimes depends on the particular context and political terrain in which it finds itself. As suggested by Rueschemeyer *et al.* (1992: 8), they have historically tended to support authoritarian regimes, particularly when fearing the threat of popular unrest.

This is not to say that middle classes, whether managerial, professional, technical or 'intellectual', have no central core of abiding interests deriving from the very nature of their social position. Although they are found as the *apparatchiks*, the experts and technicians within a range of regimes, middle classes ideally require some degree of protection from arbitrary rule, some basic civil rights, some organised framework within which skills, qualifications, credentials and expertise might be expected to prevail over political whim. So long as these basic protections are afforded within frameworks of economic growth, political security and rising living standards, it is difficult to argue that middle classes are inherently supportive of liberal and democratic reforms.

Not surprisingly, it was the question of unaccountability and non-transparency that brought liberal, middle class reformists together, rather than democratisation as such. Common to virtually all middle class critics was the belief that Pancasila had been appropriated by powerful elements within the New Order and that the so-called 'common interest' of society was nothing less than the interest of powerful officials and corporate clients. Thus, they argued for solutions that included recognition that controls upon despotism necessarily involved some form of representative politics, and guarantees of rights and freedoms for its citizens. This would make the holders of state power more accountable and render their actions more transparent.

Among the most prominent of the middle class reformist figures was Abdurrahman Wahid, leader of the Nahdlatul Ulama (NU) and a founder of the Forum Demokrasi, a loose association of liberal, social democratic and religious intellectual and political figures. Wahid was often to convey his criticism of the order that existed in cultural terms, and argued that the claims of the New Order to rule by consensus merely justified authoritarian rule that stifled creativity by encouraging conformity. This was a most direct rebuttal of the state's claims to constitute the general will of all and, indeed, a fairly clear allegation that the state really constituted the interests of the wealthy and powerful. It was also in terms of this kind of domination that he explained the withdrawal of the Muslim organisation, NU, from participation within the PPP. Such a withdrawal, he explained, was required to enable it to genuinely address the interests and aspirations of the diverse Muslim community, including the poor and workers. Accepting positions in the government, he stated, would ruin the NU because the interests of the state were not always the same as those of the people (*Kompas* 20 April 1993; *FK* 14 May 1993; *Detik* 4–10 August, 1993: 8–9).

Another noted middle class Muslim reformer was Sri Bintang Pamungkas, a former parliamentarian and ICMI member who, as mentioned earlier, would eventually fall out badly with Soeharto. Pamungkas's criticism was more directly aimed at the way that politics was organised and the glaring social disparities that it made possible. According to Pamungkas, national unity required the removal of the conglomerates, the eradication of poverty and the functioning of the MPR as the instrument of people's sovereignty' (*FK* 1 April 1993: 16). In other words, he saw the anti-democratic nature of the New Order as emerging from its social and economic base.

This head-on challenge to the organic-statist position by Pamungkas and others was most explicitly drawn out during the period of debates on 'political openness'. Perhaps the most important theme that came through was the argument that organic statism was neither a functional response to chaos nor a condition inherent to Indonesian culture, but was the institutional vehicle whereby the bureaucracy and the military imposed their political domination. Legal scholars and activists T. Mulya Lubis, Marsillam Simanjuntak and Adnan

Buyung Nasution have made extensive expositions of this theme in a number of studies (Lubis 1993; Simanjuntak 1994; Nasution, Adnan 1992).

But as already mentioned, the recurring problem was that middle class reformists had no effective organisational vehicles with which to challenge the concrete arrangements of state and society relations premised on the organic-statist vision. They were faced with the option of forming new organisations, thereby risking suppression, or joining official ones, and thereby risking the likelihood of co-optation.

Some of those who chose the former option came to set up the Wahid-led Forum Demokrasi in 1991, which was for a short time the focal point for the proliferation of critical ideas. It was filled with some of Indonesia's best-known intellectuals, including Lubis, Simandjuntak, Rachman Tolleng, and political scientist Daniel Dhakidae, all of whom were active spokesmen for democratisation in the media and public seminars. However, the organisation was never poised to play a major role in the formal political arena.

Though the Forum Demokrasi never got to the stage of widening its base among intellectuals and activists, official response to its establishment was almost immediate. Minister of the State Secretariat Moerdiono and Home Affairs Minister Rudini, for example, invoked laws that defined the approved political parties and required their adherence to Pancasila. The Forum Demokrasi, they warned, could be nothing more than a discussion group (*Kompas* 9 April 1991).

It is likely that state officials were worried that Wahid's position as head of the NU, the largest Muslim organisation in Indonesia, would eventually lend the Forum the wide mass base that its liberal intellectuals were lacking. Wahid had little choice but to confirm that the Forum would not become the fourth political party but would concentrate on encouraging democracy through discussion and study (*Media Indonesia* 3 June 1991; 13 June 1991).

The military apparatus was no less severe on the Forum. The Chief of ABRI's Social and Political Affairs Department (Kassospol), General Harsudiono Hartas, expressed concern that the Forum would confuse and mislead the public and could resort to agitation because, 'there are members of the Forum of Democracy who are still affiliated to the past'. This could be seen, he said, 'from their statements on justice, human rights and democracy' (*JP* 25 April 1992). However, he guaranteed that the Forum would not be banned so long as it did not promote change in the existing system – an effective negation of any proposals for liberal reform.

Although the Forum continued to exist, its formation emphasised the fact that middle class reformers were not free to form a party united by a common ideological commitment. It emphasised also that those wishing to take part in the formal political process were forced to join one of the existing state-sponsored and approved parties, Golkar, the PPP or PDI. And indeed, Laksamana Sukardi and Kwik Kian Gie joined the PDI and become close advisors to its leader, Megawati Soekarnoputri. NGO activist Hadimulyo, in addition, became a parliamentarian for the PPP. Many others with political ambitions were to join

Golkar. These people did so in spite of the requirement for parties to formally adhere to the conservative, organic-statist ideology constituted in Pancasila.

As we have seen, there could be no party united by a commitment to liberal, social democratic or radical ideologies in the Pancasila scheme as concocted by Soeharto. The negation of these possible alternatives clearly helped to secure the appropriation of the existing organic-statist political shell in the interests of ascendant bureaucratic-capitalist oligarchs. Notwithstanding the example of Laksamana Sukardi, Kwik and a few others, the PDI and PPP remained largely unattractive to the reformist intelligentsia, as they offered no clear route to power and influence under the system that existed.

The conservative bourgeoisie

The middle class intelligentsia has played an important role in the process of political change in industrial capitalist societies, though not always in the direction of liberal democratic reform. They have, for example, sometimes been proponents for social radicalism as well as populist fascism. The essential lack of social cohesion that characterises the middle class requires alliances with other social classes, whether they be the bourgeoisie, the petty bourgeoisie, or workers. As mentioned earlier, one of the critical problems of reformist opposition in Indonesia was that it was unable to forge such cross-class coalitions, including with the increasingly vital bourgeoisie.

But what would bring the bourgeoisie into a liberal political project with the middle class? As we have observed, a necessary ingredient is the confidence that authoritarian politics is no longer necessary to ensure its social ascendancy. Indeed, that representative politics will provide a mechanism for bourgeois political interests and enable them to impose accountability upon the state and its bureaucrats. This, argued Anderson, was a present condition in the case of Thailand:

> As the financial backers of many MPs, the banks can exert direct, independent political influence in a way that would be very difficult under a centralised, authoritarian military regime. Furthermore, as the representatives of a national electorate, the parliamentarians as a group veil bank power (and the power of big industrial and commercial conglomerates) with a new aura of legitimacy. This is a real and valuable asset. It can thus provisionally be concluded that most of the echelons of the bourgeoisie – from the millionaire bankers of Bangkok to the ambitious small entrepreneurs of the provincial towns – have decided that the parliamentary system is the system that suits them best; and that they now have the confidence to believe that they can maintain this system against all enemies.
>
> (Anderson 1990: 46)

In Indonesia, however, such a situation had not developed fully and thus the small group of middle class liberals was unable to entice the bourgeoisie into the reformist project. Throughout the Soeharto years, big business instead remained tied to the project of authoritarian corporatism, and indeed constituted a major element of the coalition of interests that was expressed as oligarchic power by the 1980s.

This is not surprising. Rueschemeyer *et al.* have argued that, historically, the bourgeoisie frequently have acted against political inclusion extending to the lower classes. Threat perception has always been important to any bourgeoisie: if they felt that their vital interests were acutely threatened by popular pressures, they invariably opposed democracy, 'and once democratic rule was installed, attempted to undermine it' (Rueschemeyer *et al.* 1992: 8).

There is little by way of public statements about the political regime or expressions of opinion about the political future from the larger Indonesian bourgeoisie, particularly those leaders of big Chinese-owned conglomerates. This silence is partly because their economic objectives have been so successfully secured in the context of rent-seeking relationships with political power-holders and also because of the social and political vulnerability of the Chinese in Indonesia. But it is not surprising that those few insights we have into the thinking of the Chinese element of the bourgeoisie show a primary concern for stability, security and economic growth, rather than democratic reform. For example, timber baron Bob Hasan, speaking in October 1993 before the Golkar Congress, was clear that the function of Golkar was primarily to provide national stability for business. To Hasan, Golkar was an effective arm of authoritarian rule and should not in any way become a representative or democratic instrument (*Kompas* 20 October 1993: 10).

However, only a small proportion of Chinese capitalists could be compared to Bob Hasan, Prajogo Pangestu or Liem Sioe Liong, either in terms of size or direct reliance on political patronage. Large corporate capitalists had become increasingly integrated into international financial circuits and markets. Sole reliance on rent-seeking for the longer term had become less viable and Chinese businesses began to move their cash out of Indonesia in the late 1980s in anticipation of difficulties that might arise from the inevitable political transition. The declining attractions of patronage systems particularly affected those who perceived themselves to be locked out of the increasing range of strategic sectors of the economy where the President's family and a few close business associates had secured dominance. However, liberal reformist sentiments were not part of the political agenda of even these elements of Chinese business. For the Chinese bourgeoisie in general, liberal or representative politics potentially opened the door for the sort of xenophobic and populist ideologies that lurk under the surface of Indonesian political life and have proven so damaging to the Chinese community in the past.

If the position of the Chinese tycoons made it unlikely for them to pronounce opinions on social and political matters, this was somewhat less the case for their *pribumi* counterparts. Significantly, the views of representatives of this element of the bourgeoisie have frequently expressed a strong populist, rather than politically liberal, bent. Long-standing *pribumi* resentment of Chinese business dominance became entangled with populist protest, aimed not only at the Chinese themselves but also at the alliances they formed with *pribumi* officials and political leaders. Ultimately, anti-Chinese sentiment was hijacked by wealthier *pribumi* business interests, not as part of a larger protest against concentration of wealth and power, but because they sought to replace the Chinese within the existing system. Golkar, ICMI and Kadin (the Chamber of Commerce and Industry) were among the institutional avenues of protest for the *pribumi* bourgeoisie.

It is important to note that the xenophobic tradition of petty bourgeois populism that they appropriated stretches back to the beginnings of the twentieth century, and that anti-Chinese sentiments and resentment of collusion between officials and Chinese business were often included within an umbrella of Islam. After independence, this petty bourgeois Islamic populism was politically overwhelmed by the more secular forces of civil and military bureaucracy that continued the colonial tradition of association with Chinese business. But despite the regular resurfacing of their resentments, in the 1950s, in the events surrounding the anti-Tanaka riots of 1974, and in periodic anti-Chinese outbursts in various rural towns, petty bourgeois populism found it difficult to organise itself effectively as a political force.

New Order-era organisations of *pribumi* business interests ranged from the anti-Chinese Pribumi Business Association, HIPPI, in which Soeharto's stepbrother, Probosutedjo, was a leading figure, to the elitist Young Entrepreneurs Association, HIPMI, founded by the young Pertamina contractors in 1972, and the state-sponsored Kadin. While these were widely regarded as mechanisms for capturing and allocating rents in the interests of their leading members, rather than as institutions representing the broader interests of *pribumi* business, they nevertheless successfully mobilised the *pribumi* issue. Calling for anti-trust legislation in the late 1980s, they also publicly advocated legislation similar to Malaysia's New Economic Policy to provide a legal basis for preference to indigenous business (*Suara Pembaruan* 16 October 1989; *Editor* 16 December 1989; *Tempo* 6 January 1990; 24 August 1991; Schwarz 1994: 125, 129, 131).

However, neither of these objectives was immediately realistic. As Ciputra, Sofjan Wanandi and other Chinese business leaders were to argue, Chinese business was structurally critical to the well-being of the Indonesian economy and would be forced to leave if the environment became too hostile (*Kompas* 2 October 1989; *Jakarta Jakarta* 30 January–5 February 1993: 87, 88; *Tempo* 20 July 1991). In any case, most of the leading *pribumi* business leaders had Chinese joint venture partners. The appeal to *pribumi* interest was further undermined by the rapid and pervasive growth of the business interests of the Soeharto children, who came to dominate their ranks and needed no formal regulatory arrangements to guarantee their positions.

Nevertheless, tension was to persist between an inner circle of Chinese conglomerates in partnership with the Soeharto family and various *pribumi* business interests. Building business alliances with conglomerates and the politico-bureaucrat families, and political alliances with Sekneg and Golkar, the so-called *pribumi* contractors described in Chapter 3 (including such figures as Aburizal Bakrie, Fadel Muhammad, Ponco Sutowo and Siswono Judohusodo) were to play an increasingly important role in economic and political life. Two prominent figures among them, Siswono Judo Husodo and Abdul Latief, were to become Cabinet Ministers.

But the honeymoon ended, as mentioned earlier, when the flow of contracts through Sekneg's Team Ten was halted in 1988 (Pangaribuan 1995: 77). Contractors began to watch with increasing resentment as certain Chinese conglomerates continued to gain access to state bank credit and mega project licenses during the period of tight money, and to allegedly transfer funds offshore. When two consortia of *pribumi* contractors bidding for large petrochemical projects were shouldered aside for Liem Sioe Liong and Prajogo Pangestu in 1991, a delegation of angry *pribumi* businessmen visited Soeharto to discuss access to projects and loans (*Tempo* 20 July 1991: 82–88).

Thus, Soeharto was forced to defend the government against charges that commitment to broader social and national goals had been abandoned in favour of 'free-fight' liberalism as critics drew links between liberal markets, individualism, increasing economic concentration and conglomeration, unemployment and growing social inequality (*Tiara* 19 June 1993: 103–110). Aware of the possible impact of such perceptions on the government's legitimacy, he was vigorously to deny that deregulation meant a shift to liberalism or a diminution in the role of the state as guardian of the national interests (Soeharto 1990: 31, 32, 34). He ordered the Indonesian Association of Economists to construct a new blueprint for economic democracy that reconciled national interests with liberal economics (ISEI 1990). At the same time, Soeharto emphasised the regime's success in poverty alleviation through targeted rural credit and investment policies aimed at ensuring availability of appropriately priced rice and other commodities.[8]

Particularly disquieting was the prospect that underlying currents of anti-Chinese xenophobia might be mobilised into a broader movement against the regime. In a highly symbolic gesture, he met with thirty-one of Indonesia's leading business figures at his private ranch in 1990, inviting them to transfer 25 per cent of their company shares to co-operatives, still the symbol of the New Order's commitment to egalitarian ideals (*Editor* 31 March 1990: 11–24). Although this produced few tangible outcomes, leading business figures were again called to a meeting in 1995 to consider the threat of a widening social gap, and to consider strategies for social programmes. One outcome of the meeting was a decree that companies with profits of more than Rp.100 million donate 2 per cent of their profits to the Yayasan Dana Sejahtera Mandiri, established by the President and ten conglomerate heads to assist in the elimination of poverty (*Indonesia Business Weekly* 4 September 1995: 9; *Asia Times* 13 February 1996: 13).

Nevertheless, it was within the framework of the regime that the *pribumi* capitalists sought politically to achieve their economic goals. Thus, they became increasingly active in such organisations as Golkar and ICMI. Business success for them rested heavily upon policies of protection and rent-seeking, access to government contracts, inclusion in joint ventures and access to state bank finance. The possibility of political parties able to seize government representing the interests of workers or even middle class liberals wanting to open the world of commerce to transparent and genuine competition was a prospect that did not fit well with their own position. What they sought was a more influential place in the prevailing system of authoritarian corporatism, and a greater role for them in the formulation of policy. And indeed, indigenous capitalist groups became an increasingly important component within the broader coalition of oligarchic interests crystallising around Soeharto, and were best represented by the burgeoning business interests of his own family.

The working class and the peasantry: challenges from below

The opportunities for cross-class reformist alliances were also curtailed by the weakness of Indonesia's subordinate classes. Both radical peasant and labour unions were smashed along with the Indonesian Communist Party in the mid-1960s, and any resurfacing of militancy within these quarters automatically invited accusations of a communist resurgence (Hadiz 1997). The authoritarian corporatism of the New Order was particularly thorough in establishing institutional and repressive mechanisms to control and demobilise lower class-based organising. The state-sanctioned FSPSI and HKTI, both formed in 1973 on the initiative of Ali Moertopo (Moertopo 1975), respectively served to pre-empt independent organising activities on the part of workers and the peasantry, rather than functioning as associations which genuinely represented their interests. It is instructive that the state has been especially inclined to utilise outright violence and coercion – the 'security approach' – when facing labour and peasant unrest, largely because of the particularly weak position of these groups.

The weakness of the working class in particular has arguably been detrimental to democratisation projects in Indonesia, in view of Rueschemeyer *et al.*'s (1992) broad comparative study of capitalist development and democracy, which suggests that democratisation has historically been linked to the emergence of strong and effective labour movements. In much the same vein, Therborn (1977) drew a distinction between the advent of bourgeois dominance over politics and the historical development of democracy in advanced industrial countries, suggesting that the latter was in large part the product of sustained pressure from working class-based organisations and political parties. These findings fit well with Anderson's argument that democracy is often confused with electoralism, the latter primarily being a framework that allows for bourgeois dominance in the political sphere without bringing perceivable change in the lives of the broader public, either in terms of social justice or real political participation (e.g. Petras 1989). It is also instructive that in Western European history, reforms guaranteeing such civil rights as freedom of speech and association as well as rule of law, and those dealing with factory acts and labour laws, often preceded the establishment of actual democratic systems.

If the working class has had such a historically critical role in democratisation projects, a crucial ally has often been the middle class – in spite of the fact that the latter's own role as a democratising agent has been ambiguous, as the above discussion on Indonesian middle class reformism suggests. Again according to Rueschemeyer *et al.*, middle class attitudes towards the political inclusion of lower classes 'is dependent on the need and possibilities for an alliance with the working class'. Thus, the middle class has been most in favour of democracy where it faced 'intransigent dominant classes and had the option of allying with a sizeable working class'. Where strong and radical working class movements found allies among the middle class, they generally managed to overcome the resistance of dominant classes to democratisation. Such alliances also 'typically softened the character of working class demands, which contributed to the stabilization of democracy' (Rueschemeyer *et al.* 1992: 283).

In view of these observations it is significant that working class organisational capacities remain low in Indonesia, and have not threatened to become the basis of stronger coalitions with more than a limited section of the middle class opposition (Hadiz 1997). Nevertheless, the development of industrial capitalism under the New Order did result in the gradual emergence of a more sizeable urban proletariat with a greater propensity to organise, as demonstrated in the growth of organising vehicles, albeit in rudimentary form, and the dramatic rise in industrial action in the 1990s. According to official statistics 350 strikes occurred in Indonesia in 1996, compared to just nineteen in 1989, though these must be treated as conservative figures.[9] Many of the new organisational vehicles involved collaboration with middle class NGO activists. The emergence of industrial workers and a labour movement in the burgeoning export-oriented industrial estates and cities introduced a new variable because – like small rural landowners, peasants, and the unemployed – they were not well incorporated within the political institutions of the New Order. These often constituted the raw material of revolutions.

Also foreboding were the ripples of discontent that had begun to spread ominously in the countryside since the late 1980s. Land disputes were to become an increasingly important issue as residential development projects, golf courses, dams and industrial estates displaced rural peasants. Often with ambiguous land titles, peasants and small landowners were increasingly to find rural farmland being bought up by wealthy Jakarta families or local propertied classes, or transformed for commercial purposed by large corporations. A series of high profile land disputes occurred in the 1980s involving peasant demonstrations, delegations to parliament and co-operation with small bands of students, NGO activists and other middle class supporters (Fauzi 1998: 115–118).

What really worried Soeharto were the risks of instability if the general anti-conglomerate mood mentioned earlier was to ignite other resentments being fuelled by the new enclosure movements in the countryside, or by unemployment and rising prices, thus producing a populist political alliance involving workers or small landowners, peasants and rural petty bourgeoisie in a backlash against predatory officials and wealthy families. It is instructive that Soeharto frequently resisted World Bank pressures to introduce market prices for such items as kerosene and cooking oil, transport, water, and electricity and irrigation services.

Though challenges emanating from the urban factory or from the rural village were never strong enough to pose an immediate, fundamental threat to the social and political order, they had the potential to produce a politically destabilising populist backlash against the interests of predatory officials and wealthy families. There is a long history of social resentment and racial antagonism in rural protest movements, and militant labour movements played a significant role in Indonesia's nationalist struggle as well as the first decades of independence (Ingleson 1986; Shiraishi 1990; Hawkins 1963). Worst of all was the prospect of populist or radical resentments coming together under the umbrella of Islam, as Soeharto must have been aware of the dynamics of populist social unrest and resistance in Egypt and Algeria.

Thus, challenges from below had to be pre-empted or contained, often quite brutally through the use of the so-called 'security approach'. In relation to labour, this approach was perhaps most vividly demonstrated in the case of Marsinah – the young, female labour leader in East Java who was kidnapped, sexually assaulted, and murdered in 1993. Though the case remains officially unsolved, it is widely believed that the local military command was intimately involved in the affair, with the aim of quelling strike action at a watch factory that she had led (YLBHI 1994). Violence and intimidation were also widely utilised in facing protest action by the peasantry, for example in Kedungombo, Central Java, with residents being accused of communist sympathies for their protests against being forcibly removed to make way for a dam.

Not surprisingly, it was the military leadership that especially favoured the security-oriented approach. Military intervention into labour matters, for example, was defended by reference to the normal workings of the doctrine of *dwifungsi*, or military dual function (e.g. Department of Foreign Affairs 1993: 20). Often at the heart of the official rhetoric was the idea of cultural relativism, which was the foundation for the formulation in 1974 of a culturally specific 'Pancasila Industrial Relations' – a model that does not recognise antagonistic relations between worker and employer. Thus, labour unrest was portrayed at the same time as the result of the workings of 'New Left' or liberal forces, alien to Indonesian culture and society.[10]

But rising worker unrest and inclination to organise independently was essentially tied to the gradual development of a more substantial industrial working class, especially in crowded new industrial zones in major urban centres. A stimulus for this development was Indonesia's shift in the mid-1980s to a more export-oriented industrialisation strategy based on labour-intensive manufactured products, itself a reaction to the fall of international oil prices.

Another dimension of the problem was increasing international pressure on Indonesia to adhere to international labour and human rights standards, itself a response to frequent outbreaks of labour unrest. Thus, the threat emerged in 1993 that Indonesia would be taken off the US list of countries receiving GSP (General System of Preferences) privileges. The programme, enacted in 1974, provides duty-free entry to the US market of eligible products from beneficiary developing countries, but only those that have taken or are taking steps to ensure internationally recognised labour rights.[11]

It was especially the AFL-CIO, the American labour federation, which had taken the lead in applying pressure on the Indonesian government. In the late 1980s it lodged petitions with the US Trade Representative's Office to have Indonesia removed from the list of countries receiving GSP facilities, alleging the infringement of international labour standards (American Embassy 1994: 43). Its actions were followed up in 1992, as two separate petitions were filed by US-based organisations to the United States Trade Representative (USTR) for the elimination of Indonesia's GSP privileges.[12] While the USTR did order that Indonesia's GSP status be reviewed for non-compliance with international labour standards (*FEER* 13 May 1993: 13), nothing concrete ever materialised from the extended period of threats. The momentary pressure did, however, help fledgling independent labour associations win additional precious room to manoeuvre, albeit temporarily.

Among the numerous new organising vehicles established in the early 1990s was the first 'independent' union of the New Order era, the Solidarity Free Trade Union, known as *Setiakawan*. Though the organisation quickly crumbled in the face of intimidation as well as internal conflict, it is significant that there was no general clampdown on the organisation, a fact that must have heartened those who would later attempt similar endeavours. Bourchier (1994) raised the possibility that the lack of overt moves to destroy Setiakawan was an outgrowth of the interests of the Benny Moerdani-led military faction to discredit Indonesia's official union – the SPSI – an organisation increasingly linked to Golkar, which was itself being gradually removed from the military's ambit by President Soeharto. Demonstrating links being forged by military officers with sections of the labour movement, Muchtar Pakpahan, leader of another independent union, the SBSI (Indonesian Prosperity Trade Union), for some time entertained the notion that he was being given 'protection' by an increasingly 'enlightened' military leadership.[13]

In spite of these possible entanglements in elite politics, neither labour nor the peasantry were emerging as a significant enough force to be fully co-opted by contending elites or to elicit the interest of a larger segment of middle class reformists in terms of developing a more broadly based coalition. Thus, they continued to be contained primarily by repressive means, even though the government did relent on such issues as low wages to placate workers (Hadiz

1997: 159–169). Most middle class leaders of the mainstream political opposition had no organic links with the labour movement or the peasantry. The consequence of this was to be seen immediately after the fall of Soeharto, in the actions and strategies taken by leading opposition groups to realise the aims of *reformasi*, as discussed in further chapters. Thus, workers in particular have continued to organise under difficult conditions – including mass unemployment due to economic crisis – in the immediate post-Soeharto period without much recourse to broader political alliances. Most importantly, the weakness of the 'challenge from below' has meant the weakness of any significant impulse in Indonesia to pursue social democratic or more radical political projects.

The authoritarian response

Intra- and extra-elite challenges to Soeharto's wider project of refashioning the existing authoritarian political shell in the interests of newly ascendant bureaucratic and capitalist families were not very successful. Soeharto was to continue to promote those with an interest in protecting the oligarchy to strategic positions in Golkar, the DPR/MPR and the military, as well as sideline those deemed as posing a threat. He was also to maintain the political disorganisation of civil society, and in fact, dealt severe blows to the political opposition on several occasions. In spite of profound economic crisis beginning in mid-1997, and mounting calls from opposition leaders and students for him to step down as Indonesia's situation became increasing dire, Soeharto was able with consummate ease to secure his ascension to yet another five-year term as President in March 1998.

Soeharto's response to the challenges to his rule, whether emanating from within or without elites, was typically to call up the issue of ideological legitimacy, to reconfirm the Pancasila model and to increase anti-liberal rhetoric. In broad terms he was to juxtapose what he saw as the destructive and confrontational nature of liberalism with Pancasila as the form of political democracy best able to harmonise with Indonesian cultural values of consensus and the family system.[14] On another tack, he portrayed liberalism as a system that sacrifices economic development for individual rights. Pancasila, on the other hand was affirmed as the form of democracy that enables political stability, social order and economic development to be appropriately balanced with freedom.[15] This point was to be a consistent theme in his attacks on the legitimacy of liberalism. In 1993, Soeharto was to repeat that 'those who were seeking to establish a Western-style democracy, rather than a democracy based on consensus … were obstructing further development' (*Indonesian Observer* 22 December 1993: 1).

Clearly, for Soeharto political life could only proceed freely once there was a common agreement upon the modes of political behaviour and the principles that would guide political and social life. In practice this meant the continued rejection of representation and opposition as mechanisms of politics, and the elimination of communism, Islamic fundamentalism, and liberalism as political philosophies. Given these conditions, the freedom to participate in political life and to express ideas only exists if it is attuned with Pancasila and does not challenge the existing state or regime (*Kompas* 29 June 1990).

The end of 'political openness'

Thus, it was no real surprise that the period of 'political openness', so easily misconstrued as a period of real democratisation, eventually came to an abrupt end. This happened in June 1994, with the closure of three press publications, *Detik*, *Editor* and *Tempo*, mentioned earlier (Hill, D. 1994: 41). With this event the hopes of the liberal intelligentsia regarding any automatic link between economic development and democracy experienced a fundamental setback.

Tempo's crime in particular involved the reporting of a conflict between Minister of Finance Ma'rie Muhammad and B.J. Habibie, already mentioned, over the latter's purchase of thirty-nine former East German warships at a highly inflated price for the Indonesian Navy. Moreover, these ships required serious overhaul and repairs, to be conducted in Habibie's plan by one of the companies under the aegis of his BPPT, and significantly, with little co-ordination with the military top brass (*Tempo* 11 June 1994). Perhaps quite appropriately, political 'openness', which originally emerged out of conflict between the military leadership and Soeharto over ABRI's institutional autonomy, came to a close with an event that provided an excellent indicator of the extent to which military influence and autonomy had in fact eroded within the regime.

With the end of political 'openness' government critics of all dispositions encountered a renewed atmosphere of repression. In April 1994, the government had in fact already reacted harshly to labour riots in Medan by arresting and eventually gaoling NGO activists and SBSI leaders, including Muchtar Pakpahan. Harassment of intellectuals critical of the government also took place at about the same time.[16] Another indicator of the changing political direction was the closing down by security forces of an NGO conference in Yogyakarta, held to discuss a proposed government bill to regulate NGOs (*Kompas* 23 September 1994).

The culmination of the renewed period of repression came with the storming of the PDI headquarters by security forces and paid thugs on 27 July 1996. This act was part of a wider aim of ousting Megawati Soekarnoputri as chairman and replacing her with a more pliant rival, the career politician Soerjadi. The significance of Megawati was that she had begun to be a rallying point for various sections of the opposition, including middle class intellectuals, various student groups, NGOs and labour groups as well (*AWSJ* 22 July 1996). There were high expectations that the PDI would do well in the parliamentary elections scheduled for 1997, and that she would even challenge Soeharto for the Presidency the following year.

The attack on the PDI headquarters occupied by Megawati supporters was the catalyst for rioting in the capital city (*JP* 28 July 1996), the worst since 1974.

Rather than admit that the outbreak of riots was the outcome of public anger at brazen acts to depose Megawati, the government chose to single out members of the small, leftist PRD (People's Democratic Party) as culprits. The unrecognised political party, through its labour arm, the PPBI, had been helping to stage labour strikes for over a year.[17] The rioting was also strangely blamed on SBSI leader Muchtar Pakpahan, who was arrested yet again only nine months after being released by the Supreme Court in connection with the Medan riots. Significantly, the activities of both the PRD/PPBI and Pakpahan were presented as proof of the resurgence of communism in Indonesia.

In reality, however, the government was hitting back at elements of the political opposition that were more vulnerable than Megawati or her closest advisors. It probably calculated that the PRD and the SBSI represented weaker elements of the opposition, and unlike the core leadership of Megawati's PDI, their punishment would not incite large-scale public protest. Interestingly, Muhammadiyah leader Amien Rais – also a major figure in ICMI until he later fell out of favour – expressed support for the government explanation of the riots. As Hefner recalls (Hefner 2000: 187), the man who would emerge as Soeharto's most vocal critic largely agreed that the rioting had been part of some devious communist-inspired plot to destabilise the country, and even defended the military action taken against Megawati's supporters at the PDI headquarters. Indeed, Muslim organisations such as KISDI (The Indonesian Committee for Islamic World Solidarity) also lent their weight to the government cause, thereby prefiguring the conflict between secular nationalist and populist Islamic forces in the immediate post-Soeharto period.

It was also significant that prior to the attack, Megawati was quickly becoming a symbol of the possibility of peaceful change from within the existing political system. The PDI was after all an official part of that system – and her supporters certainly hoped that she could utilise actually existing political institutions to organise her challenge on Soeharto. The brutal fashion of her ouster, however, must have symbolised to many the limited actual possibilities for such peaceful, internal reform. Given the Megawati experience, mounting a successful challenge on the electoral front was demonstrably out of the question. As a result, whatever hopes remained about the eventual unravelling of the New Order rested on Soeharto's eventual departure from the political stage, especially in view of his advancing age. It became clear that only this could bring opportunities for forging a new political map and new political alliances.

Soeharto's last hoorah: securing the Presidency in 1998

What the developments described earlier demonstrated was the success Soeharto enjoyed in imposing a new form of oligarchic dominance and gaining control of both the military and Golkar in the process. At the same time he proved successful in disorganising civil society and restricting politics to the institutions of the state.

Ironically, the fatal flaw lay in Soeharto's very success. The complex coalition constituting the ruling oligarchy was totally reliant on his authority and possessed no political institutions and organisations able to guarantee its social ascendancy beyond Soeharto. For Soeharto, the immediate task was to secure his nomination for the Presidency in 1998 – which required that he maintain control of both Golkar and the MPR. In the longer term, however, he needed to ensure that his successor would guarantee the interests of the oligarchy in general and of the Soeharto family in particular.

These were difficult tasks. Opponents such as General Soemitro, for example, had suggested that the time was right for a 'young civilian' to take over the helm in 1998 (*JP* 3 June 1996). Moreover, both Amien Rais and Megawati were already presenting themselves as new alternatives to Soeharto – although both lacked effective political vehicles to launch a serious challenge. Rais himself was particularly adamant that succession must take place in 1998 (Rais 1998; Soekarnoputri 1998). What no one foresaw was the devastation of the Asian economic crisis and its effect on Soeharto's grip on power.

Initially, in spite of the particular harshness of the Indonesian manifestation of the Asian Crisis, Soeharto still continued to be sufficiently in control. The months preceding the March 1998 MPR session were notable for the usual chorus of expressions of support for yet another Soeharto Presidency. Groups such as Kosgoro Youth, the *Ikatan Keluarga Besar Arief Rachman Hakim* (comprised of 1966 activists) and the KNPI all expressed their hope that Soeharto would serve another term (*JP* 1 November 1997). Earlier, noted intellectual Juwono Sudarsono, who would later be recruited into Cabinet, had suggested that Soeharto's motives in seeking to retain the Presidency were largely altruistic – he would not be satisfied until poverty was eradicated from Indonesia (*JP* 9 August 1996).[18] In reply to a Soeharto 'request' (*JP* 20 October 1997), Golkar chairman Harmoko made public assurances that it was really Soeharto that the people wanted for President. Unlike in 1993, the military, now under Soeharto close associates Feisal Tandjung and Wiranto, did not play games about re-nominating Soeharto for the Presidency (*JP* 25 October 1997; 23 December 1997; *D&R* 25 October 1997: 40). Soeharto himself was fuelling public debate about succession by suggesting at one point that he had reached an age which suited him for the role of *pandita* (sage), implying a readiness to step down in favour of a senior advisory role (*JP* 20 October 1997).

Behind the theatrical chorus of support – and public shows of humility – machinations were well under way to ensure Soeharto's re-election as well as to consolidate the ruling oligarchy's control over the institutions of state. Consequently, DPR and MPR seats occupied by Golkar were being increasingly filled by the oligarchy. Indeed, a public controversy emerged over the number of Soeharto cronies, associates, and relatives that were in the formal position to elect a President (*Ummat* 6 October 1997). Amongst many others, businessmen Fadel Muhammad, Aburizal Bakrie, Probosutedjo, Sudwikatmono, Sukamdani, Bob Hasan, Anthony Salim, Tanri Abeng, and Adiguna Sutowo were to act as regional or functional group delegates to the MPR, as were the wives of Bambang Trihadtmodjo, Feisal Tandjung, and Minister of Home Affairs Yogie Memed. Trihadtmodjo and his siblings, 'Tutut' and Siti Hedijati Prabowo, were also to take up MPR seats (*FK* 22 September 1997: 100).

To prevent any hiccups, General Wiranto, a member of ABRI, was named to head the *ad hoc* MPR committee charged with overseeing the election of a President and Vic-President. Soeharto did not want a repeat of 1988 – when the military tried to reject the appointment of Sudharmono as Vice-President, or 1993, when it promoted Try Sutrisno as Vice-President in a *fait accompli*. Furthermore, Hartono was elected chairman of another committee that was in charge of drafting state policies to be deliberated upon and endorsed by the MPR (*JP* 28 October 1997). Not surprisingly, Soeharto was yet again elected to the Presidency by the unanimous decision of the 1,000-member MPR. Thus, the pinnacle of oligarchic rule seemed to have been reached at precisely a time of rising economic hardships due to the onset of crisis and the threat of looming public disorder.

Significantly, as discussed further in later chapters, Soeharto was concurrently having a difficult time dealing with the IMF, on which Indonesia was relying for a bailout package (Robison and Rosser 1998). But Soeharto was repeatedly to snub the IMF by breaking pledges, and audaciously, the Cabinet he appointed in March 1998 after securing the Presidency was virtually free of technocrats trusted by international development organisations. Finance Minister Mar'ie Muhammad, for example – known as 'Mr Clean' – was replaced by family associate and former director general of the Tax Office, Fuad Bawazier. Business crony and golfing partner Bob Hasan was appointed Minister of Industry. Soeharto's daughter Tutut was named Minister of Social Affairs in an appointment that was widely taken as a signal that she was being groomed to eventually take over her father's position as President, and moreover, at the head of the ruling oligarchy. Speculation was already rife that she would take over Golkar later that year (*Business Week* 3 November 1997: 23). Confident of his authority, Soeharto thus appeared to be treating the much-debated issue of Presidential succession in dynastic fashion.

At the same time, Soeharto was sending out a clear signal of defiance, both to the IMF and to his domestic critics. But his actions, equally importantly, showed that the regime had, by this time, dramatically narrowed in terms of its social base. Soeharto could now only count on cronies and family members, and other individuals who were personally close to him.

This was the significance also of the appointment of B.J. Habibie as Vice-President who, as noted earlier, was unpopular with liberal economic reformers due to his penchant for expensive hi-tech projects. But Habibie's elevation to the Vice-Presidency, a heartbeat away from the ultimate position of power – at a time of grave economic crisis – was not only a snub to the IMF and economic liberals inside Indonesia. By appointing Habibie, Soeharto was also flaunting his accomplishment of complete control over the military, sections of which had opposed a Habibie Vice-Presidency five years earlier. Soeharto was truly acting as if oligarchic rule, cemented by his own personal authority, was unshakeable.

Ironically, his authority was already beginning to be seriously eroded. Soeharto himself could not have failed to recognise the rise of public discontent as Indonesia sank deeper into economic crisis. Indeed, the rupiah was to make dramatic falls after a key budget speech by Soeharto in January and following the unveiling of the new Cabinet in March. By this time, student protests and public calls for him to step down were well on the rise. In response to the threat of growing opposition emanating from civil society groups, Soeharto apparently decided to play his Islamic 'card'. Thus, the elevation of Habibie – chairman of ICMI – could be seen as an attempt by Soeharto to mobilise support from a number of high-profile organised Islamic groups, at a time of increasing political discontent.

While on the surface Soeharto's grip on power seemed as unassailable as ever, in reality the end of his long rule was finally approaching. The conditions that led to Soeharto's eventual historic downfall, on 21 May 1998, as well as the ensuing process of contestation to reconfigure power in Indonesia, are to be explored further in the chapters to follow. The major question that then emerges is whether the oligarchic interests incubated and nurtured under the rule of Soeharto can reconstitute themselves without him and maintain their political position within a new regime to be constructed.

Notes

1 See the comments of *Tempo* editor, Fikri Jufri, and *Republika* editor, Parni Hadi, on the greater freedom available to the press in these years (*JP* 7 July 1993: 1).

2 It was argued by the businessman Sofjan Wanandi that Sumitro's outburst was in response to his son Hashim's unsuccessful bid for the automobile company, Astra, which went to several Chinese business groups. Indeed, Hashim headed one of the major new conglomerates, including a range of partnerships with Soeharto children in petrochemicals and other sectors. Another Sumitro son, Prabowo, was a prominent military officer and married to a Soeharto daughter. In this sense, Sumitro's remarks might be taken less as a fundamental assault on corruption than an attempt to destabilise the prominence of Chinese conglomerates in the networks of rents.

3 See *Suara Pembaruan* 23 October 1989. How this interpretation is reconciled with the economic performance of Korea, Taiwan and Singapore in the 1970s and 1980s is unclear.

4 Also interview 1 July 1993.

5 With economists Hadi Soesastro and Mari Pangestu in the lead, the formerly conservative think-tank, CSIS (Centre for Strategic and International Studies) now became a focal point for neo-liberal critiques. Former banker, Laksamana Sukardi, and economist, Rizal Ramli, established the consultancy, Econit, a source of systematic and critical analysis of government policy and rent seeking. Economist and consultant, Sjahrir, was also prominent in the debates.

6 See Kurth's (1979) distinction between bourgeois liberalism and that of the intelligentsia.

7 There were some indications that in areas of public tendering of large infrastructure projects, the engagement with global markets could influence a move towards more open and accountable practices. The 1995 tendering exercise for telephone contracts

worth US$2 billion was reportedly conducted according to international standards of transparency, excluding bids by several well-connected companies. Last minute efforts by the Minister for Telecommunications, Joop Ave, to convince the successful consortia to include selected unsuccessful bidders were not productive (*AWSJ* 15 June 1995: 1, 2; 20 June 1995: 1, 24; *JP* 20 June 1995: 1). This apparent victory might be attributed to the then-immanent listing of Telkom shares on the New York Stock Exchange, expected to raise US$2 billion and to contribute to a more rapid disbursement of Indonesia's debt. As such, it required confidence among potential investors that Indonesian Telkom operated on the basis of international standards.

8 Professor Mubyarto, a strong advocate of rural and co-operative-based development, was included in his 1993 Cabinet as Assistant Minister for Equity Improvement and Poverty Alleviation, while the sixth five-year plan, Repelita VI, contained important provisions for poverty alleviation and local investment programmes (*JP* 15 April 1993: 1).

9 For government statistics see http://www.depnaker.go.id. For alternative sets of numbers, see Kammen 1997, which contains an excellent analysis of strike action in Indonesia

10 Interview with Sudomo, Minister of Manpower 1983–1988, 9 May 1994.

11 This is the elaboration provided by the GSP Information Centre (1993: 1). Though the GSP helped to raise Indonesian exports to the US, its real importance was more indirect. US investment levels could have been affected by the removal of the GSP facilities because American firms would no longer be covered under the so-called OPIC (Overseas Private Investment Corporation) scheme, whereby the US government insured their operations in Indonesia.

12 These were Asiawatch and the International Labor Rights Education and Research Fund.

13 Commenting on his relationship with the military, he suggested that 'in order to provide prosperity to the common people, he is obliged to co-operate with various parties' (*Detik* 4–10 May 1994: 7).

14 See, for example, Soeharto 1990: 12; *Kompas* 6 June 1989; *Suara Pembaruan* 13 May 1989.

15 See, for example, Soeharto 1990: 16; *Kompas* 21 October 1990; *Suara Pembaruan* 13 May 1989; *Suara Karya* 28 June 1989).

16 For example, George Aditjondro, a critical academic, relocated to Australia in 1995 to avoid arrest and possible imprisonment for comments made at a seminar (*Merdeka* 7 June 1995).

17 In December 1995, for example, the PPBI helped organise a strike by 14,000 workers at the giant PT Sritex textile factory in Solo, Central Java – manufacturer of Golkar uniforms – which was violently broken up by military personnel (*FK* 17 June 1996: 61). The factory was partly owned by long-time Soeharto aide, Harmoko. Just prior to the Jakarta riots, the PPBI organised well strikes in East Java, after which three of its leaders, including chairperson Dita Sari, were charged with inciting hatred towards the government (see *Surya* 9 July 1996; *Jawa Pos* 9 July 1996; *Surabaya Pos* 9 July 1996).

18 He did point out as well, however, Soeharto's need to assure a successor that would safeguard his personal interests and that of his family.

Part III

THE OLIGARCHY IN CRISIS 1997–1998

6

ECONOMIC CATASTROPHE

Even as the storm clouds of crisis gathered in 1997 and currency traders began to stalk the Indonesian rupiah, there was a widespread belief that the Indonesian economy was well insulated against such challenges. After all, the 'fundamentals' used by neo-classical economists to measure economic health were relatively sound. Indonesia ran no large budget deficits, its inflation was low and it maintained manageable current account deficits through strong export growth. In the years leading up to the crisis, most economists and the World Bank had continued to feel confident about Indonesia's economic prospects, although expressing some concerns about short-term private sector debt and flagging commitment to market reform (Hill, H. 1996: 255; Radelet 1995; World Bank 1996: xxii–xxxii, 1997c: xxi–xxxi).

As the rupiah came under attack, President Soeharto and Bank Indonesia urged calm and promised a safe passage (*JP* 18 August 1997: 1; 8 September 1997: 1). But it was not only the Indonesian government that weighed in with such assurances. The *Far Eastern Economic Review* argued that Indonesia's 'economic fundamentals and growth prospects remain sound, and investors will continue to pursue business opportunities' (*FEER* Yearbook 1997: 140). Nor did leading international financial institutions or private securities and accounting firms grasp the reality of what was happening. Lehman Brothers expressed confidence that Indonesia would ride out the currency shocks (*JP* 2 August 1997: 1). In November 1997, Timothy Condon, Vice-President of Stanley Morgan Asia, predicted that the rupiah, then at 3,900 to the US$, would settle at 3,100 and interest rates would fall (*JP* 5 November 1997: 1). Bank Indonesia Governor at the time, Soedradjad Djiwandono, recalled that World Bank and IMF officials at the joint World Bank/IMF conference held in Hong Kong in August 1997 congratulated the Indonesian government on the handling of the crisis and expressed confidence in the strategy adopted (interview 25 May 2001).

Yet, even as Soeharto stood so apparently triumphant and invulnerable, the seeds of destruction were being sown. In the end, nowhere did the Asian financial crisis of 1997 and 1998 leave such a destructive legacy as in Indonesia. An economy that had enjoyed spectacular rates of growth for over two decades found itself quickly confronted with the humiliation of rapid economic reverses as its currency entered a protracted decline that persisted for over four years. As the crisis extended into the country's financial institutions, Indonesia's public and private banks became enmeshed in a deepening and continuing crisis of bad loans that reached 85 per cent of total loans by the end of 1999, imposing on the government a crippling recapitalisation bill of US$87 billion, or around 82 per cent of GDP (*AWSJ* 11–12 June 1999: 3). A beleaguered government was to preside over a budget devoted increasingly to servicing debt, relying upon continuing injections of foreign financial assistance to shore up its deficits.[1]

Why did an economy that had appeared so vigorous suddenly collapse and why did a political regime that had been regarded as almost invulnerable disintegrate so rapidly? Were we witnessing, as the IMF had argued, the end of a system of state-led capitalism that had become dysfunctional and outmoded in a world of global capital markets, forced to surrender to the inexorable advance of liberal markets?[2] Or was this simply a run on the bank; a crisis of global capital markets and financial systems? For those neo-liberal economists within the World Bank and Western universities who had enthusiastically endorsed Indonesia's economic management over the past three decades, the IMF thesis undermined the very basis of their own position (Jayasuriya and Rosser 1999). Its emphasis on cronyism and corruption was, in this view, highly suspect; after all, Indonesia had experienced decades of strong economic growth under the same system. Instead, they were generally to support the argument that Indonesia was a basically sound economy that became victim to panic and flight in global financial markets.[3]

But the question remained, when other countries had been hit by the same turmoil, why did the attacks on Indonesia's currency wreak such particular devastation? Here, they were to argue that Indonesia's continuing plunge into the abyss was the result of failure to manage the crisis and the inappropriateness of the government's macroeconomic policy responses. With their insistence on fiscal austerity and clumsy attempts to impose structural changes, especially the early decisions on bank closure, both the IMF and the government were blamed as confidence collapsed and the withdrawal of liquidity from the economy exacerbated further economic contraction (Hill, H. 2000a; McLeod 1998: 41; Garnaut 1998: 2, 3). However, there was some retreat from the earlier insistence that such issues as corruption and cronyism were incidental. Even the World Bank was unable to continue denying the problem of predatory raids on its own funding programmes. While charges made by American political scientist, Jeffrey Winters, that up to 30 per cent of World Bank funds was being lost to corruption were initially denied, the Bank eventually admitted that 20–30 per cent of the Indonesian budget was being lost to corruption and that a similar percentage was probably being diverted from its own programmes (World Bank 1998b, cited in Schwarz 1999: 316).

Politics, too, was drawn into the equation. Andrew MacIntyre, for example, has argued that the very tight insulation of the regime from outside influences enabled it to ignore reformist entreaties to make policy changes early in the crisis

(MacIntyre 1999, 2001). At the same time, the collapse of state authority was seen to open the doors to policy chaos, making it impossible for effective and cohesive responses in the longer term. Whereas Soeharto's New Order had been viewed, for all its shortcomings, as an economically responsible and effective regime, at least in its golden years, its collapse was seen to lead, among other things, to a flirtation with populist redistribution policies and attempts to try to break the old monopolies by legislating for competition rather than getting on with the real tasks of deregulation. Such chaos frightened off much needed foreign investment (Hill, H. 2000a: 131–132, 1999c). Even corruption became dysfunctional as the predictability formerly guaranteed by central control over its operation descended into uncertainty (McLeod 2000a).

There is little doubt that the various Indonesian governments after the crisis encountered great difficulty in controlling things. But did this 'failed state' allow a fundamentally healthy economic regime to unravel? And was the task now a technical one of rebuilding it? We propose that crisis eroded the financial and political cement that had papered over a profoundly flawed and vulnerable system wholly reliant upon the protection of a corrupt and authoritarian system of state power and an unregulated engagement with volatile global capital markets. When these frameworks evaporated, the oligarchy was exposed and the very circumstances that had allowed the rapid economic growth of the past decade and the spectacular rise of corporate empires now provided the mechanism for Indonesia's descent from financial crisis into a deeper economic and political morass. Indonesia's political leaders resisted neo-liberal pressures for policy change and institutional reform not because they were insulated from social pressures but precisely because the regime constituted in a crudely instrumental way the interests that surrounded a vast and sprawling politico-business oligarchy. Their resistance was in response to the fundamental threats posed to the very social order in which their ascendancy was embedded. In its essence, this was not a problem about putting back together a now-failed state, but a conflict between coalitions of state and social power about policy and institutions and, ultimately about the way social power was to be organised.

Thus, in the period between July 1997 and the fall of Soeharto in May 1998, the political landscape was altered fundamentally as the political regime, floundering in the face of deepening fiscal and economic crisis and mounting social tension, was forced progressively to give way to IMF structural adjustment programmes. No longer able to roll over its short-term debt or to call on the government to preserve its monopolies, keep its critics at bay or bail it out, the oligarchy faltered. Major corporate interests and politico-business families were forced into insolvency with impossible levels of debt, struggling to protect their assets by capital flight and forced into protracted rearguard actions against creditors. At the same time, reformist liberal forces, now assembled around the IMF, were able to impose upon a beleaguered and bankrupted government demands that it abandon its monopolies, reform its institutions and open its public offices to audit and scrutiny. While power remained, however disaggregated, in the hands of forces whose interests were embedded in the old system of power relations, they could frustrate the reform agenda but at the same time they could no longer rule in the same way.

The impact of the crisis

A speculative attack on the rupiah following the collapse of the Thai baht triggered a rush for dollars by Indonesia's domestic private sector corporations to cover their predominately short-term and unhedged foreign debts (World Bank 1998b: 1.3, 1.5; *AWSJ* 31 December 1997; *FK* 8 September 1997; 12 January 1998). Within six months, driven in part by this rush, the rupiah was to slump more than 80 per cent in value. Attempts in December 1997 to persuade creditors to roll over the private sector's short-term debts were unsuccessful and by late January it was apparent that most of Indonesia's conglomerates were technically bankrupt. At the same time, the rupiah's collapse also generated a fiscal crisis for the Indonesian government. As revenue sources contracted and demands for social sector subsidies increased, the government was forced to a budget deficit of 8.5 per cent of GDP by mid-July 1998, leaving it with little option but to reschedule its foreign debt commitments and seek increased amounts of foreign aid.

These were not the only problems. In an attempt to stem capital flight and maintain liquidity, domestic banks raised lending rates from around 15–16 per cent before the crisis to 60–65 per cent by mid-1998. This in turn contributed to a dramatic increase in non-performing loans within the country's banking system from around 9 per cent of total credit outstanding prior to the crisis to an estimated 50 per cent by mid-1998 (World Bank 1997c: 128; *Infobank* July 1998; *AWSJ* 20 April 1998). Not only was inflation expected to reach up to 80 per cent by the end of 1998, the number of unemployed Indonesians at the end of February had risen to an officially estimated 27.8 million, up from 13.1 million at the end of 1997 (*JP* 26 March 1998). Most observers now expected the economy to shrink by between 10 and 20 per cent of GDP during 1998 alone (World Bank 1998b: 1; *Asiaweek* 17 July 1998). In early July, the Central Bureau of Statistics announced that the number of Indonesians living in poverty had surged to 79.4 million, or about 40 per cent of the population (*JP* 3 July 1998).

It was now acknowledged, even among the most optimistic neo-classical economists, that it would take years for the Indonesian economy to recover (Pangestu 1998). More important, the crisis signalled that the era of authoritarian rule and oligarch dominance might now be at an end. Not only were powerful politico-business families and Chinese-Indonesian corporate conglomerates shattered by the crisis, the structural transformation of Indonesian capitalism had become a fundamental objective of the IMF to be imposed within the terms of its US$41.5 billion rescue package. As we shall see, successive IMF reform packages required closure of insolvent financial institutions, elimination of government and private monopolies, reductions in tariffs and export subsidies, development of greater transparency in government and introduction of new regulatory frameworks and audits to deal with bankruptcy and public and corporate governance.[4] For the IMF, the crisis came as a 'blessing in disguise', providing the opportunity to sweep away the distortions untouchable in good economic times (*AWSJ* 13 November 1997: 7).

Fracturing the corporate and politico-business oligarchy

Most neo-liberal economists had been enthusiastic about the prospects for the Indonesian economy in the years leading up to the crisis, despite nagging concerns about the slowing of the reform agenda (Pangestu 1989; Soesastro 1989; Prawiro 1990). While it was recognised that growing private sector debt, often in the form of short-term, unhedged debt, raised new sorts of exchange risk to borrowers, the levels were considered to be within Indonesia's capacity to repay and, in any case, sustaining investment rather than consumption (Radelet 1995; World Bank 1991a: 52, 53, 1995: 24–26; Nasution, Anwar 1995).[5] Although the World Bank had recognised the increasing scope and audacity of the predatory raids of well-connected business on the state, it was generally felt that corruption was not a fatal cost to the larger process of economic growth. At one level, its solution was seen to lie in the erosion of the source of rents themselves through continued deregulation (*Asiaweek* 26 February 1999: 53; *FEER* 16 May 1996: 40; *Australian Financial Review* 2 July 1997: 14). At another level, the problem could be tackled through institutional engineering, including stricter rules to curtail off-budget spending, protect property rights, control infrastructure contracting and regulate banks (World Bank 1995: vii, viii, xv–xvii).

Yet private sector debt and corruption were not significant simply in terms of their immediate costs to growth. They reflected a specific system of power relations antithetical to liberal reform and sustainable only so long as continued access to debt rollover from commercial lenders and the protective intervention of state power remained. Belated recognition of these problems now began to emerge within the neo-liberal camp. Formerly optimistic supporters of Indonesia's banking reforms, David Cole and Betty Slade, now argued that:

> the preconditions and the playing out of the crisis in Indonesia were essentially political. The failure to put together an effective initial response and the ultimate severity of the Indonesian financial crisis must be understood and explained as the consequence of the lengthy process of politicisation of economic and financial activity within Indonesia, and the concomitant erosion of any effective prudential supervision over financial institutions by regulatory authorities coupled with imprudent behaviour by many foreign lenders and investors.
>
> (Cole and Slade 1996)[6]

Neo-liberals were increasingly to distance themselves at least from the latter period of the Soeharto regime where, it was argued, the increasingly unconstrained activities of the Soeharto children had led to a growing divergence from responsible macro-economic policy and to the eclipse of the technocrats (McLeod 2000a: 108; Liddle 1999: 25; Hill, H. 2000a: 132).

In fact, while neo-liberal economists frequently suggested that the extent of losses from corruption, bad debt and government collusion with business interests had been hidden by strong growth during the 1980s and 1990s, these facts were always well known.[7] The caution with which neo-liberals treated these issues stemmed in part from those close associations formed in the Cold War with a regime willing to renounce populist agendas and because of the Washington consensus conviction that macro-economic fundamentals were, in any case, the true indicators of healthy economic performance. Amongst Indonesia's own technocrats and reformist advocates the views about the regime had never been as sanguine. Such figures as Anwar Nasution, Mar'ie Muhammad, Hadi Soesastro and, of course, Sjahrir, always understood and feared the dark side of the Soeharto regime. They were under no illusions that Bank Indonesia had ever been able properly to regulate the private banking sector. By early 1998, World Bank Head, even Wolfenson, and other officials were admitting that the Bank's earlier assessments of the Indonesian economy had glossed over warning signs, perhaps because of a concern to maintain good relations with Jakarta (*AWSJ* 5 February 1998: 4).

Nowhere were these inherent problems more significant than in the banking sector, where non-performing loans were problematic before the crisis at US$12.95 billion, or 10.75 per cent of total bank loans (*JP* 10 September 1996). The collapse of the Sutowo family's Bank Pacific in March 1997, the result of excessive property investments, represented only the tip of an iceberg in private banking failures stretching back into the mid-1980s and discussed in Chapter 3. Indonesia's banks remained beset by bad loans and questionable lending practices, and the debate about whether some might need to be liquidated began as early as March in 1997 (Rosser 2002: 51–84; *JP* 7 March 1997: 4; *AWSJ* 5 March 1997: 1). Huge losses continued to be piled onto a haemorrhaging state apparatus by mark ups, overpricing and other illegal arrangements in allocating contracts and distributorships, irregularities in spending and losses through bad debts in the state banking system. The State Audit Agency ranked the Finance Ministry itself among the most corrupt government departments (*JP* 13 June 1996: 1; 19 June 1996: 1).

At the heart of the matter were the huge debts that now accrued from the previously unconstrained flows of investment and borrowing by the Indonesian private sector. Attempts by officials to constrain rising debt or to reform the state banks had proven futile in the face of the powerful corporate and political coalitions that had straddled private and state power in the pre-crisis period. Foreign financial institutions had also enthusiastically supported the huge debt flows. Indonesia's private sector corporations now enjoyed undreamt-of access to huge funds from global capital markets and flows of credit that had grown globally from negligible levels in the 1950s to more than US$6 trillion at the time of the crisis (Perraton *et al.* 1997: 265). These were, however, volatile and potentially destructive. As John Plunder has pointed out, it had become increasingly difficult to impose controls on the free flow of capital. 'Any attempt to clamp down on imprudent financial practices,' he argued, 'risks driving business and jobs away to other more lightly regulated centres' (cited in Gill and Law 1988: 188). Because of the extremely mobile nature of portfolio and commercial bank investment they were liable to dramatic swings in sentiment that resulted in widespread panic among investors and a scramble for the exit (Eichengreen 1997: 379; Winters 1997; Radelet and Sachs 1998).

Spreading quickly into the very heart of Indonesia's commercial world, debt engulfed the major business groups and politico-business families that had enjoyed over a decade of unchecked expansion into booming and newly deregulated sectors of the economy and access to the largesse of global capital markets. With loans for land and property purchases growing at over 35 per cent annually, an increasingly concerned government placed a prohibition on all loans for land purchases in July 1998 (*AWSJ* 8 July 1997: 1, 13; 10 July 1997: 1, 4). However, by the time of the Asian economic crisis, the World Bank estimated Indonesia's private sector external debt at US$72 billion with repayments of US$20.8 in largely unhedged loans due by the end of 1997 (World Bank 1998b: 1.9, 2.3). Other estimates were gloomier. Analysts at Indosuez estimated the corporate sector debt at US$140 billion of which US$83 billion was undeclared offshore borrowing. According to de Koning, economist at ABN/AMRO Bank in Jakarta, total repayment of private sector debt and interest in 1998 would be US$59.8 billion (cited in Soesastro and Basri 1998: 37). Even at an exchange rate of Rp.5000/US$, Soesastro and Basri estimated that about 50 per cent of all listed companies had a dollar debt to asset ratio of around 50 per cent (1998: 37, 38). Highly exposed to short-term and largely unhedged loans, they were paralysed by a debt crisis of massive proportions and by early 1998 most had defaulted on their loans (World Bank 1998b: 1.9, 2.3).

None of the high profile companies that had emerged in the previous decade escaped the carnage.[8] The largest of the conglomerates, that of Liem Sioe Liong, saw its Bank Central Asia taken over by the government after panic by depositors, while his giant Indofood noodle maker, with reported debts of US$1 billion, had to suffer the removal of the wheat subsidy that was formerly at the heart of the operation (*AWSJ* 18 May 1998: 1; 24 May 1998: 1, 9; 24 August 1998). The Soeharto family was also in deep trouble. Carrying huge loans for investments in a range of major power generation and petrochemical projects, automobile and telecommunication ventures, the family debt was estimated at around $4 billion in 1999.[9] Bambang Trihatmodjo alone owed $2.5 billion to the state banks, and Tommy $600 million (*FEER* 13 May 1999: 10–14). Tommy Soeharto owed IBRA over Rp.5.5 trillion, while Bimantara debt totalled Rp.3.5 trillion and Bambang Trihadtmodjo's debt stood at Rp.1.4 trillion (*Kontan Online*

13 March 2000). Suddenly, the Cendana companies were no longer immune to the demands of creditors and were besieged by government and foreign creditors and embroiled in an ongoing scramble to restructure.[10]

They now confronted their former foreign creditors who held billions of dollars in debt from Indonesia's previously buoyant business groups. The government, too, became a major creditor. As we have seen in the previous chapter, the government had allocated huge levels of discretionary credit through state banks to well-connected business groups whose viability rested upon the sorts of state 'guarantees' it was no longer able to deliver. In the private banking sector, too, banks had been raided to supply cheap credit for the corporate groups of which they were a part to the extent that intra-group loans often exceeded the assets of the banks. They had also been plundered to provide the means of a capital flight in the months following the crisis. By December 1998, non-performing loans for state banks were estimated to be Rp.150 trillion or 50–60 per cent of outstanding loans while in the private banks they were Rp.175 trillion or 80 per cent (*JP* 23 December 1998: 5; World Bank 1997c: 128; *Infobank* July 1998; *AWSJ* 20 July 1998).

An important split was now driven into the business/state relationship. As the banking system collapsed, the state was forced to fund a huge programme of recapitalisation. According to Finance Minister Bambang Subianto, the costs of recapitalisation had surged to an estimated Rp.406 trillion in May 1999, rising to Rp.550 trillion or US$82 billion in July of that year. By the end of 1999, Standard and Poor's estimated non-performing loans in Indonesia's banking system would reach between 75 per cent and 85 per cent of total loans, and the costs of recapitalisation Rp.600 trillion (US$87 billion) or around 82 per cent of GDP.[11] It now became a priority for the government to recoup the debts as the costs of recapitalisation consumed an increasing part of the state budget and threatened fiscal crisis. The private sector debtors (often the bank owners themselves in the case of private banks) were now placed under increasing pressure to hand over billions of dollars in assets and to dismantle much of their corporate empires.

Resolving the debt problem was now a primary concern of the IMF and it built into its Letters of Intent requirements that Indonesia introduce new bankruptcy legislation and establish a commercial court especially to deal with the attempts to recover debt and unlock assets. The IMF also required that the government establish an agency to recapitalise the banking system (IBRA) that was eventually to have the power to close banks and seize assets of debtors. Both moves were complemented with requirements for extensive auditing of government corporations, Ministries and agencies, and the establishment of a supervisory presence within the state banks. Among the main IMF targets were the pervasive off-budget funds held by most of the state agencies. Audits of Bulog, Bank Indonesia, Pertamina, PLN and other state corporations were set in train (Government of Indonesia *Letters of Intent* (hereafter *LOI*) 4 July 2000).

No longer able to prevent legal action against themselves or to escape scrutiny of their activities to the same extent, members of Indonesia's politico-business oligarchy now required a quite different set of strategies if they were to survive. Most immediately, they were forced to liquidate as much of their equity as possible, moving the cash out of rupiahs and out of Indonesia, resulting in an outflow of capital in late 1997 and 1998. They were forced also to default on loans. Most of Indonesia's private business groups gained valuable breathing space by simply stopping their debt servicing. Refusing to participate in debt renegotiation schemes and stalling on handing over assets to IBRA, they hoped to dilute their debt problems without the need for structural change, intending either to ignore or restructure them on their terms as time dragged out and, hopefully, the rupiah rose again in value. Early efforts by the government to entice corporate debtors into restructuring through an agreement with international banks to roll over US$60 billion in debt provided little attraction while the value of the rupiah remained so low (*JP* 5 June 1998: 1; 20 August 1998: 8). While this may have appeared a clever tactic, it nevertheless came at a high cost. Access to credit was now severely constrained.

Foreign creditors and the government closed in on debtors, initially assuming that they could enforce repayment by taking them to court where necessary. They began a series of attempts to secure bankruptcy judgements against local debtors. Among the first were the World Bank's International Finance Corporation, Amex and PT Ing. As we shall see in the following chapter, despite the introduction of new bankruptcy laws and the establishment of a commercial court in August 1998, the courts were not to be an effective channel for debt resolution. The rampant corruption of the judiciary meant that they were susceptible to bribery and willing to throw out cases brought by even the largest international corporations. By mid-1999, only sixteen out of fifty-five cases had been successful (*AWSJ* 7–8 May 1999: 1, 8). Only thirty bankruptcy cases had been registered by June 1999 compared with several thousand in Korea (World Bank 1999: 2.4). Conglomerates and the politico-business families avoided prosecution and kept their assets out of the hands of creditors, at least in the period to May 1998, able to continue trading while insolvent with impunity and free of debt servicing. Nevertheless, so deeply had Indonesia's corporate world become enmeshed in global financial markets and with international equity partners that it was inevitable that some accounting of the debt be made. It was this negotiation that was to consume them for the coming years.

Fracturing the framework of 'political capitalism'

As the crisis bit deeper in the period between July 1997 and the eventual fall of the government in May 1998, Soeharto and his allies fought an increasingly desperate rearguard action to keep intact the policies and institutions that had sustained the system of 'political capitalism' and their own economic ascendancy.[12] In a desperate attempt to protect its sliding currency, the Indonesian government called in the IMF on 8 October 1997, and a $US43 billion rescue package was announced on 31 October. In return, the government agreed initially to a range of reforms involving fiscal austerity and further trade deregulation (Government of Indonesia *LOI* October 1997–2001; *JP* 1 November 1997).

It appeared that Soeharto had expected the mere announcement of IMF assistance to restore confidence and avoid the need to undertake fundamental structural change. On the contrary, the IMF had more fundamental reforms in mind. Whether attempts to stop the haemorrhaging through capital controls would have worked or not, the IMF's advice to float the rupiah in July 1997 and its subsequent opposition to the currency board proposal nevertheless denied Soeharto a potential way out that might have left intact the very fabric of the regime. For economist Steve Hanke Soeharto's currency board advisor, this was clear evidence that both the IMF and the US government were involved in nothing less than a deliberate programme of regime change. No longer useful to them in the post-Cold War era, Soeharto just had to go (Hanke 2003).

Whether the IMF was aiming so precisely at the fall of Soeharto or not, it clearly intended that the price of its assistance required nothing less than a fundamental restructuring of institutions and policies that would break the nexus of oligarchic and state power and dispossess much of Indonesia's politico-business elite. It demanded the postponement of several large industrial projects, the abolition of state trading monopolies in various agricultural products, reductions in tariffs, the liquidation of insolvent banks, and the adoption of tighter fiscal measures, including the removal of a range of subsidies. Other demands aimed directly at the very power structure included the elimination of taxation privileges for the controversial national car project, the abolition of the clove trading monopoly, and the scrapping of a controversial jet airplane project being pursued by the state-owned aircraft manufacturer, IPTN (Government of Indonesia *LOI* 1997–2001).

When it became apparent that the currency was not stabilising and that the IMF was intending to carry out a protracted programme of reform, Soeharto quickly revealed that he had little intention of implementing many of the conditions of the packages. Within a few days of the initial package being announced, Soeharto had reinstated fifteen large and expensive infrastructure projects. Five of these involved the construction of power plants in Java at a time when electricity on that island was already in oversupply. Needless to say, business groups associated with the Soeharto family were the main beneficiaries of this policy reversal (*JP* 8 November 1997; *AWSJ* 7–8 November 1997; *FEER* 4 December 1997).

There was resistance to early attempts at banking reform. When the Minister of Finance, Mar'ie Muhammad, announced on 1 November 1997 that sixteen ailing private domestic banks, including three partially owned by members of the Soeharto family, would be liquidated in accordance with the IMF package, the Soeharto family reacted angrily. These were actions they had never before experienced. Contemplating the closure of his Bank Andromeda, Bambang

Trihatmodjo accused the Finance Minister of plotting against the Soeharto family and threatened to take the government to court. Whilst Trihatmodjo was eventually persuaded to drop his court case, it appears that he did so only because he was guaranteed that he would later be allowed to re-establish Bank Andromeda under a different name (*JP* 26 November 1997; *AWSJ* 13 November, 1997). Similar efforts by Probosutedjo to keep his bank alive through complicated legal manoeuvres confirmed a general view that the family still considered itself immune from demands for structural reform (*JP* 31 December 1997; *AWSJ* 6 January 1998).

The government also resisted demands for fiscal austerity. Running a budget surplus meant cutting into large state-funded projects critical to Indonesia's corporate moguls. Perhaps more important, the rapid spread of poverty in the regions and urban centres made current levels of spending on items such as food and fuel subsidies even more essential, especially where social unrest was possible. Yet, failure to embrace fiscal austerity in the January 1998 budget and to base its estimates on realistic predictions of currency movements was taken as further evidence by financial markets that the Indonesian government was not serious about fiscal reform. As a consequence, the rupiah collapsed to more than 10,000 to the US dollar, driving companies with large foreign debts into technical insolvency and provoking a wave of panic buying as Indonesians attempted to stock up on commodities before inevitable price rises kicked in. By the next day, most basic food products had vanished from supermarket shelves and traditional markets. The collapse on the Jakarta Stock Exchange was just as severe. By 9 January the market index had fallen to 340 points, after having been more than 700 points six months earlier (*Australian Financial Review* 8 January 1998; 10–11 January 1998; *JP* 10 January 1998; 12 January 1998; *AWSJ* 9–10 January 1998).

However, by mid-January, Indonesia's increasingly desperate situation allowed the IMF to secure a far greater range of reform commitments in its second package of negotiated reforms. Among the new reforms were: greater independence for the central bank; the withdrawal of taxation privileges for the national car project; the elimination of cement, paper and plywood cartels; the withdrawal of credit privileges as well as budgetary and extra-budgetary support for IPTN; the removal of restrictions on investment in the retail sector; the introduction of revisions to the budget; the elimination of the clove monopoly; the abolition of state trading monopolies in flour, sugar, soybeans and other basic commodities; and the phased elimination of subsidies for fuel and electricity (*AWSJ* 16–17 January 1998; *JP* 16 January 1998).

Nevertheless, full implementation would continue to be frustrated by resistance from the major politico-business families and conglomerates. For instance, although the government officially abolished the plywood cartel, Apkindo, it appeared that the cartel continued to exercise authority over exporters through its control of plywood shipping. Timber companies found it difficult to operate outside, and continued to pay fees and adhere to its pricing policies (*AWSJ* 10

February 1998). The government also put in place a range of disingenuous devices to keep the national car project and clove monopoly alive (*Australian Financial Review* 26 February 1998; *AWSJ* 24 February 1998).

By now, the IMF's main concern was switched from Soeharto's attempts to subvert and undermine individual reform measures to his controversial plan to form a currency board to manage the value of the rupiah. Despite the introduction of the second IMF package, the value of the currency continued to fall, at one point trading at around 17,000 to the US dollar. Within this context, Soeharto was convinced that the only way to prevent what he saw as currency speculators trying to destroy the Indonesian economy was by fixing 'a certain exchange rate' in order to bolster the currency and assist local companies with hefty foreign debts. On 11 February, the Minister of Finance announced that the government was 'preparing steps toward the setting up of a currency board system, including legislation to support the board' (*Australian Financial Review* 11 February 1998; *Sydney Morning Herald* 12 February 1998; *AWSJ* 10 February 1998).

To a beleaguered Soeharto, a currency board was attractive because it offered a possible way out of the dilemma without the radical structural surgery demanded by the IMF. Cynics suggested that a currency board, even if established for only a short time, might also allow well-connected business groups to cover their foreign debts and Soeharto to be re-elected as President in March. Given the involvement of Soeharto's eldest daughter and the prominent businessman, Peter Gontha, in arranging the visit of Professor Steve Hanke to organise the board, such suspicions appeared to have substance (*AWSJ* 10 February 1998; *Australian Financial Review* 11 February 1998).

Soeharto pressed ahead with the proposal despite heated opposition from leading figures within the IMF, including Camdessus and Fischer (*AWSJ* 17 February 1998; *JP* 14 February 1998; *Suara Pembaruan* 11 March 1998). His dismissal on 17 February of the Governor of the central bank, Soedradjad Djiwandono, was a move believed by many observers to be in response to Djiwandono's opposition to the currency board proposal. In early March, Soeharto appealed to the IMF and foreign governments to help Indonesia find 'a more appropriate alternative' to the existing IMF programme. What Indonesia needed, he proposed, was an 'IMF Plus' programme which, besides the IMF's reforms, would also include measures specifically designed to stabilise the rupiah, including the adoption of a currency board system (*AWSJ* 2 March 1998; 16 February 1998; 18 February 1998; *JP* 18 February1998).

In response, the IMF began to turn the screws on a government confronted by fiscal disaster and mounting debt. On 7 March, it announced that it would delay the second $US3 billion tranche of its bailout package, and on 10 March, the World Bank and the Asian Development Bank followed suit, withholding $US1 billion and $US1.5 billion respectively (*Bisnis Indonesia* 8 March 1998; *Suara Pembaruan* 11 March 1998). By now, Soeharto was clearly in a desperate situation. On 8 March, he declared that the IMF package could not be implemented because it was 'unconstitutional', arguing that, 'The IMF package will impose a liberal economy, which is not in line with Article 33 of the Constitution' (*Straits Times* 9 March 1998). Frustration with the political price of the reforms was growing within Soeharto's inner circle. His daughter, Siti Hardiyanti Rukmana, was quoted as saying: 'If the funds sacrifice and degrade our nation's dignity, we do not want them' (*Asiaweek* 20 March 1998).

Effectively forced to choose between continued resistance to the IMF or political suicide, Soeharto predictably opted for the former when he retreated to the most trusted and closest of his political, family and business associates in selecting his new Cabinet and his Vice-President. This was an act of defiance and, perhaps, hubris. He nominated as Vice-President his long time supporter, B.J. Habibie, in spite of well-known antipathy towards him within the international finance community and in the IMF. When he announced the composition of his new Cabinet a few days later, gone were the technocrats who previously had controlled such portfolios as finance and trade and industry. In their place, as mentioned earlier, were people personally close to Soeharto. Bob Hasan, who controlled a private business empire based upon a range of trading and forestry cartels, became Minister for Trade. Other Ministers included Siti Hardiyanti Rukmana (Tutut), Soeharto's former tax chief, Fuad Bawazier, and the businessman, Tanri Abeng, who was given the State Enterprises Ministry (*FK* 6 April 1998; *FEER* 26 March 1998). It was clear that, despite the intensifying economic difficulties, Soeharto remained politically strong.

At this stage, the Indonesian government also began to press the IMF over its harsh programme of fiscal austerity. It claimed that continuing reduction of government subsidies on basic commodities was posing a severe threat to political and social stability as Indonesians found it increasingly difficult to maintain their livelihoods in circumstances of rising prices and unemployment (*AWSJ* 16 February 1998; *JP* 20 March 1998; *FEER* 26 March 1998). This was a claim given weight as outbreaks of rioting and looting targeted ethnic Chinese in a range of cities and towns, including Jakarta itself. Although facilitated by military involvement, the riots were driven by long-held resentment of Chinese business interests amongst indigenous Indonesians because of their dominant economic position and suspicions that they were hoarding food and deliberately inflating prices.

A more flexible approach on the part of the IMF was confirmed in the third round of negotiations between the IMF and the Indonesian government that began on 17 March (*AWSJ* 18 March 1998). These approved the Indonesian government's plan to continue subsidies on imports of basic commodities and to keep Bulog in existence (*JP* 21 March 1998). As a result of this concession, when a third IMF agreement was signed in early April, the IMF was also forced to agree to a budget deficit of 3.2 per cent of GDP (*South China Morning Post* 8 April 1998). At the same time, the IMF also retreated from its earlier warning to the Indonesian government not to take part in debt rescheduling negotiations for fear of encouraging local companies to ask for a government bailout rather than repay their debts. Nevertheless, the Indonesian government was, at the same time, forced to agree to the divestiture of state-owned shares in six listed companies, privatisation of seven other state-owned enterprises within twelve months, an end to the allocation of monopoly privileges, the introduction of a new bankruptcy law and commercial court, and reductions in foreign ownership and trade restrictions in areas such as wholesale trade and palm oil exports (*JP* 13 April 1998; *AWSJ* 13 April 1998). Perhaps most significantly, the new Finance Minister, Fuad Bawazier, announced on 20 March that the government was no longer seriously pursuing the currency board proposal.

By this time, it was apparent that power was shifting in the government and this was reflected in the negotiations with the IMF. On 22 April the government announced that the first deadlines for the implementation of reforms agreed to under the third package had been met. Most significantly, these reforms included the removal of the ban on palm oil exports and its replacement with an export tax of 40 per cent, the firmest indication so far that the authority of such figures as the new Minister for Trade and Industry, Bob Hasan, was in eclipse. It was an impression reinforced by a declaration from Co-ordinating Minister for Economics, Ginandjar, that Indonesia would 'fully adhere' to the IMF package and that the government would become increasingly transparent (*AWSJ* 15 April 1998; 23 April 1998; 1–2 May 1998; *JP* 23 April 1998). As we shall see, however, the removal of concerted resistance only opened the door to a new phase of frustration for the IMF.

The contagion: economic crisis becomes political crisis

That the IMF was able progressively to impose its will signalled that below the surface things were changing. Deepening fiscal and debt crisis gave the government little room to manoeuvre against reformist pressures for policy and institutional change. It also meant that the very cohesion and authority of the politico-business oligarchy was being fractured and the highly integrated state-business relationships that had sustained it were dissolving. Unable to rely on the intervention of a centralised state, the priorities of the conglomerates and the families now became those of capital flight and isolating their remaining assets from attempts to restructure debt. It was estimated by one source that around US$10 billion was sent overseas by Indonesian citizens between July 1997 and June 1998.[13]

Not only was the state no longer in any position to protect the politico-business families or the conglomerates, it was increasingly forced to move against them; to seize their assets and enforce debt repayment in order to protect its own fiscal position and in a vain hope that foreign capital inflows might begin again. As IMF-sponsored supervisory groups entered the state banks and corporations and ministries, the authority of reformist state officials was enhanced. Other political reformers, outside the state apparatus, saw the IMF reforms as at least an opportunity to bring down the conglomerates and the Soeharto family, even if they shared little sympathy with the market ideology that underlay them. Student protesters and middle class opponents of the regime supported moves for institutional change aimed at undermining cronyism, and the implicit IMF view that regime change was a precondition for reform. Thus, the door was opened to a wide assembly of critics, reviving the fortunes of populist politics and its calls for equity before growth, redistribution of wealth to co-operatives and protection of indigenous Indonesians against the incursions of Chinese and foreign investors.[14]

Nevertheless, until almost the very end, Soeharto appeared to be in control of the political apparatus at least. As we have seen, his last Cabinet, appointed in March 1998, was a demonstration of his confidence in the face of a fragmented opposition. It included, in an unprecedented way, many of his closest personal associates. If he could retain the levers of power through the dark days of economic crisis, conceivably the same power relations and the same faces might prevail in the long term. Yet, despite the appearance of invulnerability on the surface, the foundations of the political regime itself were being eroded in the wake of the crisis as Soeharto proved increasingly unable to control events. Not only were opponents marshalling themselves with increasing impunity outside the state, within its very walls, its own apparatchiks were faced with the task of preserving intact the apparatus of administrative oligarchy. Ironically, the very person who had created the New Order and stood at its apex for so long had now become its greatest liability. While the crisis had fractured and disorganised the oligarchy, it was, at the same time, providing a clear message that its long-term survival required a new political shell. This will be the subject of the next chapter.

Notes

1 The government was forced to run an 8.5 per cent budget deficit for 1998, leaving it little option but to reschedule foreign debts and seek increased amounts of foreign aid. In late July of that year, a fiscal disaster was averted only when Western creditors pledged US$7.9 billion in loans and grants to the government (*AWSJ* 31 July–1 August 1998: 1; *JP* 21 July 1998: 4). More importantly, in the longer term, debt servicing was to consume an increasing proportion of the routine budget. In the 2000 state budget, APBN, and in the planned budget for 2001, RAPBN, debt servicing stood at just over 30 per cent of outlays.

2 The First Deputy Managing Director of the IMF argued that the main cause of the crisis had been the inadequacies of the East Asian model of development. 'Crony capitalism delivered for a long time in Asia but the interlocking nexus of banks, governments, corporations became quite rotten and it is rotten in the countries in crisis and they have to be reformed' (comments made to the Australian Broadcasting Commission's *Four Corners* programme, Australia, 11 May 1998, cited in Jayasuriya and Rosser 1999: 6).

3 The Head of the World Bank in Jakarta, Dennis De Tray, chastened by criticisms of the World Bank's persistently upbeat assessments of Indonesia's prospects before the crisis, now attributed the crisis to weak institutions, difficult cultural values and the fact that funds managers operated on the basis of movements in margins rather than

the fundamentals of economies (*JP* 14 April 1999). In other words, markets over-reacted in an irrational way.

4 The main sources for details of the IMF packages and the progressive agreements with the Indonesian government come from the various *Letters of Intent* (*LOI*) of the Government of Indonesia to the IMF and Memoranda of Economic and Financial Policies. These were found on the IMF website (http://www.imf.org/external/np/loi). The *Jakarta Post* also offered these documents through its website, *Jakarta Post.Com* (http://www.thejakartapost.com/special/imf_loi_4.asp). Colin Johnson provides a detailed summary and analysis of four IMF agreements (1998).

5 Generally, however, it was felt that this risk could be handled, and that its risk was less given that it was being directed into investment rather than consumption (World Bank 1996: 1.28).

6 Cole and Slade went further, now suddenly finding that in the 1990s, the 'Soeharto connection' had become

> 'guarantee' or collateral underlying the viability of many enterprises and financial institutions, most obviously in banking and securities markets. Any financial regulator who attempted to apply prudential rules to such connected financial institutions or transactions, for example Bank Indonesia Managing Director Binhadi in 1992, or Director General Matriono at the Ministry of Finance in 1996, was removed from his position. ... Foreign investors were as much influenced by this 'guarantee' as were domestic players.
>
> (Cole and Slade 1998: 65)

7 See, for example, Stanley Fischer's statement about Indonesia's hidden weaknesses, in *JP* 12 November 1997: 1.

8 Although a clear picture is difficult to assemble, by the time the initial confusion had cleared, the Salim group was reported to have debts of US$5 billion, while Eka Tjipta Wijaya's Sinar Mas Group confronted a massive debt of US$9.1 billion, the legacy of huge borrowings for ambitious forestry, pulp and paper ventures. The Soeharto family's Cendana group was reported to have debts of US$4 billion, while the Bakrie group owed US$1.05 billion. The petrochemical venture, Chandra Asri, alone owed US$1.1 billion, while Astra owed US$935 million (*Financial Times* 6 October 1997; *JP* 28 August 1998: 8; 25 December 1997; 1 January 1998: 15; 28 October 1998; *AWSJ* 23 September 1998: 1, 5; 23 October 1998: 3; 16 August 1999: 1, 8).

9 Assets of the Soeharto family corporate interests, collectively known as the Cendana group, were estimated by the Indonesian Business Data Centre to be worth US$17.5 billion or Rp.200 trillion in June 1998, including $3.16 billion equity in joint ventures with foreign partners (*JP* 2 June 1998: 1).

10 Bank Bumi Daya settled a Rp.250 billion debt of SCTV by taking over 52.5 per cent of equity from Sudwikatmono in 1998 (*JP* 24 November 1998: 12). IBRA seized 14 hectares of land owned by Siti Hardiyanti Rukmana to cover costs of US$100 million debt to Bapindo. Tommy Soeharto was forced to seek refuge in bankruptcy as his Sempati Airlines collapsed in 1999 with debts of US$220 million (*JP* 8 July 1999: 8; 8 October 1999: 8). Attorney General Ghalib estimated Tommy Soeharto's PT Timor Putra Nasional to have cost Indonesia US$1.05 billion in foregone import duties and taxes and US$500 million in syndicated state and private bank loans. The government demanded repayment of US$425 million in exempted import duties for 39,000 cars (*JP* 8 December 1998: 1). Tommy's container terminal at Tanjung Priok was taken over by the government for failure to pay US$82 million and Rp.79.3 billion in debts (*JP* 12 May 1999: 8). Most important, Soeharto family companies lost their

entire holdings in PT Astra and Chandra Asri as the government moved in to settle the huge debts of these corporate giants.

11 See *AWSJ* 11–12 June 1999: 3. In contrast, the costs of recapitalisation in Thailand, South Korea and Malaysia are estimated at 35 per cent, 29 per cent and 22 per cent respectively. This compares with a figure of 15 per cent for the 1994 crisis in Mexico and 41 per cent in Chile during the 1981–1985 crisis. Levels of non-performing loans were also lower in other Asian economies, estimated to be 55 per cent in Thailand, 50 per cent in South Korea and 35 per cent in Malaysia (*AWSJ* 22 October 1998: 1). These figures compared with 11 per cent and 16 per cent in Chile and Mexico respectively (*ASWJ* 22 October 1998: 1, citing data from Barclays Bank and the IMF).

12 The most comprehensive analyses of the complex and bitter conflicts that accompanied the progressive economic collapse of the Soeharto regime in the face of the crisis and the more concrete interventions of the IMF can be found in Robison and Rosser 2000; Rosser 2002: 171–182. The following section draws heavily on this work.

13 See the estimates of ABN AMRO's C.J. de Koning in *JP* 10 August 1998: 4.

14 As we shall see in the following chapter, the growing influence of populism was briefly to enter the very heart of the government when long-time populist politician, Adi Sasono, was appointed Minister for Co-operatives under Habibie and given around Rp.20 trillion (US$2.9 billion) to allocate to co-operatives. Sasono came with the support of several indigenous businessmen eager to participate in the redistribution of the assets of Chinese conglomerate assets (*FK* 21 September 1998: 76–80; *FEER* 3 December 1998: 14–16; *JP* 11 August 1998: 1).

7

POLITICAL UNRAVELLING

Soeharto's fall

The crisis of the Thai baht in mid-1997 precipitated a chain of events that would eventually lead to Soeharto's downfall after more than three decades of autocratic rule. Though Indonesia was plunged into the turmoil of deep economic crisis, Soeharto initially appeared so politically invulnerable that in March 1998 he managed to orchestrate a seemingly routine, unanimous re-election by the MPR to yet another five-year term in office. But in reality the legitimacy of the Soeharto government was already being progressively undermined as it failed to halt the progressive unravelling of Indonesia's economy and the collapse of its corporate and financial institutions. Domestically and internationally, Soeharto himself was increasingly being perceived as a cause of Indonesia's economic problems rather than the solution to them. He would finally be forced out in May 1998, but in the absence of a coherent opposition, the remnants of his regime were able to doggedly cling to power. By November they had successfully ensured that they would not be swept away by reform – but instead would reconstitute within the institutions of Indonesia's new democracy.

But why did Soeharto fall after he had staved off repeated challenges in the past with relative ease? Crouch (2000) suggests that Soeharto's downfall was significantly due to advancing age, which had diminished his capacities for leadership. According to Crouch, Soeharto's ability to identify and solve problems at a time of sudden acute economic crisis was impaired. Crouch's approach is one that rejects the predictive value of 'grand theories' of political change and advocates a voluntarism that hinges on the importance of elite 'crafting' and on political will. There are implicit similarities to the 'transitions' arguments of O'Donnell and his colleagues in this way.

To Crouch, such an approach is advantageous, for it allows an understanding of how authoritarian regimes collapse and how 'democracy' can emerge in situations where structural conditions do not seem propitious – such as in Indonesia after Soeharto, where the middle class is relatively small and the working class weak. Thus, important in his explanation of the dramatic changes in Indonesia since Soeharto's fall are such factors as military paralysis due to 'shock' at its inability to control events after riots shook Jakarta in May 1998.

Liddle (1999) offers an explanation for Soeharto's downfall that is equally voluntarist. Like Crouch, he suggests that Soeharto's old age was a major factor – and identifies a series of wrong moves and bad judgements he made from the beginning of the economic crisis as critical in bringing down the regime. Since these went against the advice of international development organisations such as the IMF, and given Soeharto's suggested past ability to recognise good economic policy even when he didn't like it, Liddle regards this behaviour as 'aberrant' (Liddle 1999: 22–23). Thus, he concludes that Soeharto's downfall at the age of seventy-six was due to a growing inability 'to distinguish between the interests of his family and his cronies and those of the nation' (Liddle 1999: 25).

A different understanding of Soeharto's fall and of subsequent developments in Indonesia underpins the analysis that is offered in this book. Rather than the state of mind of key actors, it emphasises the fatal flaws and sources of tension within the Soeharto regime that accelerated its disintegration in the context of economic crisis. In this chapter and more fully in chapter 9, an analysis of the reconstitution of politics in Indonesia after Soeharto is offered which looks at elite struggles and negotiations to construct a new regime, but in the context of the constellation of social power that is a legacy of three decades of New Order rule. It is not the political will of actors that is emphasised in this analysis, but their underlying interests and the way these are expressed in the forging of new alliances and coalitions.

On this basis, Soeharto's fall in May 1998 was not primarily a factor of advancing age or inability to rationally identify the objective interests of the nation as a whole. More important than impaired judgement was that Soeharto, for the first time, was faced with a crisis he could not control. He was confronted with a bankrupt state, faltering giant conglomerates, burgeoning debts and the withdrawal of investors. On previous occasions, Soeharto had been rescued by oil money, or had been able to secure the support of Indonesia's propertied classes and corps of state officials, partly by manipulating social conflict. Equally important was the fact that the economic reforms demanded by the IMF to deal with the crisis would have seriously threatened the entrenched interests and position of the oligarchy that had consolidated under the New Order. Conceding to these demands would have been tantamount to admitting an inability to safeguard the conditions for rule by oligarchy and risking its unravelling.

Thus, Soeharto's ultimate failure to cling to power after so masterfully outmanoeuvring critics and opponents for so long involved the interplay of a number of factors. Most obviously, Soeharto's inability to steer Indonesia out of economic crisis had quickly revealed the limits of his powers. With the high growth rates gone and not expected to return in the immediate future, the darker side of his authoritarian rule became subject to greater critical scrutiny. Though it would be a mistake to explain Soeharto's ability to stay in power solely by reference to the workings of an extensive apparatus of control that had severely disorganised civil society, it is important to realise as well that the legitimacy of his rule was clearly overestimated by those who had routinely pointed to his regime's ability to produce economic growth. The superficial nature of this legitimacy was quickly revealed by the rapid appearance of student demonstrators, yuppie stockbrokers, housewives and industrial workers on the streets, especially of Jakarta, as soon as it appeared that the regime was beginning to unravel.

It would be a mistake as well, however, to attribute Soeharto's fall solely to these politics of the streets. An equally important factor in his downfall was Soeharto's abandonment at a critical juncture by key elements within his regime. Most prominent among these were his key economic ministers, as well as associates in Golkar, parliament and the military. Such an accelerated disintegration of the regime would not have been possible if it had not already embodied some fatal flaws. The economic crisis exacerbated these flaws and weakened the coherence of the complex alliance of officials and business that extended through Indonesian society and underpinned the New Order.

A fatal flaw was ultimately Soeharto's inability and unwillingness in the past to organise a viable social base of power outside of the apparatus of the state. Golkar, for example, in spite of his victories over the military, was never actually transformed into an effective party of the rich and powerful, and therefore, the maintenance of oligarchic rule remained dependent on Soeharto remaining in office. When Soeharto's own position became vulnerable, elements of the very oligarchy he had nurtured, and which he had resolutely protected against IMF conditions of assistance, decided that their own survival could only be assured by reorganising themselves within a new regime and with new allies. In other words, Soeharto had become redundant.

Thus, ironically, Soeharto had become a liability to the social order he had built, consolidated, and protected. This fact was underlined by an increasingly assertive people's uprising and the loss of the confidence of international capital – symbolised by the IMF's little-disguised disgust at Soeharto's attempts to protect various projects and banks it had targeted in the conditions outlined for assistance.[1] For the first time, the support of the Western industrial powers for the Soeharto regime was ambivalent at best. Even for the social interests embedded in the New Order, Soeharto now had to go.

But what of the success of the increasingly bold student and people's protests against Soeharto? Did these represent the emergence of new agendas and alliances of social interests? Fatal flaws existed here as well. Although the student movement in particular was crucial in mobilising popular protests against Soeharto in his last days, in reality, a new, solid coalition of interests, ready, not only to overthrow the regime, but also to impose a clear agenda of reform, remained sorely lacking to the end. There was no liberal party or social democratic coalition, for example, to fill this void. This was to show especially in the jostling and repositioning that would characterise the struggle to construct a new regime after Soeharto had left the political stage that he dominated for more than three decades.

A growing social base for opposition

Though political opposition remained internally fragmented and lacking a clear agenda of reform, its social base had quickly broadened during the last months of Soeharto's rule. This in itself was a significant development. No longer confined to bands of NGO and student activists, former officials of the regime, yuppie professionals, housewives, workers and urban slum dwellers, were all to be involved in protest action against Soeharto. That such open and lively protest could not be quelled easily was an indicator of the degree to which things were unravelling under the surface of calm control that Soeharto tried to project until near the very last days of his regime. That he could not easily sweep aside these protests indicated that he was in trouble, and thus opponents sensed a regime in distress.

As mentioned earlier, students were among the key actors in the chronicle of Soeharto's fall. Although their crucial part in the events of May 1998 was best exemplified by the dramatic five-day take-over of Parliament House, in reality student action had been slowly gathering pace from 1997, mainly taking place in and around regional campuses. Media attention was to focus on student action, however, when a series of demonstrations took place at the campuses of the University of Indonesia (UI) in February 1998 – just prior to the start of the MPR session that would re-elect Soeharto for the last time (Aspinall 1999: 215).

Whether or not stimulated by developments in Jakarta, a sudden increase of larger-scale student demonstrations was then to take place on campuses across Java as well as in Sumatra, Bali and Sulawesi, right until the end of the MPR session on 11 March that year. While students typically protested against *korupsi, kolusi dan nepotisme* (KKN) and called for *reformasi total,* the main galvanising element was clearly Soeharto's unopposed re-election to the Presidency. The mood of the times was best expressed by the spectacle of the burning of Soeharto's effigy during a demonstration in March at Yogyakarta's Gadjah Mada University (Aspinall 1999: 215; Ecip 1998: 43–46) which – given the frequent brutality of the security apparatus – was then regarded as a particularly daring act of defiance.

But student protests were nothing new to Indonesia – 1974 and 1978 witnessed major crackdowns on student activity, and demonstrations would take place intermittently from the late 1980s around a number of social and political issues. However, the distinct character of the new unrest in 1997–1998 was becoming increasingly apparent as the broad base of public support for student opposition gradually grew and the capacity of the regime to control events declined. Whereas student movements previously operated in a vacuum, after the crisis they became plugged into a wider anti-regime uprising.

Thus, student actions became progressively bolder. Expressions of support among the political elite, tacit as well as explicit, clearly encouraged students not to relent even in an atmosphere where the kidnapping and torture of known activists by the military had become well known (*Kompas* 18 July 1998; 5

December 1998). As mentioned earlier, Amien Rais, who had fallen from favour with Soeharto and the ICMI mainstream, was by this time frequently at the forefront of calls for Soeharto to step down, and was beginning to emerge as a key *reformasi* leader.[2] Similarly, encouragement from former Soeharto associates and ministers such as Ali Sadikin, Kemal Idris, Emil Salim, Subroto, Sarwono Kusumaatmadja and Edi Sudrajat – discarded by Soeharto at one time or another – also buoyed student activists.[3]

Nevertheless, students clearly realised that they were riding a wave of support that extended beyond political elites and intellectuals. Ordinary people, whose standards of living had severely declined since the onset of economic crisis, now felt able to come onto the streets. The main reason was clearly that prices for basic commodities had skyrocketed since mid-1997, but especially after a dramatic plunge of the rupiah's value in January 1998. Hyper-inflation in the midst of widespread unemployment due to mass sackings clearly created an environment conducive to mass unrest, especially in the urban centres. By March 1998, 300,000 people had lost their jobs in the garments and textiles sector alone, while 1 million workers had been retrenched in property and construction, and a further 50,000 in the banking sector (*JP* 26 March 1998). In the months immediately after Soeharto's fall, 40 per cent of Indonesians were reported by the Central Bureau of Statistics to be living below the poverty line compared to 11 per cent in 1996 (*JP* 3 July 1998). In this context, industrial workers, initially quiet due to the particular precariousness of their employment status, were to be involved by early May in outbreaks of strike action, especially in the industrialised areas around Jakarta.

Likewise, the urban middle class, including its previously more pliant elements, was displaying an uncharacteristic political restlessness as the security of its lifestyle became increasingly threatened by retrenchment and dwindling real incomes. No longer confident of its relative affluence, it became haunted by the growing spectre of unemployment, especially as management level workers now became among the first to be retrenched when banks collapsed or construction and real estate development firms abruptly folded.[4] Middle class demonstrators increased in number as new homeowners became economically crippled by interest rates on home loans that spiralled in some cases to 50 per cent per annum.[5] Significantly, middle class housewives were widely reported as having provided students with food and other logistical support during the dramatically charged take-over of Parliament House on 18–22 May 1998 (Budianta 2003).

That students played such a central role in the fall of Soeharto was indicative of the disorganisation of civil society during the New Order. In the absence of effective social organisations representing labour and other interests, and given the weakness of political parties, students filled the void – though it should be emphasised that the students themselves were often badly fractured. Clearly many of the students spearheading protest actions were personally jolted by the effects of the crisis as dreams of secure middle class lifestyles faded with the promise of scarce employment opportunities in the immediate future. In the regions in particular many student activists would have had lower middle class backgrounds and certainly would have struggled to keep up with the rising cost of basic commodities, transportation and books.

In Jakarta, the most militant grouping was Forum Kota, a loose association of students centred largely on private universities and colleges in the Jabotabek (Jakarta–Bogor–Tangerang–Bekasi) area, a number of which were regarded as second-tier. Interestingly, these students often expressed a vociferous rejection of what they regarded as the elitism and moderation of the student bodies of more established state universities like UI, from which government technocrats were commonly recruited.[6] Indeed, UI, and other state universities like ITB (Bandung Institute of Technology) had played a central role in the rise of Soeharto's New Order in 1965–1966, virulently leading the charge against the old left. Another major grouping, KAMMI, was an association of Islamic-based student organisations, which espoused a Muslim brand of populism, but rejected the overtly left-wing tendencies of other students linked to the then-banned PRD. The fractures within the student movement were a key factor in its inability to influence the direction of political change once Soeharto had fallen. In other words, Soeharto's fall eliminated the basis for a broad-based, cohesive, student movement.

Tensions within the regime

Though student demonstrations, by themselves, were unlikely to lead anywhere without being translated into organised popular movements, they may have convinced some within the regime that survival was not possible without resorting to violence. Thus, events leading to Soeharto's fall involved the fatal shooting of students at Jakarta's Trisakti University on 12 May 1998 by security forces. They also involved mass looting and violence (particularly against ethnic Chinese), especially in such areas as the Chinese district of Kota and its surroundings in Jakarta, as crowds wantonly razed buildings and other property. The worst of this took place on 13 and 14 May: numerous cases of rape against ethnic Chinese women were reported in the city at the same time that the vast suburbs enveloping Jakarta were terrorised by rampaging groups of provocateurs and looters. While the breakdown in law and order occurred primarily in the capital and its vicinity, Solo in Central Java was also the scene of a similar kind of destructive frenzy. Rather than simply demonstrating growing public frustration and desperation, these episodes masked increasingly serious tensions within the regime itself as the crisis worsened.

It has been argued by critics of the Soeharto regime that the outbreak of violence and lawlessness in May 1998 could only have taken place with the connivance of some sections of the security forces (*Kompas* 4 November 1998; 25 February 1999; 10 October 1999). In the past, Soeharto critics had also blamed military involvement for outbreaks of violence against Chinese or Christians, suggesting that they were intended to send messages to Indonesia's middle class, and to foreign investors, that stability could only be guaranteed by Soeharto rule. Outbreaks of disorder also seemed to conveniently channel public frustration into nationalist and anti-Chinese feeling.[7]

While the truth behind the shooting and rioting in Jakarta may never be completely known, there appears to have been manoeuvring involved by Lieutenant General Prabowo Subianto, the ambitious Soeharto son-in-law and then-commander of the elite army unit, Kostrad. Through Kostrad troops, and the Kopassus units that he formerly commanded, together with the Jakarta garrison commanded by close ally General Sjafrie Sjamsuddin, Prabowo had effective control of security in the capital. The interventions of the military were undoubtedly crucial to the unfolding of events that led to Soeharto's downfall, but the motives of the individuals involved remain murky. For example, there is dispute about whether Prabowo acted independently during the tense days and nights of mid-May 1998, or under the instructions of Soeharto.[8] Also, if the riots of May were orchestrated, were they partly calculated, as suggested by some, to embarrass and discredit Prabowo's main rival for control over the military, armed forces commander and Minister of Defence General Wiranto?[9]

Even if intended to confirm the need for order and stability, and by consequence elicit support for the government, especially among the middle class and propertied, the violence also sent other messages. If the riots were an attempt to channel public anger through religious and racial conduits and thus divert attention from the regime, this clearly backfired. The outbreak of violence was only to strengthen the perception that the regime – so focused on maintaining order in its official rhetoric – was indeed quickly unravelling.

However, as mentioned earlier, Soeharto's fall was not solely attributable to the politics of the streets and the breakdown in law and order that followed. It was directly related to such events as the sudden declaration, on 20 May, by his economic ministers – led by Ginandjar Kartasasmita – that they could no longer perform their duties under him (Ecip 1998: 123–125; Walters 1999). That Ginandjar should have played such a key role is very significant, given the prominent standing of the Kartasasmitas in the ranks of New Order oligarchic families.[10] A survivor, Ginandjar was to re-emerge as Deputy Chairman of the MPR only a year-and-a-half later.

The abandonment of Soeharto by his economic ministers was a relatively low-key affair, in spite of its importance. More dramatically playing the collective role of Brutus were Harmoko, Sjarwan Hamid, and Abdul Gafur, DPR/MPR leaders and all close Soeharto associates. Earlier, in a widely televised press conference on 18 May, the ultimate 'yes-man' Harmoko – who only months earlier had made public assurances that Soeharto was the genuine choice of the people – had called on Soeharto to step down (Ecip 1998: 83–85). This was the initial act that rang alarm bells for others within the regime who would later queue to jump off Soeharto's Titanic.

But even abandonment by someone so previously dependant on his patronage as Harmoko did not at first deter Soeharto from clinging on to power. Reaching outside the regime, Soeharto reacted by inviting a group of Muslim leaders to a consultation, on 19 May, after which he announced the intention to establish a 'reform committee' and hold fresh elections – going so far as to declare that he would not accept another nomination for the Presidency (Ecip 1998: 101–102; Richburg 1999: 76). That the plan was endorsed by Abdurrahman Wahid – who also urged students to call off their protests – gave an early indication that opposition elites were becoming uncomfortable about developments that, it now became apparent, they would not easily control. At no time during the height of student unrest in May 1998 did Megawati Soekarnoputri make a clear statement of support for the student movement. Even Amien Rais, who had been the opposition leader most identified with the apparent 'people's power' movement, was forced to call off a public 'show of force' on the day prior to Soeharto's resignation, reportedly due to threats of a Tiananmen-style suppression (Ecip 1998: 118–119).

It appeared that opposition leaders still considered that their ambitions could be achieved from *within* the regime, even if Soeharto eventually had to go. In this respect, rather than as rank outsiders seeking to break down the regime, it may be useful to view such figures as Abdurrahman Wahid, Amien Rais and Megawati Soekarnoputri – each with experience as leaders of mass organisations or parties with a formal place and function within the New Order[11] – as insiders who had each been pushed to the margins of the regime at one time or another. Their primary interest was to change the regime without losing control to more radical and populist forces.

Significantly, students rejected Soeharto's overtures and continued to occupy Parliament House, in spite of the pleas of Wahid. This demonstrated a growing rift between some sections of the student movement and the mainstream opposition leaders. Ironically, the Forum Demokrasi, which Wahid also nominally led, rejected Soeharto's concessions as well, declaring in no uncertain terms that 'the *resignation* of President Soeharto and Vice President Habibie from their positions' was the 'necessary *precondition* for *reformasi*' (Forum Demokrasi 1998). In an unusual intervention, the rector of Yogyakarta's venerable Gadjah Mada University declared that 'President Soeharto should withdraw from his position as soon as possible in accordance to the wishes of the people' (Gadjah Mada University 1998). The fact that such a statement could be made and reported further demonstrated the extent to which the regime had lost many of its punitive powers as it disintegrated.

The statements above were quite typical of reactions among much of the politically active intelligentsia in their expression of support for student efforts and for the idea of transitional national leadership to be formed immediately. Indeed, as discussed further below, the establishment of a transitional 'presidium' or 'people's committee' was to continue to be part of student demands long after Soeharto had resigned, for they regarded it as the appropriate transitional instrument through which democratic elections could eventually be held in Indonesia.

Thus, Soeharto's offer of the carrot did not appease the majority of student demonstrators and did not put an end to the politics of the streets or for demands for him to go. It would appear that Soeharto finally decided that the game was up when Ginandjar and his group of ministers notified him of their unwillingness to enter a reshuffled Cabinet and when Wiranto, among the most loyal of his lieutenants, reportedly advised him of the futility or high cost of clinging to power (Ecip 1998: 205; Walters 1999: 81). Soeharto's dire situation was made increasingly clear when only three out of forty-five people contacted to sit on the planned 'reform committee' reportedly accepted his offer, showing the lack of confidence among elites in Soeharto's ability to hold things together for long (Ecip 1998: 127; Richburg 1999: 80).

Significantly, Wiranto had previously quashed opposition hopes that he would play the role of an Indonesia Juan Ponce Enrile or Fidel Ramos when he reiterated – on the evening of 18 May – the military's backing of Soeharto (Ecip 1998: 86–89; 206; Richburg 1999: 79; Jenkins 1998). However, now faced with an increasingly confident popular uprising, and without international support, Soeharto had to come to terms with one brutal fact: the representatives of two crucial components of his regime – its 'bodyguards' and its economic policy-makers – were no longer to be relied upon.

Made to realise that he was now regarded as a liability to the social order he had built and maintained, and abandoned by close allies, there was little option for Soeharto but to devise a way out without opening the door for a wider conflagration. He was to attempt this escape by transferring powers to the recently appointed Vice-President, B.J. Habibie. A political figure previously completely dependent on Soeharto's patronage, few would have given Habibie much of a chance to survive. As we shall see, however, Habibie was able to retain the Presidency for seventeen months and, moreover, preside over the beginning of the process of constructing a new, post-Soeharto, political regime.

After thirty-two years in power, Soeharto's resignation signalled the end of an era in Indonesian political history. To some, it may have appeared that anything was now possible with Soeharto gone, and that there were few reasons to doubt that the fall of the dictator would usher in a new era of transparent and accountable democratic governance in Indonesia. They would have taken heart from the fact that the framework of power over which Soeharto had presided appeared unviable almost as soon as he had departed from the political stage. There seemed to be little chance that Habibie could survive within the framework he had inherited. He too would be caught in the inexorable unravelling of the regime.

Habibie's challenge

The politics of gradual reform

But Habibie proved more resilient than most observers had expected and indeed, presided over an interregnum that, crucially, allowed for the regrouping of old New Order forces, albeit in new alliances and in a new political environment. In the absence of organised popular pressure, and with the eventual – perhaps inevitable – abating of student pressure, the course of *reformasi* was to be determined through negotiations between the surviving remnants of the New Order regime, over which Habibie struggled to exert authority, and the largely middle class opposition elites who had been for years on the fringes of political power. The latter, as we shall observe, would consistently opt for the route of gradual political reform as outlined by new President Habibie due to their own interest in the non-disruption of the social order.

This should not have been entirely unexpected. As already demonstrated by Abdurrahman Wahid's calls for students to withdraw from the streets just prior to Soeharto's resignation, opposition elites were not interested in encouraging mass movements or the alternative student agenda of radical reform. The latter course would have effectively dispensed with all semblances of constitutionality and order, and unleashed a process over which they would have had little control.

Habibie himself faced a difficult dilemma. On the one hand, he had to ensure his own political survival and, arguably, as much as possible demonstrate his capacity to protect much of the coalition of interests that had been nurtured under the New Order. On the other hand, this was, ironically, not possible without constructing a new political framework that would necessarily open up the political process to new social forces and actors. Thus, Habibie was to rely increasingly on the appeal of Islamic populism while allowing ICMI figures such as Arnold Baramuli, head of the Supreme Advisory Council, Adi Sasono, Minister of Co-operatives, and parliamentarian Marwah Daud Ibrahim, greater influence over Golkar. The solution for Habibie then was to attempt to retain Golkar's political machinery as a base of power, but to marry this more closely with the state-centred Islamic populism that Soeharto had already cultivated.

Habibie's answer was also to cast himself as a reformist who was capable of reading the sign of the times and stronger aspirations for democracy.[12] This role involved contriving a process of gradual reform, clearly in the hope that its end result could ultimately be controlled. However, as we shall see more fully in Chapter 9, he lacked the authority over the institutions of power to fully engineer such a process. Indeed, in an environment in which politics had inevitably become more fluid following the fall of Soeharto, this was immensely difficult to achieve. What was ultimately necessary for Habibie was to democratise, but in such a way as to ensure that old interests continued to dominate a ruling party. Thus, control over the bureaucracy could be retained and reforms, for example in the judicial and legal system, which might endanger these forces, could be repelled.

But Habibie was evidently constrained by the fact that Soeharto – the master of retaining power – had failed to put in place a mechanism for stable political succession, and thus any immediate successor's legitimacy within and outside the regime was always likely to be seriously questioned. In the end, like Soeharto,

Habibie's fate was to be regarded as a liability to the survival of the interests he needed to effectively protect in order to survive, as the spectre of mass politics temporarily resurfaced in October 1999, as the MPR met to elect a President following elections the previous June. Thus, he too was to be abandoned as elites, old and newly ascendant, began a new round of jostling and re-positioning in the struggle to reconfigure power and construct a new political regime.

Clinging to power

While Habibie took over the reins of government after Soeharto resigned, he so evidently lacked a firm base from which to maintain political power that bold demands were quickly put forward by opposition groups regarding the political prerequisites of overcoming the economic crisis. Essentially, these were the immediate resignation of Habibie and the complete overhaul of the New Order political system. Because Soeharto had been the lynchpin that kept everything together, opponents expected that the system he put in place would surely fall apart now that he had departed from the political stage, leading to expectations of a quick transition to democratic rule. Importantly, however, apart from the students, much of the opposition centred on middle class figures that had been in the margins of the New Order. Nurtured to varying degrees by the regime they came to oppose, whether they intended to transform it or merely to open it up to new forces was always a major question, notwithstanding the frequently tough rhetoric employed.

Opposition demands for change were given the opportunity for continuing vigorous advancement given that Soeharto's fall did not immediately improve Indonesia's desperate economic situation. Habibie's legitimacy and capabilities were also publicly questioned while investors remained fearful of a repeat of the mass rioting in May. Moreover, the rupiah remained weak. Basic commodities continued to be both increasingly expensive and scarce.

Even the moderate Emil Salim was suggesting that parliamentary elections be held by late 1998 (rather than mid-1999, as envisaged by the Habibie government), in order to more quickly produce a credible, legitimate government that could oversee the economic recovery process. An economist and former minister who by this time was leading a group of intellectuals and NGO activists under the banner of a loose organisation called *Gema Madani*,[13] Salim viewed Indonesia's continuing economic problems as stemming 'from the lack of belief in, and the poor economic performance of, the government'. Moreover, he argued that 'in order to overcome this economic crisis' the term of the Habibie government must be 'shortened via a general election' (Salim, E. 1998). Echoing the views of many economic commentators, Salim held that the Habibie government lacked the popular legitimacy to make the sweeping economic reforms needed. Failure to install a more legitimate government, he warned in no uncertain terms, would result in sheer economic disaster.[14]

In spite of such pressure, the Habibie government was able to design and implement its own, albeit more protracted, plan of political transition, which as we shall see, involved an MPR Extraordinary Session in November 1998, fresh parliamentary elections in mid-1999, and another MPR session to elect a President later in the year. The plan also involved the controversial drafting and passage of a set of new laws on politics to replace three of the five draconian 1985 laws on parties and mass organisations. Henceforth, the attention of opponents was diverted away to a large extent from calling for Habibie's resignation to jostling for position in the new democracy that was now more or less guaranteed.

Earlier, Habibie had also been able to quash inter-elite threats to his newfound position. Most significantly, he was able to stave off a challenge by General Edi Sudrajat for control over Golkar during a Special Congress in July 1998, by securing the ascension of long-serving Soeharto cabinet member Akbar Tandjung as chairman. Tandung would replace Harmoko, the former Soeharto loyalist turned defector. It was widely believed that Sudrajat had plans to use Golkar as an instrument to remove Habibie from the Presidency by calling for an extraordinary MPR session through the party. The next step was apparently to install former Vice-President Try Sutrisno in place of Habibie (Crouch 1999: 131–132).

General Sutrisno was obviously calculated to be acceptable to a wide range of groups: although regarded as bland and unimaginative, he had the odd combination of association with the Moerdani clique as well as credentials as a devout Muslim. The Moerdani connection seemed to guarantee support from significant sections of the military apparatus, which would see him as a guarantor of the military's institutional interests. On the other hand, an ability to project Islamic sympathies would help him carve out a role as a 'man of the people' and allow him to challenge the support Habibie was receiving from various Muslim groupings. Ironically, Sutrisno has forever been plagued by the reputation he garnered as the military commander responsible for the shooting of Muslim demonstrators in the port area of Tanjung Priok in the 1980s.

Essentially, the failed Sudrajat ploy should be read as an opportunistic attempt to regain influence by the forces within the military that had been sidelined from Golkar earlier by Soeharto (see Chapter 4). It should be recalled that this clique of military officers had been vigorously opposed to Habibie's authority over strategic enterprises, which infringed on the military's economic and political interests.

Habibie's position was ultimately saved by an accommodation struck with General Wiranto, which involved the elimination of Prabowo from the political scene. This was in spite of the fact that Soeharto's son-in-law was commonly regarded as the benefactor and patron of the most vocally pro-Habibie Islamic groups. Regarded by Wiranto as dangerously ambitious, Prabowo was forced to give up his position as head of Kostrad and then to enter into early retirement. In return for Habibie's support for the removal of Prabowo,[15] Wiranto apparently guaranteed the ascension of Tandjung to the Golkar chairmanship by virtue of his influence over the party's regional delegates, many of whom were military men. Instructively, press reports suggested that the military chief of social and political affairs at the time, General Mardiyanto, had personally telephoned regional military commanders to instruct them to influence Golkar regional chapters to vote for Akbar Tandjung (*JP* 15 July 1998).

With his position at least temporarily secure, Habibie was in a better position to promote an electoral process that he clearly hoped would facilitate the reconstitution of the alliance of interests, albeit in modified form, which had held the New Order together. However, as mentioned before, Habibie could not control the process of political reorganisation, for he did not hold sway over state institutions in the way that his predecessor did. Within Golkar for example, his authority was constantly challenged by Tandjung – who owed his position partly to Habibie's power broking – as well as politicians such as Marzuki Darusman, who opposed the Islamic populist appeals of the ICMI-related figures within the party. In fact, rather ironically, Tandjung would be a key player in Habibie's ultimate downfall.

Securely in office for the moment, an attempt was initially made to establish new political laws that would have almost guaranteed the production of a DPR/MPR that was likely to reinstate Habibie in the Presidency, while managing to maintain the façade of reform (Hadiz 1999). Prolonged public debate on these laws (see King 2000), however, eventually produced a set of legislation that would govern an electoral process that was much more fluid and open-ended than evidently intended by Habibie. Extensive changes were eventually made on the original blueprint produced by the team of 'seven experts', linked to the Ministry of Home Affairs, and recruited to establish the political framework for reform. This was in spite of the fact that the legislature was overwhelmingly dominated by Habibie's own political vehicle, Golkar. Rather than a gift from an enlightened leader, the democratic reforms during the short Habibie period were concessions that had to be made because of a lack of control over parliament and Golkar, as well as the increasingly unpredictable behaviour of the military.

For the time being, however, Habibie was still able to take advantage of the fact that, although resurgent, political opposition remained fragmented, poorly organised, and confused about the direction it wanted to pursue. This undoubtedly prolonged Habibie's Presidential tenure and was a natural legacy of the disorganisation of civil society so clinically pursued by Soeharto for three decades.

The fragmented opposition

There were two main elements of the anti-Habibie forces in May 1998 and the months that immediately followed, each with their own often mutually antagonistic components. First was a range of sometimes mutually competing student groupings, and second, an array of mostly *ad hoc* associations coalescing around middle class politicians, many of which later established political party vehicles. No cohesive coalition quickly emerged, however, with a clear agenda of reform, or of action to push for the more immediate and thorough unravelling of the New Order framework. This was essentially because a central dividing line remained over the extent and speed with which the old regime should be overturned. The most radical of student activists insisted on a thorough reform process largely based outside of the existing institutions, and which would largely exclude the old elites. Mainstream political leaders, however, were always more likely to pursue the path of reform from within the old institutions, which would necessarily include alliances with old forces.

Endeavours in the months immediately after Soeharto's resignation to form some kind of 'people's committee' or transitional 'presidium', comprising noted opposition figures and student and people's representatives, never took off. The idea was probably partly inspired by the *ad hoc* groupings established mainly in Jakarta just before Soeharto's fall – such as the 'People's Council', the 'People's Working Committee' and the 'National Reform Movement Presidium' – which typically comprised intellectuals and activists, but also former government and military officials (*FEER* 28 May 1998: 16–17). That such a presidium never materialised, uniting opposition forces during this critical juncture in Indonesian history, revealed the absence of unanimous support for reform initiated from outside of existing state institutions.

In any case, immediate radical political change could not have been achieved without mass mobilisations. In the face of the state's still-formidable security apparatus, the most important leaders of the *reformasi* movement were clearly disinclined to take this route. Apart from demonstrating well-founded fears of the New Order's still potent arm of repression, this disinclination also displayed long-standing elite and middle class wariness of the unleashing of uncontrollable (especially urban lower class) social forces. It also reflected a lack of confidence in the capacity of existing organisational vehicles to mount a frontal challenge even to an already severely weakened New Order framework, particularly without setting off unintended consequences.

While Abudurrahman Wahid, Amien Rais and Megawati Soekarnoputri were adopting a non-confrontational approach, militant student groups rejected the notion that the Habibie government –which they viewed as a mere extension of Soeharto's – was capable of carrying out free and fair elections. They did not accept that Habibie – such a close associate of Soeharto in the past – could be entrusted with presiding over political transition. According to student propaganda (e.g. Forum Kota 1998), only a 'people's committee' could oversee perhaps a longer period of transition before elections were held, during which parliament and the upper echelons of the government were first to be cleansed of New Order influences. Significantly, the students also displayed their distrust of political elites who emerged within an environment which: 'bred a rotten mentality and culture that afflicted almost all government agencies and even the People at large. Right and wrong were confused, the boundaries between good and bad were blurred, and morality and ethics were increasingly ill defined' (Forum Kota 1998).

Rather than accept even a transitional Habibie Presidency, on 28 May 1998 the Forum Kota called for the establishment of a temporary 'Indonesian People's Committee', which was to be 'vested with supreme executive and legislative powers' and which would also serve as an interim cabinet. The members of the committee were to be 'drawn from various groups in order to produce a leadership representative of the entire Indonesian population of 210 million people'. Most significantly, the committee's job, according to these students, would be to 'annul the five political laws ... compile a list of amendments to articles of the 1945 Constitution ... as well as compile new political laws and carry out fresh general elections to elect a new DPR/MPR' (Forum Kota 1998).

The Forum Kota students also envisaged that this new DPR/MPR would be 'free of collusion, corruption and nepotism' and 'capable of fighting for the aspirations of the people'. It would be an instrument to construct a government that is 'clean, committed to human rights and able to guarantee the pure and consistent implementation of the Pancasila and the1945 Constitution' (Forum Kota 1998).

It was evident that the students looked for leadership from outside their ranks, in spite of the distrust of political elites shown in their tough rhetoric. Unable to run the state themselves, they were clearly waiting for, even relying on, support and co-operation from civilian politicians that were relatively untainted by association with the Soeharto regime. Perhaps demonstrating some vagueness in political strategy, students, however, were typically disinclined to name specifically who such persons could be.[16]

In any case, it later became clear that the most salient elements of the mainstream opposition that had coalesced around such figures as Amien Rais, Megawati and Abdurrahman Wahid were on a completely differently route. Together with the tenth Sultan of Yogyakarta,[17] they made up the so-called Ciganjur group – named after Wahid's place of residence, where a meeting took place between them prior to the November 1998 MPR Extraordinary Session, which was tasked to formalise the political transition plan worked out by the Habibie government. While many student groups continued to press for the idea of a transitional presidium, it was clearly not the preferred option of opposition elites by this time.

Although the meeting only occurred due to student coercion – Wahid and Rais were long-standing bitter rivals in the Muslim political community – the results of the Ciganjur meeting were clearly disappointing to the more militant elements of the student movement. These had hoped that the Ciganjur meeting could settle whatever differences existed among opposition elites and that the four would agree to take a leading role in a transitional presidium which students would go to the streets to demand. The meeting highlighted a central dilemma of the opposition. They were held together by a common desire to get rid of Soeharto and a stated intent to eliminate corruption and arbitrary power. But a common organisation or vision remained sorely lacking.

Nevertheless, the Ciganjur group did produce a statement outlining the elements of a gradual, accommodating reform process (*Kompas* 11 November

1998), but one which also fell short of condemning the Extraordinary MPR session that the students considered illegitimate. It also expressed a readiness to accept the military's 'dual function' for another six years, which contrasted sharply with student demands for its immediate removal. The only point that could have encouraged student activists was the Ciganjur group's plea for the disbanding of a civilian militia made up of the poor and unemployed – the *Pam Swakarsa* – which the army had established to stave off student protests (*Kompas* 12 November 1998).[18] Given such a stance, it was not surprising that Wahid, Megawati and Rais all indirectly refused to provide their support for students when they attempted another take-over of Parliament House (see Budiman 1998) to thwart Habibie's plan for gradual reform. It was during this attempt that a series of bloody and deadly clashes with the military ensued (*Kompas* 20 November 1998).

As mentioned earlier, part of the reason that Habibie was able to survive in spite of inheriting a teetering New Order framework lay in his foes' lack of interest in developing a wider base of support within civil society. Clearly this was inimical to the objectives of the moderate opposition. Indeed, a striking feature of the opposition movement was its relatively limited base among sections of the urban middle class. The continued exclusion of other sections of society from the strategies of the *reformasi* movement demonstrates that the legacy of New Order 'floating mass' politics – in which direct political participation by the poor was considered potentially destabilising – was inherited by salient opposition forces. Indeed, mobilising mass support was essentially contradictory to their leadership's commitment to gradualism and non-disturbance of the social order.

All of this was apparent in the public statements made by opposition leaders during such a critical time. For example, Wahid was to warn that: 'Social revolution is a condition in which the people revolt against everything, and there is no government or ruler. In a social revolution ... what each person wants is unknown and every one does as he pleases' (*Kompas* 11 February 1999). Such statements betrayed a fear of politics beyond the confines of elite negotiations, as much as they displayed a fear of anarchy. But these views were prevalently expressed on both sides of the political divide. General Wiranto had earlier expressed the same sentiment when he said on the occasion of an Indonesian Muslim Congress that: 'If we [the nation] cannot manage the reform movement well, it will turn into a revolution ... and no revolution has solutions [to the problems it generates] as there will always be a power which turns anarchic and which will get rid of anything in its way' (*JP* 5 November 1998). The similarity of views is hardly surprising, however, as the political careers of virtually all the key leaders of the opposition were embedded in Soeharto's long rule, during which the skills of backroom dealings required more honing than the ability to lead mass-based movements. Hence, the path to power for them lay in exploiting or extending existing alliances and in co-operating with Habibie's political transition plans. Significantly, the upper echelons of the parties, old and new, led by

Megawati, Wahid and Rais, were to be well-populated by New Order elites such as retired generals, business people of varying stature, ICMI activists, politicians and rural *ulama*.

Their conservatism was well illustrated in the rejection of Megawati as President by some of Wahid's followers in 1999, on account of Islam's supposed rejection of a female President (*Jawa Pos* 29 April 1999), Megawati's strident defence of Indonesia's claims on East Timor (Harsono 1999), and Rais' equally strident anti-communism.[19] Upon becoming President in October 1999, Wahid was reported to have dismissed student action as being the work of 'hooligans', thus stressing his by then long-held displeasure with those attempting to influence political events from the outside (*Expresso* October 23 1999).

But the student groups themselves were open to criticism that they were exclusive, and that they were merely the middle class children of the New Order. Indeed, they evidently inherited many of the key elements of the political culture that grew out of the system they reviled. For example, many of the student groups were adamant, during the crucial months following the fall of Soeharto, about portraying themselves as a 'moral', rather than 'political' force. In their definition, this meant fighting for an idealised 'common good', while negating self-interest. Interestingly, the juxtaposition of 'moral' and 'political forces' harks back to the mid-1960s, when anti-Soekarno student groups backed by the army attacked the 'divisive' and 'self-interested' array of political parties that many then saw as being a root cause of Indonesia's economic and political crisis.[20] The emphasis in student rhetoric on maintaining the 'unity' of the Indonesian people also displayed some of the features of the organic state ideology propagated by Soeharto – for it meant the negation of the idea of political movements based on class or group interests.

Most importantly, unlike student activists in South Korea, Thailand or the Philippines, student groups in Indonesia in 1998 largely ruled out alliances with workers and the urban poor because the latter's struggles were perceived as being 'social-economic' and 'self-interested'.[21] Thus, lacking support from middle class opposition groups, including student organisations, workers largely developed new organising vehicles – in spite of the real constraints posed by rampant unemployment (Hadiz 1998; Ford 2000) outside of the mainstream *reformasi* movement.[22] Thus, notwithstanding their differences, there was, ironically, at least one important similarity in the outlook of some of the student groups and that of opposition elites: both rejected the idea of alliances with lower class-based movements.

Some student organisations also decried the quick establishment of new political parties – regarding them as expressing political opportunism rather than genuine *reformasi* intentions. Reminiscent of early New Order attacks on political parties, students protested that they only divided the Indonesian people in their common struggle against the vestiges of Soeharto's authoritarianism.[23]

It should not be suggested, however, that there was ever one, united student movement, in spite of this rhetoric of unity. As mentioned earlier, the students themselves were internally fragmented with various groups in Jakarta alone supporting or opposing Habibie's initial elevation as President on 21 May 1998. The Islamic group KAMMI (*Kesatuan Aksi Mahasiswa Muslim Indonesia* – Indonesian Muslim Students' Action Front) accepted the legitimacy of Habibie, while the militant Forum Kota rejected him quickly. Another organisation, the FKSMJ (*Forum Komunikasi Senat Mahasiswa Jakarta* – Communication Forum of Jakarta Student Senates), made up of the more formal senate bodies of universities in the Jakarta area, also rejected Habibie, though it was generally considered more moderate than the Forum Kota. It is significant that these varied organisations rarely co-operated with each other after Soeharto's fall, though many were briefly united in opposing the November 1998 Extraordinary Session of the MPR that was crucial to Habibie's scenario of transition.[24]

This lack of willingness to pursue the possibility of cross-class alliances might have cost the students dearly. Indeed, support for students from labour groups and the urban poor had been visible in both the key moments of May and November 1998. It had been ordinary poor people in Jakarta, for example, that stopped the rampaging Pam Swakarsa civilian militia from physically attacking students during the latter month's session of the MPR (*Kompas* 23 November 1998).

Maintaining the 'purity' of their struggle, students would quickly lose what influence they had over the direction of political change as others staked a stronger claim on *reformasi* and the contest to construct a new post-Soeharto political regime. Such a contest, as we shall see, was to be one dominated by those that had matured within the New Order's extensive system of patronage, albeit only on the fringes, and this was greatly to influence the character of the new regime being constructed.

November 1998 was thus an important turning point in the contest to construct a new political regime after Soeharto. Habibie was successful in holding an Extraordinary Session of the MPR, in spite of opposition from a number of student groups. It should be recalled that, to the consternation of students, this was the same MPR that had re-elected Soeharto to the Presidency just eight months earlier. Not surprisingly, this Extraordinary Session primarily served to provide legitimacy to the Habibie Presidency and ensure that his plan for gradual reform would be pursued rather than the more radical route favoured by the students. It established a more protracted time frame for a possible change of government – stipulating general elections to be held in mid-1999. Importantly, it also placated the military by assuring them that the number of seats they automatically occupied in the national and regional parliaments would be only gradually reduced, in contrast to student demands for an immediate military withdrawal from political life.[25] Habibie's plan was effectively endorsed by then-opposition elites, who had decided to pursue the path of gradual reform.

Such developments displayed some interesting parallels to those that followed the fall of Soekarno in the 1960s. Like their predecessors, the 1998 student movement came out as heroes in popular renderings of events constructed by those emerging victorious from political struggles. However, both the student movements of the 1960s and 1998 were effectively banished from the political stage as soon as the real contenders for political power had determined that they no longer had any role to play. Effectively, the failure of the student movement to develop alliances with groups such as urban workers – among the natural beneficiaries of the *reformasi* total that was advocated – meant that a radical agenda was to be absent in the struggles that would shape the contours of Indonesia's new democracy.

Notes

1 It is significant also that the US Secretary of State, Madeleine Albright, representing the concerns of the advanced industrial countries of the West, also eventually asked Soeharto to resign (Ecip 1998: 129).
2 Rais had been expelled from ICMI's Board of Experts for incurring Soeharto's wrath due to comments made on Busang and Freeport. See 'Amien Rais Terganjal Rumput Kering?', posted by apakabar@clark.net, 23 February 1997, from *Tempo Interaktif*, edition 52/01.
3 Salim made an unusual independent bid for the Vice-Presidency in 1998. Though he had little chance of winning – there was no reason for Soeharto to pick him over Habibie – his candidacy attracted support from a wide range of intellectuals and activists due to his reputation for political integrity.
4 Thus, at least two banking and financial sector unions were set up by middle class employees of banks and financial institutions, outside of the structures of the official state-sanctioned union. This was a new development, as most of the middle class had previously been uninterested in union activities – which were associated with less skilled, blue-collar workers.
5 See, for example, Letter to the Editor, *Kompas* 12 May 1998.
6 Discussion with Forum Kota activist, 7 September 1998.
7 See, for example, YLBHI 1994: 3, which suggests that labour riots in Medan in 1994, which involved the murder of an ethnic Chinese businessman, served to provide the military with legitimacy to continue intervention and repression in the labour area. YLBHI, an NGO, also suggested that the riots reinforced the perception among Chinese business that they required military protection.
8 Soeharto himself had left the country for Egypt on 9 May in order to attend a meeting of G-15 countries, and only returned after the outbreak of rioting
9 Forrester (1998) is one among many who puts forward this argument.
10 It will be recalled that Ginandjar Kartasasmita played a key role in the Sekneg group's attempts in the early 1980s to promote *pribumi* businesses.
11 NU leader Wahid was once a member of the MPR, Megawati of the DPR (and head of the PDI), while Rais was of course once a key figure in ICMI.
12 See the interview with Habibie (*New York Times* 3 July 1998)
13 Literally meaning 'Echo of Civil Society'.
14 This he argued at a seminar on general elections held at the University of Indonesia on 21 July 1998.
15 Tensions were thus growing between the President and Prabowo – who was acting increasingly like a rogue figure, according to Habibie himself (*New York Times* 3 June 1998; *Kompas* 17 February 1999). A desperate Prabowo apparently threatened him on the day he became President, demanding a promotion. By this time Prabowo would have been very much aware that he was in jeopardy due to Wiranto's consolidation of power. Indeed, Prabowo was said to have arrived at the Presidential palace the following day with armed troops, protesting his removal from Kostrad – part of a bargain between Habibie and Wiranto.

16 Various discussions with student activists, including 7 September 1998. It was pointed out that students feared more serious conflict between the different groups should they be more specific about whom they supported to run the country.
17 Though a Golkar functionary, the Sultan has acquired reformist credentials and is extremely popular in the Central Javanese heartland.
18 The establishment of the Pam Swakarsa was itself a significant development, for it was effectively the precursor for the subsequent rise of a number of paramilitary groups or civilian militia. Consisting of thugs, martial arts experts, and the urban and rural underemployed, the Pam Swakarsa rampaged through Jakarta in November 1998 in an attempt to help quell student opposition to the MPR proceedings. Other groups like Pemuda Pancasila and Pemuda Marga, long associated with Golkar, and often acting as the Soeharto regime's civilian goons, joined the Pam Swakarsa in 'safeguarding' Parliament House, as did an organisation called Furkon, formed by followers of the slain leader of the Tanjung Priok riot, Amir Biki. For at least a short while they were even joined by Ansor, a youth group associated with the NU (Young 1999: 90).
19 He once warned of the re-emergence of *Nasakom* ideas (the Soekarno doctrine which fused nationalism, religion and communism), as evidenced by a rise in left-wing ideas. Rais was understood to be referring to student ideas about the establishment of a people's committee or presidium (*SiaR* 15 September 1998). He also once suggested in a television interview that the Christian labour leader Muchtar Pakpahan could be a communist, after he allegedly made derogatory statements about ICMI. Pakpahan had created an outrage when he was reported by a Muslim magazine to have suggested that ICMI and Muslim activists were behind the May 1998 riots in Jakarta (see *Republika* 29 June 1998).
20 This perspective was included in Arief Budiman's account of the 1966 student movement in his presentation at the conference 'Democracy in Indonesia? The Crisis and Beyond', Monash University and the University of Melbourne, Melbourne, 13–14 December 1998. Budiman, who may have coined the term 'moral force', nostalgically recalled that he was inspired by an old Hollywood Western, 'Shane', in which the hero protected a town from outlaws, but refused to stay on to receive any reward from the helpless but grateful townsfolk. Likewise student activists are a 'moral force' because they seek no personal political gain.
21 Various discussions with students at the Faculty of Social and Political Sciences, University of Indonesia, November 1998.
22 There are some exceptions. Kobar (Komite Buruh Untuk Reformasi) combines worker and radical student activists, mostly in the Jakarta area.
23 Discussion with Forum Kota activist, 7 September 1998.
24 The highly unstructured Forum Kota eventually developed internal rifts of its own, out of which emerged new organisations like Famred (*Front Aksi Mahasiswa Untuk Reformasi dan Demokrasi* – Students Action Front for Reform and Democracy) and the Jakarta Front. Other players included students grouped in 'NGOs' such as Pijar and Aldera, combining students of various private universities, and Kobar (*Komite Buruh Untuk Aksi Reformasi* – Workers Committee for Reform Action), which was exceptional because it combined workers with students. Many Kobar activists have links with the labour and student wings of the formerly banned People's Democratic Party, the PRD. The University of Indonesia's Forum Salemba represented a moderate wing of the student movement, quickly accepting, for example, the products of the November MPR extraordinary session. Another grouping partly emerging out of UI was the more radical Komrad (*Komite Mahasiswa dan Rakyat Untuk Demokrasi* – Student's and People's Committee for Democracy).
25 See Majelis Permusyawaratan Rakyat Republik Indonesia (1998).

Part IV

OLIGARCHY RECONSTITUTED

8

REORGANISING ECONOMIC POWER

With Soeharto gone, and the IMF-sponsored reform programme seemingly in place, it seemed that the floodgates of change would now open. Formerly ascendant political and business interests found themselves politically fractured and paralysed by huge debt burdens. Hounded by domestic and overseas creditors, vilified in the media and dragged before the courts, the conglomerates and the old politico-business families were in retreat. No longer able to call on a centralised, authoritarian state to protect their interests and to dispense the rents they had relied on, they now faced the prospect of bankruptcy and even imprisonment. The collapse of Indonesia's banking system and the distress of its leading business groups offered reformers an opportunity, not only to impose new policies and institutional arrangements, but to restructure power in the corporate world.

In a series of protracted and acrimonious confrontations, many of the largest private banks were closed or nationalised. Formerly untouchable business oligarchies were forced to hand over billions of dollars in debt-restructuring programmes, assets were seized and sold while many of the state contracts and monopolies that had provided such lucrative sources of super profits were cancelled. The government introduced legislation that sought to supplant the old gate-keeping conduits with new and accountable systems of governance in public institutions. More immediately, there were growing calls for former officials and business figures of the Soeharto era to be investigated, arrested and charged with corruption.

Despite what appeared to be a desperate situation, the interests that had underpinned the Soeharto order, including many of the figures that had been dominant then in business and politics, managed to survive and to reorganise their economic power. They did this, not by fundamentally transforming and adjusting to competition in a new liberal market economy and within transparent and legally framed regulatory institutions, but by forcing the state to assume much of the cost of debt and economic restructuring, and by striking new political alliances. In the end, the state was never able to impose systemic reform because it remained ultimately in the hands of political interests whose survival relied ultimately on preserving many of the central institutions of the old regime.

In a remarkable metamorphosis, the oligarchy of authoritarian rule now became the diffuse and confused oligarchy of money politics, as patronage networks and mechanisms for the allocation of public power and wealth were reassembled within the new arena of parties and parliament. In an accommodation that took place within an increasingly decentralised and diffused system of power, those who came as reformers were often inexorably drawn into the new amalgam of predatory capitalism and democratic politics. The flood of investors and global institutions expected to buy up cheap assets and transform the institutions of corporate life in Indonesia never materialised, and the anticipated end of the Chinese family conglomerate at the hands of efficient modern corporations remained an ambiguous dream.

The political terrain of reform and resistance

The replacement of Soeharto as President by Habibie in May 1998 was greeted with widespread cynicism. After all, Habibie had been at the very centre of the old power arrangements and carried the legacy of the old regime. Not only had he been a leading proponent of strategic industry policy, presiding over an extensive apparatus of state industries, the Habibie family was also a major, second tier politico-business family, embedded in partnerships with the Soeharto family business and with large Chinese corporate groups. If anyone could usher in an era of market reform, clearly Habibie was an unlikely candidate. Both the government and the administrative apparatus, its ministries, the judiciary and the state corporate sector, continued to be pervaded by figures previously instrumental in the organisation of oligarchic power under Soeharto. No less a figure than Ginandjar Kartasasmita, that consummate Soeharto-era apparatchik, became Habibie's Co-ordinating Minister for Economics.

By now, however, the reform process had established a momentum of its own and Habibie was to some extent unable to resist being drawn into the vortex. Presiding over an economy in retreat and with its budget under pressure from growing debt-servicing obligations, there was little option except to agree to the agendas for policy changes and reform of the state apparatus contained in the 'Letters of Intent' devised between the IMF and reformist technocratic elements within the state bureaucracy. Whatever the outcome, the mere fact that leading figures within the old business and political oligarchy were now being dragged before IBRA to account for their debts and to answer charges of corruption in the courts represented a major shift in the way power operated. Nor could Habibie any longer ignore the mood within new political parties and student groups and from within the middle class itself. This demanded reprisals against Soeharto and his associates, and the dismantling of the old regime and its institutions. As elections loomed it was clear that his survival depended at least in part on laying a claim to the reformist high ground.

Nevertheless, Habibie's commitment to reform appeared suspect as attempts to prosecute Soeharto and other leading figures of the old regime on charges of corruption came to nothing amid a sea of allegations about collusion and corruption in the judiciary and as major business figures continued to treat with contempt attempts by IBRA to negotiate the recapitalisation of banks and the restructuring of debts. Indeed, Habibie was soon to be implicated in a major scandal when it was revealed that US$80 million in public funds allocated by Bank Indonesia (BI) for the recapitalisation of the private Bank Bali had been diverted into the coffers of Golkar.[1] It seemed that the old web of money politics remained as sticky and entangling as ever.

When Abdurrahman Wahid was elected President in October 1999, it initially appeared that the reform agenda had finally triumphed, although perhaps not in the way the IMF might have intended. Although he did not command the majority in parliament, in the early stages of his government he presided over a buoyant reformist mood within parliament and amongst the broader population. Himself an outspoken, if sometimes calculating, critic of the Soeharto regime, Wahid appeared to bring none of the baggage of the Soeharto era with him. His early appointments to Cabinet and the higher echelons of the bureaucracy confirmed for many that a new broom was sweeping through. Gone were many of the figures critical in organising the oligarchic interest within the Cabinets and government departments under Soeharto and Habibie, among them Ginandjar Kartasasmita, former Bulog Head, Beddu Amang, former Finance Minister and Bulog heavyweight, Fuad Bawazier, and Habibie himself.

Many of the new Ministers and Directors General of Departments came to office with strong reformist credentials, with Laksamana Sukardi (Minister for State Enterprises and Investment) and Kwik Kian Gie (Co-ordinating Minister for Economics), the most high profile reformers appointed to the first Wahid cabinet. In the second Wahid Cabinet, the appointment of Rizal Ramli as Co-ordinating Minister of Economics brought to power one of the Soeharto regime's most persistent critics. Perhaps the most dramatic appointment was that of Justice Minister, Baharuddin Lopa, who brought promise of energetic and robust reforms to an area plagued by inaction and reluctance to pursue corruptors before his strange and premature death. The appointment of Kuntoro Mangkusubroto, who became Minister for Energy and Mines and, later, Director of PLN, and Cacuk Sudarijanto to head IBRA in 1999, brought back bureaucrats pushed aside after disputes with the Soeharto family in previous years. These represented only some of the senior officials who emerged across the bureaucracy, apparently prepared to push the reform agenda.

There were good reasons for thinking the reformers might prevail. Hundreds of millions of dollars were injected by the World Bank, the IMF and international aid agencies into programmes of institutional reform that promised greater efficiency, accountability and transparency. State agencies were to be subject to regular audit, resulting in a process of scrutiny and public reporting not witnessed before. Formerly supine public agencies such as the Supreme Audit Agency (BPK) and the Government Financial and Development Comptroller (BPKP) assumed a new authority. Together with international accounting firms they conducted far-reaching audits of government ministries and agencies. Widely disseminated in the public arena, these reports detailed what most had suspected. Private groups also emerged to track down and make public cases of corruption and corporate fraud, just as similar groups had done in the field of human rights.[2] At the same time, newly released from the stifling control of the Soeharto regime, a burgeoning media widely and enthusiastically reported the ongoing saga of assault on the citadels of *korupsi, kolusi dan nepotisme* (KKN).

Despite the flood of reformist legislation and the new enthusiasm for change in many quarters, reform proved to be difficult and inconclusive, not at all a simple process of learning how to manage the economy in a more rational way. Why did the reform agenda falter? Indonesian supporters of reform looked to decisive and courageous leaders while neo-liberal reformers expected that the prospects for opportunists and predators would be curtailed by newly imposed institutional arrangements and policies of deregulation. In the end neither the attempts to address regulatory failure through institution building, nor submitting the bureaucracy to public scrutiny, were able to deliver the expected transformation in public governance.

That the mere presence of regulatory institutions or trained personnel would not in itself guarantee reform was a point made by Natasha Hamilton-Hart in her analysis of Indonesia's banking reforms. Nor was the resistance of officials embedded in the bureaucracy a guarantee that reform would fail. It was, instead, the incapacity of governments to pursue reform that mattered, 'the government has rarely been able to deploy either authority or resources in a disciplined manner for public purposes' (Hamilton-Hart 2000: 109). The observation that institutions or resources in themselves guarantee nothing is an important insight but one that begs critical questions. What is the magical ingredient that provides the will or capacity to act? We return to our proposition that institutions are created and enforced in the process of social conflict. That an effective regulatory state has not emerged in Indonesia, we argue, is precisely because those complex interests, the beneficiaries of the system of distributive administrative oligarchy, proved to be more resilient and pervasive than expected and able to reorganise their power and insinuate themselves successfully into the new economic and political regimes.

In three landmark struggles, these interests were able to avoid bankruptcy and dispossession despite huge debt and widespread default, they survived confiscation of monopolies and attempts to reform the gate-keeping institutions that were their lifeblood, and they resisted the demands for their prosecution and imprisonment.

How oligarchy reorganised: the politics of debt

The sheer magnitude of private sector debt and the collapse of the private banking sector provided Indonesia's reformers a unique opportunity, not only to purge the oligarchy and to recast economic and corporate power in Indonesia, but to reform the very institutions within which state–business relations were forged. For their part the major business groups and politico-business families attempted to hang onto their key assets and to force others, creditors or the government, to bear the costs of their huge debts. In this task they quickly confronted foreign creditors who initially adopted a hard line, in many cases taking their Indonesian debtors to court in Indonesia and demanding full disclosure as a precondition for debt restructuring and forgiveness (*JP* 3 November 1998: 1).[3]

As the attempts to prosecute through Indonesian courts failed one by one, it soon became clear that pursuing debt through the courts was an ineffectual strategy, not only because judgements invariably went in favour of the debtors. The difficulties of tracking down assets even in the event of a positive judgement meant that creditors would most likely end up empty-handed (*AWSJ* 11 August 1998: 1, 8; 9 February 1999: 3; 26–27 March 1999: 3; *JP* 15 September 1998: 1; 25 November 1998: 12; 8 December 1998: 4, 7). Foreign creditors now turned to negotiation outside the courts, although they continued to be frustrated by obstructive tactics and the opaqueness of the financial arrangements of Indonesian companies.[4] By the end of 2001, it was estimated that only half (around US$31billion) of external corporate debt remained non-current in meeting payments falling due (IMF 2002: 67). Able to decide the pace, timing and sequence of the restructuring of their international debts, the conglomerates and politico-business families faced their greatest challenges within Indonesia in a protracted struggle with the government over their debt in the private banking system.

The implications of bank recapitalisation

An agreement between the IMF and the Indonesian government provided for a Banking Reconstruction Agency (IBRA) to be established in January 1998, with the task of closing weak banks with serious structural problems and recapitalising those considered to be basically sound but requiring an injection of funds. In a drawn out process punctuated with accusations of corruption and favour, IBRA progressively closed, recapitalised or nationalised banks. While officials initially expected that the number of banks would be reduced to twenty or thirty, this figure was never achieved. Of the 160 private banks that had existed in July 1997, by the end of 1998, sixty-six had been closed and twelve had been taken over by the state, of which four had been recapitalised, and another eight jointly recapitalised by the state and the owners. Between 30 June 1997 and 31 March 1999, the market share of private banks declined from 45 to 15 per cent.[5]

While some dubious banks were effectively bailed out by the state, there were, nevertheless, some significant casualties. Among those banks closed were Sjamsul Nursalim's BDNI and Bob Hasan's Bank Umum Nasional, two of the most important private banks of the pre-crisis era. Liem Sioe Liong lost all but 6 per cent of his share in Indonesia's largest and most profitable private bank, Bank Central Asia (BCA), when it was subjected to a run on deposits in May 1998, forcing the government to place it too under supervision. However, expectations of a vibrant new banking system where foreign banks now played a critical role were not realised. The attempt by Standard Chartered to take over Bank Bali in 1999 failed in the face of an employee revolt, and it was only after several inconclusive attempts to overcome opposition within the parliament and indifference in global markets that a sale to foreign interests of a majority ownership in a substantial private Indonesian bank, BCA, was completed.

For the most part, surviving banks struggled on under their old ownership propped up by the state which, by 2000, owned 75 per cent of the liabilities in the banking system and had Rp.600 trillion of assets under its control, including non-performing loans. Written up with non-marketable government bonds, the surviving banks presided over fragile and over-valued asset bases and a shortage of credit-worthy borrowers, wary of lending for fear of overstepping CAR limits, and preferring to make their money by investing in Bank Indonesia SBI promissory notes (*JP* 16 October 2000: 10; 18 October 2000: 12; Dick 2001: 21–23; Pangestu 2001: 167, 168).[6]

But the problem was not simply about recapitalising banks. Not only did reformers fail to resuscitate the banking system and to infuse new blood into the industry, the old structural problems remained. As former Chairman of the Indonesian Capital Market Supervisory Agency (and later Head of IBRA), I Putu Gede Arya Suta, observed, recapitalisation was little less than an attempt to, 'reconstruct an obsolete and defective structure' (Suta 2000b: 4). He argued, 'instead of rigorously separating ownership of private banks from borrowers and putting banks on a professional basis, there is an eagerness to return banks to their original owners – as long as they can raise capital to repay government loans' (Suta 2000a: 4).[7] In other words, all the old problems of private banks owned within larger corporate groups remained with all the pitfalls of intra-group lending and the task of imposing legal lending limits. In the meantime, the real problem for the bank owners was not that of retaining the banks themselves, but preventing their debts to the banks from destroying their larger business empires.

Resisting debt restructuring

The rescue and recapitalisation of Indonesia's private banks was to cost the government over US$85 billion by late 1999.[8] In their attempts to recoup this outlay by restructuring debt and seizing and selling the assets of debtor companies, the government found itself involved in a struggle over the very survival of Indonesia's largest conglomerates. Ironically, the beleaguered conglomerates received an early advantage when the government decided, in the dying days of the Soeharto regime, that BI would guarantee depositors and meet inter-bank commitments in accordance with the Banking Law of 1992. By August 1998,

RP.146 trillion (around US$11 billion at the time) in Bank Indonesia Liquidity Assistance (BLBI) had been injected into the banking system (*AWSJ* 17 August 1998: 1, 4). Of this sum, Rp.100 trillion went to five of the biggest private banks.[9] In January 2000, a report of the Supreme Audit Agency (BPK) revealed that some Rp.80.25 trillion of BLBI had been allocated in violation of BI's own prudential requirements to banks that did not meet capital adequacy requirements and had transgressed legal lending limits (*JP* 20 January 2000: 1; *Prospektif* 6 December 1999: 8–22).[10]

Not only were banks able to access government largesse virtually without condition, state auditors also reported that almost all of these funds were used by their recipients to pay other debt and for investments in other companies within the group and overseas, as well as for foreign exchange speculation.[11] In other words, these loans enabled an easier escape route for bank owners whose major concern was not propping up banks no longer useful to them but trying to protect their corporate empires and avoid paying debts. Even as they accepted BLBI, owners and employees of the banks were withdrawing their own funds *en masse* (*Kontan Online* 15 February 1999). In their attempts to recover the BLBI, the government confronted a private banking system where the owners of banks were also their biggest debtors.[12] In this situation, the banks, which had been established mainly as cash counters to fund other companies within the group, were now virtually expendable, valuable only to raid in the rush to get funds offshore or as dumping grounds for debt.[13] The priority was clearly to hang onto their larger corporate assets. With bad loans concentrated in the hands of the largest and most influential conglomerates and politico-business families, the government faced a formidable political task in enforcing debt restructuring.[14]

Early efforts by IBRA to recover assets were met with contempt by many of these corporate bosses, who often failed to turn up for meetings with IBRA negotiators. Such displays of arrogance initially heightened popular anti-conglomerate sentiment. Various political figures in the volatile Habibie era rode this mood with calls for expropriation of their assets.[15] Continuing lack of progress in pursuing these recalcitrant debtors prompted claims that the government was doing little more than warehousing the debt of prominent business figures.[16] It was only after a series of sometimes acrimonious meetings with leading debtors that IBRA was able to conclude Master Settlement Acquisitions Agreements (MSAA) with four of the largest debtors that would, according to IBRA head, Glen Yusuf, cover the Rp.80 trillion in loans outstanding.[17] But dissatisfaction with these agreements was widespread. Even Habibie initially refused to sign the government's agreement with Liem Sioe Liong, arguing that if his companies were worth so much, why did he not just sell them and give the government the cash? He set a deadline of one year for Liem to come up with the money. Ironically, it was IMF Asia Pacific Director, Hubert Neiss, who came to the rescue, writing to Habibie in October 1998, requesting that the deadline be extended to four years. It was the IMF view that forcing Indonesia's conglomerates into fire sales would have broader economic consequences, inhibiting their return to Indonesia and undermining their capacity for recovery (*AWSJ* 13 January 2000: 1, 10).

When it came to negotiating the MSAA there was little doubt that the conglomerates had the upper hand. The opaque nature of corporate governance prevented a clear picture of corporate assets. Indeed, some senior IBRA officials continued to see some of the enduring problems in asset recovery coming from these original MSAA. Allowing Liem to hand over a complex package of assets from 105 companies made due diligence appraisals a nightmare. When surrendering just three companies, for example, Indofood, Bogasari and Indasair, would have covered all debt.[18]

Another key advantage for the conglomerates was their ability to insulate highly successful businesses within Indonesia and overseas from assets negotiations. While their banks and highly geared infrastructure projects were weighed down with debt, other Indonesian-based enterprises within the same groups, in forestry, foodstuffs and agriculture, flourished, providing cash flow for groups such as Sinar Mas and Salim.[19] It should also be remembered that many of these groups had moved considerable funds offshore in the early 1990s. Liem Sioe Liong's First Pacific, based in Hong Kong and cash rich from the US$1.8 billion sale of Hagermeyer, used these funds to buy up new assets, including a US$700 million stake in the Philippine Long Distance Telephone Company. It was also to purchase a 40 per cent stake in Indofood, one of the Salim group's flagships in Indonesia. The deal enabled Salim to keep a grip on Indofood, shift the controlling stake offshore and provide funds for paying debt to overseas creditors (*JP* 28 December 1998: 8; *Warta Ekonomi.Com* 14 January 1999; *FK* 11 January 1999: 61; *AWSJ* 23 June 1999: 1, 4).[20]

IBRA was also hamstrung by the judicial system, where it fared no better than foreign creditors in efforts to pursue debtors through the courts. As we shall see in more detail later in the chapter, in the fluid and opaque environment that still prevailed, legal judgements were subjected to political pressure and bribery. The new commercial and competition laws, designed in haste after the fall of Soeharto, proved to be easily manipulated in favour of entrenched commercial interests in the less predictable environment of democratic Indonesia (Lindsey 2000: 283). It was no surprise that IBRA continued to suffer significant defeats in the bankruptcy courts until well into the year 2000 (*JP* 22 May 1999: 8; 5 April 2000: 1; 28 April 2000: 10; *Jakatra Post.Com* 28 July 2000).

The continued reluctance of big business figures to co-operate with IBRA remained an important political issue as prominent identities thumbed their noses at the government in ongoing cat and mouse games in the courts. None were more blatant than the efforts of Sjamsul Nursalim, who continued to resist settlement of his Rp.28.5 trillion debt (*JP* 25 May 2000: 1; 4 July 2000: 1). These battles were treated prominently in the Jakarta press, playing to the nationalist and anti-conglomerate passions unleashed in the fall of the Soeharto dictatorship. Such perceived arrogance was to become increasingly resented when it was revealed that conglomerates and other debtors had deceived the government in asset negotiations. To cover a debt of Rp.47.7 trillion, for example, the Salim group handed over assets of Rp.53 trillion that were subsequently valued at Rp.24 trillion. Assets originally claimed to be worth Rp.27.4 trillion pledged (but not actually handed over) by Sjamsul Nursalim to cover a BDNI debt of Rp.28.4 trillion were subsequently valued at Rp.9.4 trillion (*Panji Masjarakat* 19 August 1998: 16–22; *JP* 8 August 1998: 1; *Tempo* 11 June 2000: 130; *Jakarta Post.Com* 25 July 2000).[21]

It was not only popular sentiment that was outraged. Claiming that these asset agreements now imposed a potential total loss of Rp.80 trillion on the government, Economics Minister, Kwik Kian Gie, announced that the government's Financial Sector Policy Committee (FSCP) was proposing to cancel the agreements (*JP* 25 July 2000: 12). This new hard line was not just a knee jerk reaction to the behaviour of the conglomerates and other debtors. Pressures to step up the pace and scale of asset recovery and dispersal were increased by looming fiscal crisis. Interest payments on Rp.624 trillion (around US$80 billion) in Treasury Bonds issued to recapitalise banks and reimburse creditors and depositors were expected to reach Rp.60 trillion in 2001 and Rp.130 trillion in 2002. Servicing the interest on these bonds in the state budget of 1999/2000 was Rp.34 trillion, but was expected to rise to Rp.60 trillion in 2001. IBRA now came under intense pressure to plug these yawning budget gaps. From its debt recovery programmes and asset sales, it was required to deliver Rp.17 trillion by 31 March 2000, and another Rp.18.9 trillion by December 2000 (Pardede 1999: 26; *JP* 15 July 1999: 1; 6 March 2000; 14 March 2000: 8, 4).

Wahid's appointment of Cacuk Sudarijanto as the new IBRA chief in January 2000 was widely perceived as a move to address problems of both asset recovery and disposal. Under Cacuk, the government took an increasingly hard line against recalcitrant debtors, targeting some of the most powerful business interests – confiscating land and filing bankruptcy proceedings.[22] Appointed later in the year, new Economics Minister, Rizal Ramli, a long-time critic of the conglomerates, announced that bank owners would be required to surrender additional assets if the amounts originally pledged were not sufficient. He demanded that Salim and others cede their 'cash cow' assets to repay BLBI. Almost immediately, IBRA set about renegotiating MSAA with the major conglomerates (*JP* 20 September 2000: 1; 10 October 2000: 1).

Yet the determination of the government to secure maximum asset return was complicated by other factors. Chair of the National Business Development Council, businessman Sofjan Wanandi, argued that setting aside the MSAA would damage confidence in legal certainty and inhibit a revival in investment (*JP* 26 July 2000: 12). More important, though, the government was now forced to recognise the importance of quick sales and to accept that the price of forcing the pace might be asset recovery levels as low as 30–40 per cent (World Bank 2000: 14; *JP* 3 August 2000: 4). Promising to restructure 35 per cent of debts by June, Cacuk began negotiating debt-restructuring agreements with a number of leading debtors. By September 2000, it was claimed that Rp.62.3 trillion in debt belonging to the top twenty-one debtors was in a late stage of resolution (*AWSJ*

10 May 2000: 4; *Tempo* 30 April 2000: 102, 103; 16 July 2000: 105; *JP* 5 July 2000: 11; World Bank 2000: 14).

It soon became clear that negotiated debt restructuring was sinking into the murky waters of relationships between powerful business groups and the new power centres. Wahid himself had stepped in to resolve a restructuring package for struggling petrochemical giant, Chandra Asri. Under the proposed agreement, Marubeni took a 20 per cent share by converting into equity US$100 million of the US$723 million it was owed, while IBRA converted US$413 million of its loans into a 31 per cent equity stake. Inexplicably, Prajogo Pangestu was to retain 49 per cent equity in the company.[23] Even more contentious was the case of Texmaco, a large, diversified industrial group with assets of $4.5 billion and over 150,000 workers that found itself with more than US$3 billion in debts after the crisis struck. With a record of continuing difficulty despite several costly government rescue attempts, the controversial company, now owing Rp.16.97 trillion (US$1.8 billion) to IBRA, was one of several companies included in debt restructuring deals worth US$3.7 billion approved by the FSCP.[24] Public cynicism about the processes were reflected in *Tempo*'s speculation about whether Tommy Soeharto might not now ask for credit to revive the Timor car project (26 March 2000: 98, 99).

Both Texmaco and Chandra Asri created a dilemma for the IMF and the World Bank. While supporting the need for speed in restructuring, they expressed alarm at some of the deals being done and concern at the opaque nature of the negotiations (*AWSJ* 24 May 2000: 1, 7; *Kontan Online* 15 May 2000; *Tempo* 21 May 2000: 102, 103). Committing funds to equity in an asset such as Chandra Asri, widely regarded as having little capacity to compete in the global petrochemical market, raised a storm of objections from within the IMF and from a range of Indonesian economists, politicians and former officials (*JP* 7 October 2000: 1). In a letter to Rizal, jointly signed by IMF Jakarta representative, John Dodsworth, and World Bank Indonesia Country Director, Mark Baird, warned that the Texmaco deal could be costly for future Indonesian budgets, and urged a second opinion from international consultants. Economist, Sri Mulyani, was more blunt, casting doubt on the quality of the assets and warning of future burdens on the public purse (*JP* 7 October 2000: 1; 5 October 2000: 1; 9 October 2000: 1; 10 October 2000: 4).

In the end, it became clear that the government possessed neither the capacity nor the will to enforce a rigorous programme of debt restructuring. Although it was claimed that 74 per cent of the outstanding debt to its largest debtors was now under MOU, only 17 per cent of agreements were actually being implemented (Dick 2001: 23, 24). According to the IMF 'asset recovery rates have been quite low ... total recoveries will fall well short of off setting the total cost of ... recapitalising the banking system', and 'after all assets have been disposed of, Indonesia will be left with a large public debt stock as a legacy of the 1997–98 crisis' (IMF 2002: 29,34) The government's decision to increase repayment time from four years to ten years in early 2002, and to reduce interest rates to 7 per cent was *de facto* recognition of the need to speed up the process. It was a move Economics Minister Kwik claimed would cost the government over US$46 billion (*Laksamana.net* 25 January 2002).

The politics of asset sales

Despite the difficulties, by mid-1999 IBRA controlled assets with a nominal value of Rp.546 trillion, or 57 per cent of GDP, including non-performing loans, equity investments in nationalised and recapitalised banks, and claims on former bank owners' assets.[25] Their rapid sale was critical, not only to prop up Indonesia's flagging fiscal position but also to turn inactive resources into productive assets and to regenerate investment and economic growth.[26] But the sale of assets was also potentially significant in reshaping the very configuration of business in Indonesia. It had been widely predicted, not least within the international business community, that the entry of powerful and efficient international corporate institutions to buy up cheap assets would bring to an end the era of the Chinese family conglomerate and the predatory arrangements that defined the relations between state and business (*AWSJ* 11 August 1998: 1, 7, 8). Amongst many Indonesian observers, too, there were expectations that the reorganisation and sale of assets would result in the replacement of the old style of rent-seekers with sleek new business interests emerging in the booming export manufacturing and resource sector, unconnected to the old networks of power and patronage.[27]

However, attempts to sell assets met with opposition from several quarters, not least populists and economic nationalists within parliament and the parties who regarded the increasing entry of foreign investors as a potential sell-out of Indonesian interests. Hence, we find that in the Habibie interregnum, the Minister for Co-operatives, Adi Sasono, had demanded that IBRA assets and the proceeds of privatisation be redistributed to co-operatives as the basis for a new populist revival (*JP* 16 December 1998: 1; Cameron 1999: 30). Spokespersons within the largely indigenous business associations, Hipmi and Kadin, called for the assets to be used to bolster indigenous business (*Suara Pembaruan* 28 August 1998; 18 September 1998; *JP* 16 December 1998: 12).

Such public demands sounded alarm bells within the IMF, whose officials began to express fears that the task of getting maximum value for assets might be overridden if IBRA was turned into an instrument for populist agendas intent on reconstructing an economy focused on employment intensive industry and indigenous ownership (*Reuters* 4 July 1999). However, fears that assets would be distributed amongst populist rent-seekers proved unfounded. Co-ordinating Minister for Economics, Kwik Kian Gie, confirmed the government position that it had no intention of using IBRA to redistribute assets to small business interests (*JP* 1 December 1999: 1; 3 December 1999: 4). But neither were nationalist fears of a sell-off to foreign interests to prove real. Ironically, foreign investors were increasingly reluctant to re-enter Indonesia.

As the fiscal pressures increased, Kwik Kian Gie confirmed that the Indonesian government was now faced with resorting to asset fire sales to meet its Rp.18.9 trillion obligations to the budget (*Business Times Online* 28 June 2000). The sale of PT Astra in 1999 to Singapore interests, Cycle and Carriage, represented the first significant breakthrough for IBRA. With the Rp.3 trillion (US$370 million) sale of Salim's plantation interests to the Malaysian Guthrie group in November 2000, IBRA brought to Rp.8 trillion the value of Salim assets it had sold in the past year (*AWSJ* 27 November 2000: 1, 7). Yet the overall pace of asset disposal was slow. By September 2000, IBRA had generated only Rp.35 trillion in revenues (World Bank 2000: 13). In 2001, it restructured only 30 per cent of its NPLs and sold only 8 per cent of its total assets compared to the sale of 70 per cent and 45 per cent of assets respectively by the same restructuring agencies in Thailand and South Korea (Siregar 2001: 296). By the end of 2002, the IMF reported that only Rp.17.1 trillion of the Rp.87.7 trillion in assets pledged under the MSAA agreements had been sold. Almost all, Rp.16.3 trillion, were assets pledged and handed over by Liem Sioe Liong (IMF 2002: 47).

In the view of the World Bank, the slow pace of sales was attributable in part to an unrealistic insistence by Indonesian policy-makers on recouping full value on asset disposal.[28] It is true that within the parliament and in the political parties, there was a widespread reluctance to see 'national' assets sold off cheaply and their control pass into foreign hands. The government had encountered strong opposition to its amendments to the banking law of 1992 allowing sale of Indonesian banks to foreigners (*JP* 2 September 1999; 15 September 1999). Legislators opposed the planned sale of IBRA holdings in Bank Central Asia and Bank Niaga, arguing that continued economic recovery and strong performance by the two banks would enable a better return. Despite the fact that the DPR agreed to a 51 per cent disposal of BCA shortly after President Megawati came to power, there remained strong feelings that the sale price was too low for a bank with a 33.4 per cent CAR, into which the government had injected Rp.60 trillion in bonds (Siregar 2001: 296).

The slow pace of sales was also connected to the growing political issue of who would dominate the new corporate Indonesia. The longer assets remained unsold, the stronger became the position of Indonesia's beleaguered corporate debtors. With foreign investors showing little interest, they increasingly appeared to be the only game in town. There was widespread cynicism that nationalism often masked more sinister interests. Fears that the old owners would now repurchase their assets at a discount prompted demands by some legislators that the Salim group not be allowed to re-enter BCA (*JP* 15 August 2000: 2; 6 October 2000: 1). Rumours that the government was contemplating selling assets back to debtors had been circulating for some time and were fuelled in early 2000 by reports that Anthony Liem planned to repurchase part of the group's assets held by the IBRA holding company, Holdiko (*AWSJ* 13 January 2000: 1, 10; *Prospektif* 13 March 2000: 6–23). The choice became increasingly one of justice or cash, and the latter was attractive to a government wanting to get investment going again. As Nomura Securities' Hong Goei Siauw argued, in the existing situation, no-one wants to outlay US$2.5–US$3 billion to buy IBRA assets especially when they know the condition of these assets and Indonesia's circumstances. The re-entry of high profile conglomerates, in such circumstances, would act as a magnet to foreign investors (*Tempo Online* 27 August 2000).

Co-ordinating Minister for Economics, Kwik Kian Gie, resigned in August 2000. Kwik had launched a stunning criticism of the government position. It was, he argued, now flirting with unscrupulous tycoons and giving concessions for banking crimes. It had failed absolutely to take legal action against any of the conglomerates. Under no circumstances, he proposed, should they be permitted to buy back their assets as was now being considered by Cacuk (*JP* 16 September 2000: 1). Despite the fact that Rizal had introduced a ban on former owners bidding for their assets, allegations of lack of transparency in MOUs in the divestment process and speculation about collusion in sales were to surround the sale of interests in the Jakarta Ring Road project, the sale of BCA to the US Investment Company, Farallon, and the sale of IBRA's 72.63 per cent stake in PT Indomobil to a consortia led by Trimegah Securities.[29] In the latter case, the head of the Business Competition and Control Commission team monitoring the sale, Sutrisno Iwantono, alleged the sale was below its market price, recovering only 29 per cent of the original asset value of US$205 million. It was a deal widely believed to allow Liem back into Indomobil (*Laksamana.net* 25 January 2002).

In the end it appeared that only 20 per cent of assets pledged to IBRA would be both secured and sold. The FSCP appointed the establishing holding companies to manage unsold assets after 2004. To an important extent, business had largely forced the government to carry the burden of much of its debt. As the IMF noted, 'care will be needed to ensure that the holding companies do not simply become a warehouse for assets that are deemed either too politically contentious, or otherwise difficult to sell' (IMF 2002: 45). Nevertheless, despite their success in frustrating many of the government's attempts to force a restructuring of debt, there is no doubt that almost all of the major business groups had suffered serious losses in the aftermath of the crisis.[30] Furthermore, holding onto their assets in the face of the government's attempts to recoup the costs of rescuing the banks was only one of the problems faced by Indonesia's business tycoons. They fought on several fronts, under threat, at the same time, from attempts to confiscate many of their most lucrative monopolies and to dismantle the very institutions that had provided the rivers of wealth from the state.

Unplugging the predatory system: confiscating monopolies and reforming gate-keeping institutions

As we have seen in Chapter 3, most of the large business groups that emerged during the Soeharto era had been based initially upon monopolies and contracts allocated generally without public tender. In the aftermath of the crisis, these became the targets of reformers, determined not only that the cronies would be dispossessed but that the very institutions that made collusive and discretionary allocation of rents possible in the first place would also be removed. Ironically, Soeharto had gained much political mileage in his early years in power from the dispossession and public prosecution of former Soekarno business figures such as Jusuf Muda Dalam and Markham (Robison 1986: 91). It might have seemed logical that Indonesia's new rulers would see the advantages of appealing to public demands for blood and at the same time freeing up lucrative business opportunities for themselves.

However, the conglomerates and the big politico-business families of the Soeharto era proved less vulnerable than their predecessors. Spreading beyond trade monopolies and state contracts into sectors such as property, manufacture, resource processing and media that are not so susceptible to confiscation and cancellation, their corporate interests were also much larger and more complex, often including public share offerings and partnerships with foreign investors outside the immediate reach of any arbitrary cancellation of projects or monopolies. With a substantial part of their assets offshore the larger business groups, at least, were able to insulate important elements of their empires. Because their roots within the apparatus of the state and within Jakarta's political and economic elites were deep and pervasive and their accommodation with new power and money brokers quickly secured, the political break with the past was ambiguous and uncertain. If confiscation was to prove inconclusive, the task of institutional reform was even more difficult. Here, the reformers faced officials and political leaders who, no less than their predecessors, quickly realised the importance of the patronage networks and extra-budgetary income potentially generated by gate-keeping institutions.

The politics of confiscation

The vulnerability of entrenched business oligarchies to confiscation seemed quickly to be confirmed as many monopolies and contracts were revoked or cancelled. For example, the termination of the clove monopoly, BPPC (by Presidential Decree 21/1998), effectively ended Tommy Soeharto's grip on the industry. His national car project, PT Timor Putra Nasional was dealt a deadly blow when import and luxury tax exemptions that sustained it were cancelled and it was required to repay taxes on cars imported from Korea since 1996 (*JP* 1 October 1998: 8). Similarly, the abolition of Bob Hasan's plywood marketing and export monopoly, Apkindo, under the IMF package of 15 January 1998, led ultimately to the unravelling of his control over Indonesia's plywood producers and exporters (*JP* 7 February 1998: 1). Perhaps the most important step was the ending of Indonesia's most enduring source of trade monopolies in 1997 and 1998, when Bulog lost its exclusive right to import and distribute a range of commodities, including rice, sugar, cooking oil, soybeans, wheat and sugar.

These had long provided the Soeharto family and Liem Sioe Liong with lucrative sources of cash flow.[31]

A review of the state sector in 1999 by the Ministry of State-owned Enterprises identified 173 projects that had been established, often without public tender and involving a range of illegal practices, including over-pricing of contracts, illegal granting of credit, tax and import facilities, the sale of state assets at unjustifiable prices and compensation for unnecessary intermediary activities (Abeng 2001: 73). These practices were to be found in contracts for supply and construction, distributorships and import monopolies allocated by public sector agencies, concentrated in road construction, power generation, in the construction and operation of ports and in the state airline, Garuda. Of these, seventy-nine were cancelled and the remainder restructured, transferred to the state audit or private buyers, or continued under a co-operative contract (Abeng 2001: 74).[32]

It was the assault on contracts allocated by the State Oil Company, Pertamina, which was perhaps the most significant. Its control over the allocation of drilling and exploration leases, contracts for supply and construction, crude oil and fuel distributorships, and insurance arrangements, had enabled extensive private business interests to consolidate within this most pervasive and powerful of all the Indonesian state gate-keeping empires. A series of legal and financial audits in 1998 by the Attorney General and the Government Finance and Development Comptroller, BPKP, revealed that 159 companies, most of which were associated with the Soeharto family, had received contracts through dubious means (*JP* 12 October 1998: 12). In 1998, they were the main losers in moves by the Energy Minister, Kuntoro, and Pertamina Director, Soegianto, to progressively rescind, cancel and renegotiate contracts and to abolish monopolies. Their monopolies in fuel and oil imports through Permindo and PT Perta were ended, and several key contracts to build refineries and LNG plants, provide insurance and transport LNG and oil were cancelled or renegotiated (*JP* 10 June 1998: 1; 23 September 1998: 8). By the end of 1998, Pertamina claimed to have scrapped contracts with a total of thirty-two firms and made a saving of US$100 million (*JP* 10 October 1998: 8).[33]

But there were important limits to the confiscation process. In particular, the vexed question of property rights constrained the authority of reforming ministers and officials. This was especially the case in the forestry industry where one of the largest arbitrary allocations of rents had taken place. As the Minister of Forests and Plantations, Muslimin Nasution, noted in early 1999, it was difficult to expropriate forestry concessions or introduce a new regime of allocating logging concessions, as the pie had already been carved up and the government did not (officially) intend to open new areas (*JP* 12 April 1999: 11). While a progressive resources rent tax was introduced to replace the much-criticised royalty arrangements, extracting much higher rents from the big concessionaires, the major players remained entrenched (*JP* 29 April 1998: 8; 16 July 1998: 1; 26 December 1998: 8; 12 April 1999: 11). While the government proposed to review licenses and refuse renewal of concessions where holders were involved in poor management and misuse of reforestation funds, in the end only 1.36 million hectares were affected by these measures – a fraction of the total (*JP* 9 July 1999:1).

It was in the power generation industry, however, that the limits of confiscation and renegotiation were most decisively revealed. As we have seen in Chapter 3, the state electricity commission, PLN, had entered agreements with private power-generating companies requiring it to buy electricity in US dollars at prices higher than the prices it was permitted to charge public consumers, and in circumstances of an existing over-capacity (*FK* 29 December 1997: 12–21; *JP* 19 February 1998: 8). In a situation where it was now clearly unable to pay the existing dollar contracts or to absorb the new providers yet to come on stream, PLN Director Marsudi took a bold stance. He announced that PLN would continue to buy from existing producers at pre-crisis US dollar rates, and cancelled the contract with Sudwikatmono's PT Cikarang Listrindo and the contracts of several other companies still in the construction phase. He also threatened to review the contracts of the giant Paiton I and II projects in which Hashim Djojohadikusumo and Bambang Trihatmodjo were the central figures (*JP* 8 June 1998: 1).

Such moves outraged the foreign partners who turned to their governments and to international courts for redress. Even though it was clear that these partners had in many cases entered contracts issued without transparent public tender and in deals with Soeharto family members that involved overpricing, payment of the equity of local partners and the offloading of costs onto PLN, foreign governments were willing to offer their support. This was not least because their import and export banks and insurers had provided over US$4 billion in financing and guarantees to big US, Japanese and German companies. As lawsuits were traded, the Indonesian government was eventually forced to retreat from the battle and to try and seek some relief in renegotiating prices (*FEER* 21 October 1999: 63–64; *AWSJ* 9 March 2000: 1).

Nevertheless, many formerly well-connected business groups lost lucrative rents in this period, particularly the Soeharto family who not only lost important contracts, distributorships and brokerage monopolies with Pertamina, Bulog and Garuda, but also faced newly emerging regional authorities intent on securing some of these prizes. But did this really mean the end of the system of state-sponsored predatory allocations or even the demise of the Soehartos and other rent holders? At one level, it may be argued that many of the contracts and monopolies terminated had in any case ceased to operate effectively as the private monopoly-holders ran out of cash and the government was no longer able to underwrite the expansions, particularly in infrastructure development. Confiscation just confirmed the already obvious.

The real question is whether the strategic gate-keeping institutions were dismantled or simply in hibernation, ready to be reactivated as the Indonesian economy was revived and new power-holders installed themselves. For example, as oil trading companies began to swarm once again around Pertamina in the Singapore oil trading market, with oil imports and exports reaching US$300 million per month in 2000, the reported involvement of figures formerly associated with the Cendana oil distribution and export monopolies, Perta and Permindo, and their continuing relationships with Pertamina officials, cast ominous shadows (*Tempo Online* 12–18 February 2000).

The politics of gate-keeping institutions

It had long been recognised by the World Bank that gate-keeping institutions such as Pertamina, Bulog and the host of ministries and agencies through which contracts and monopolies were allocated functioned with minimal accountability and often outside any predictable and enforceable legal framework. While the favoured solution to this problem was seen to lie in the privatisation of public authority and increased reliance on market forces, thereby removing the very need for state sector involvement in allocation of access and guarantees, the Bank increasingly recognised that institutions might also be reformed by imposing mechanisms of accountability in accounting practices, auditing, transparent and arm's length procedures in procurement, clear rules and disclosure within a predictable legal framework (World Bank 1995: ii, iii; 29, 30, 60–62).

Thus, the IMF and reformers in the Indonesian bureaucracy embraced policy and institutional change with enthusiasm in the early post-Soeharto years. Formal and informal restrictive marketing arrangements in cement, paper and plywood were targeted for elimination by June 1998, and import and marketing monopolies and price controls, except for those on rice, sugar and cloves, were to be phased out by the end of 2000. Clove marketing was opened to the private sector, and Bulog's function was redefined as one of importing and distributing rice to maintain price stability in competition with private traders. It was required to adopt a system of open tenders, intended to eliminate the opportunity for the sorts of trading monopolies formerly allocated in closed agreements (Government of Indonesia *LOI* 31 October 1997–2001; 15 January 1998; *JP* 4 November 1997: 1; 24 December 1998: 8).[34]

Efforts also began to strip Pertamina of its extensive and extraordinary powers. Legislation initially introduced into parliament in 1998, and again in 2000, provided for the transfer of its authority over the allocation of licenses for exploration and production to a Mining Control Board under the Minister for Mining and Energy. Its authority over downstream distribution and marketing was shifted to a Management Board (BP) reporting to the President. Pertamina would become like any other contractor, giving 20 per cent of production to BPKP (*Kontan Online* 5 June 2000). But the process proved difficult. For all its problems, Pertamina had represented for many Indonesians national control over the country's most important natural resources. There were fears that Pertamina would not survive under the new arrangements, the move to a free market in oil resulting only in the replacement of a Pertamina monopoly with a private oil company oligopoly.[35] It was this nationalist sentiment, argued by some to mask more cynical vested interests, that contributed to the defeat of the initial reform legislation in a highly volatile and populist parliament in December 1998 (*FEER* 24 December 1998: 42–44).

Nevertheless, the legislation was eventually passed and on 6 August 2002 a new agency, Balak, formally took over the task of supervising and regulating the upstream sector of Indonesia's oil industry, including the signing of explorations and production contracts (*JP* 6 August 2002: 1). There was no doubt that disentangling exploration and production from other elements of the Pertamina octopus would clearly reduce the potential for the vast systemic collusion that had made Pertamina such a formidable dispenser of rents. The question was, whether such institutional engineering would now be enough to enforce accountability and transparency in its operations or whether Balak would now simply operate in the same business of rents at a more specific level.

The IMF adopted the same strategy of using institutional engineering in an attempt to undermine and eliminate predatory state–business relations in the banking and finance sector. Both the IMF and World Bank placed a high priority on the reform of Bank Indonesia and the state banks, institutions that had been so instrumental in the allocation of the floods of discretionary credit that had made Indonesia's banks and corporate entities so vulnerable when the crisis hit (World Bank 1999: 2.3). This was to be achieved by separating the banks from political and predatory influences. Under the provisions of the new central banking law introduced in May 1999, the Governor of the Central Bank was to be independent of the government and would no longer sit in Cabinet. BI's function in channelling subsidised liquidity loans to individual borrowers was abolished and its authority to supervise banks transferred to a new independent institution to be established by the end of 2002 (McLeod 1999: 148, 149). It was an attempt to protect the markets from politics by insulating rational decision-makers.

Attempting to deal with the problem by making formal changes to institutional rules and procedures had been tried before. After all, Bank Indonesia had presided over an era of gross violations of legal lending limits and capital adequacy requirements by private bank owners in blatant disregard of prudential requirements, despite the fact it already received regular reports from the banks. It had directly channelled large discretionary allocations of liquidity loans to Soeharto family business interests despite clear regulations that excluded such loans from the liquidity credit programme. Reformers in the government and the World Bank were well aware of the problem. In the decade before the crisis, efforts to strengthen Bank Indonesia had included the appointment of large numbers of well-qualified, foreign-educated staff (Hamilton-Hart 2000: 123). Credit supervisory teams had been installed at BI and a foreign bank brought in to supervise the state bank, Bapindo (World Bank 1995: 60). But this 'strengthening' strategy did little to prevent the gross excess and speculation in private and state banks, nor was it to prevent the disastrous and highly illegal allocation of BLBI to banks in 1997 discussed earlier.

Less than a year after the new banking bill had been introduced, both President Wahid and the parliament began backtracking on the issue of bank 'independence'. Amendments to the Banking Bill, debated in parliament, included measures to force BI to report regularly to parliament and giving the parliament the right to nominate the Board of Governors (*Straits Times* 20 December 2000). These moves, complicated by the prosecution of BI Governor, Sjahril Sabirin, were seen by the IMF as an attempt to perpetuate the capture of the bank by political interests and to compromise its capacity to make 'technical' decisions. It was this issue that led to a withholding of a US$400 million tranche to Indonesia in 2000.

But the assumption by the IMF that establishing the independence of the Governor from parliamentary control and insulating the bank's technocrats from the demands of vested interests and distributional coalitions outside would be enough to ensure abstracted and 'rational' decision-making was flawed. On the contrary, the bank's record suggested that such autonomy might equally insulate predators within the bank from external accountability and enhance their capacity to engage in relations with political interests outside. Indeed, the bank had been characterised by economist, Anwar Nasution, later to become Deputy Governor under the Wahid Presidency, as a 'den of thieves' (*JP* 26 June 2000: 10). Although portrayed as a disingenuous cover for extending his own influence in the bank, President Wahid's argument that a purge of existing officials was a precondition for real reform was nevertheless a reasonable proposition (*JP* 13 June 2000: 1; 26 June 2000: 10; 5 August 2000: 1, 4).

Institutional engineering without the creation of accompanying political authority in the state apparatus for reformist forces had its limits. The importance of removing 'fixers', cronies and members of powerful politico-business families from the boards of state companies and senior management positions as a precondition to the effective reform of the state apparatus was widely recognised (Abeng 2001: 90–91). But the problem was deeper. Reform ultimately depended upon the outcome of political conflicts elsewhere. Whoever occupied the boards and management, reform of the gate-keeping institutions would remain difficult so long as political power more generally relied upon control of patronage and non-budget funds. As we shall see, the transition from authoritarian to parliamentary rule did little to diminish the importance of this imperative. Indeed, efforts by the IMF, the World Bank and other agencies to impose technical reform and 'good governance' in well-funded programmes now had to contend with the counter-pressures coming from the descent into a more fragmented, frenetic and unconstrained form of money-politics.

In these circumstances, attempts to reform the gate-keeping institutions proved more difficult than expected. Audits by Arthur Andersen and the Government Finance and Development Comptroller, BPKP revealed that non-budgetary disbursements from Bulog had continued to be significant throughout the 1990s and into the period since the fall of Soeharto.[36] Ministries and agencies reported the existence of off-budget funds of around Rp.7.7 trillion in June 2000 (US$860 million) (Government of Indonesia 2000: 8). While there was uncertainty over whether things were getting worse or better, or whether the auditors could get the full story, continuing audits by BPK and BPKP showed high levels of leakage, corruption and contravention of regulations in the state apparatus.[37] The continuing audacity of these processes was most dramatically illustrated to the World Bank itself when it was revealed that Rp.8 trillion, or almost half the World Bank funds made available for social safety net allocation in the fiscal year 1998/1999, had been misappropriated (*JP* 24 April 1999: 4).

A series of episodes in which funds from Bulog were diverted to political parties dragged both Habibie and Wahid and DPR Speaker, Akbar Tandjung, into corruption scandals that confirmed parliamentary politics remained just as reliant on these extra-legal sources of political funding and patronage.[38] IBRA, now the premier gate-keeping institution and presiding over huge resources and powers, became a prize for interests fighting over the terms of recapitalisation, debt restructuring and asset sales. Controversies surrounding IBRA's debt restructuring deals worth US$3.7 billion owed by four of the biggest conglomerates fuelled public perceptions that predatory interests were at work in framing these agreements. Questions were raised about relationships between Ramli, Wahid and Marimutu Sinivasan of Texmaco, the way IBRA's negotiations were held, and the appointment of several members to its board who were close to businessmen in debt to IBRA and to political parties (*Tempo* 16 July 2000: 104). The *Jakarta Post* commented in relation to the debt restructuring deals that:

> In addition to the instability of its management in the first seven months of its operations, the agency has been held hostage to political battles between various influential groups … We are utterly disappointed with the Oversight Committee of IBRA, which seems to serve only as a perfunctory body.
>
> (*Jakarta Post* 17 October 2000: 4)

At another level, growing deflation of power obstructed real institutional reform as increasingly fragmented political interests fought to establish their control over various gate-keeping institutions. The ongoing scandal of disappearing fertiliser stocks and the collapse of the distribution system of the state-owned producer and distributor of fertiliser, PT Pupuk Sriwijaya (Pusri) was, for example, widely seen as part of a contest for control of Pusri involving the Minister for Trade and Industry, Rini M.M. Soewandi, and bureaucrats within the department linked with local mafia who increasingly controlled the distribution of fertiliser and the purchase of rice crops at the grassroots level. Commentators noted that:

> Soewandi's problems in the Department of Trade and Industry were mirrored in a number of other ministries, where ministerial staff have repeatedly come to blows with bureaucrats from the Department of State Apparatus (PAN), many of whom are closely linked to Soeharto era power structures.
>
> (*Laksamana.net* 20 May 2002)

Elsewhere, under pressure as complaints mounted from business against the continuing corruption in the Department of Customs and Excise, Director General, Permana Agung, complained of a lack of authority to curb corrupt senior officials. He also claimed:

> there is an attempt by politicians to turn this directorate into their cash cow for [financing their campaigns in] the coming general election 2004 … Believe me, right now I have great difficulty in trusting people. There are so many vested interests around here. And the most dangerous pressure is internal.
>
> (*Jakarta Post.Com* 17 May 2002; 27 May 2002)

Problems of enforcement are perhaps most clearly illustrated in the forestry industry where the government quickly lost control of a logging industry that descended into the hands of political cliques and officials operating independently of central control. The World Bank estimated in 2000 that illegal logging was a bigger business than legal logging, imposing a US$650 million loss each year in royalty and Reforestation Fund payments. Big money and powerful individuals, including the military, the police and the provincial bureaucracy, were the backers of these illegal forestry businesses (World Bank 2000: 40). Devolution of power to the regions did not assist the problem. Forestry Secretary General, Suripto, presented a picture of an industry out of control with regional officials selling forestry concessions, the alleged involvement of a member of parliament and the illegal import of heavy logging equipment (*JP* 14 August 2000: 3; 4 October 2000: 10).

It was becoming more clear that attempting to transform Indonesia's economic regimes through institutional means without any accompanying political revolution was leading simply to the hijacking of new institutions by old forces. Yet, in the looming issue of corruption, the reformers seemed to possess a unique opportunity to decimate the entrenched oligarchies in the period immediately after the fall of Soeharto. How this opportunity was missed provides us with an insight into the resilience of the oligarchies and the fatal political weakness of reformers.

Resisting prosecution: the politics of corruption

Indonesia's political and business oligarchy faced an immediate threat in growing popular demands for the prosecution of corruptors. In the heady days of street politics in the months before and after the fall of Soeharto, their notorious and increasingly rapacious activities, long a major focus for the anger of students and other opponents of the regime, now assumed a central place in the popular politics of Jakarta. It was no surprise that the eradication of corruption was quickly appropriated in the public rhetoric of the governments of Habibie and Wahid. There appeared to be no difficulty in assembling prosecutions. Auditors and state ministries had easily identified hundreds of cases where the establishment of monopolies, contracts and concessions had involved corruption, collusion and nepotism. Leading business figures had openly admitted flouting the legal lending limits applying to bank credit and to misusing the BLBI. As we shall see, various agencies, including the Attorney General's Office and the government's Joint Team to Eradicate Corruption, as well as a range of parliamentary commissions, undertook extensive investigations of individuals and companies suspected of corruption, collusion and nepotism (KKN).

With the media free of many of the controls of the Soeharto era, scandals were brought to the surface in rapid succession. Reading about the sordid and bizarre details of corruption and collusion that drew into its net most of the public figures of the Soeharto era became a major form of public voyeurism. The Bank Bali scandal had sparked a public outcry that would have been impossible in the days of Soeharto, leading to an investigation and to the arrest of Djoko S. Chandra, whose firm, PT Era Giat Prima, had been at the heart of the diversion, and drawing in other casualties.[39] But it was the successful prosecution of the Soeharto family that was the litmus test for rule of law and the real target of reformers.

The long campaign to deal with the Soehartos began when the government assumed operational control of the ubiquitous *yayasan* in November 1998. A Presidential Decree of March 2000 finally transferred authority to disburse funds from office holders, including such figures as Sudharmono, as well as Soeharto himself, and enabled the government to take over Rp.4 trillion (US$540 million) in assets (*JP* 26 November 1998: 1; 7 March 2000: 1). But prosecuting Soeharto and his family proved more difficult than cutting back his business interests. An initial and fairly insipid attempt was abandoned by Habibie but later revived by Wahid in August 2000, when Soeharto was charged with embezzling US$571 from his *yayasan* and funneling it into his own businesses.

The net appeared to close around other Soeharto family members when Tommy Soeharto was arrested and charged with corruption relating to an illegal land-swap deal with Bulog that cost the government an estimated US$13.4 million (*JP* 13 March 1999: 1; 27 March 1999: 1; 31 August 1999: 1). Siti Hardiyanti Rukmana, Sudwikatmono and Probosutedjo were all questioned as witnesses in cases where their companies had been involved in the defrauding of

Pertamina and the Reforestation Fund.[40] Bob Hasan was arrested and imprisoned pending trial on charges of corruption relating to the misuse of US$84 million in export promotion funds provided by members of his plywood export co-operative, Apkindo (*JP* 25 July 2000).

Moves to prosecute corruptors appeared to gather steam as numerous former key officials were investigated and some detained and arrested. Among these, Ginandjar Kartasasmita, former Minister of Mines and Energy under Soeharto and Economic Co-ordinating Minister under Habibie, was charged with involvement in the illegal transfer of US$24.8 million from Pertamina to the private company, PT Ustraindo Petro Gas, and faced questioning over the mark up of costs at Pertamina's Balongan refinery involving a potential loss to the government of US$800 million. He also faced accusations of irregularities in Bakrie's purchase of 10 per cent shares in the mining giant, PT Freeport (*Tempo* 26 February 2001: 14, 15). Others who were the subject of investigation included former Bulog Chief, Bustanil Arifin, Finance Minister, Ali Wardhana, and former Pertamina Director, Faisal Abda'oe.[41] The DPR Commissions for Foreign Affairs and for Finance and the State Budget also investigated allegations of a Rp.189 billion scandal involving the Army Strategic Reserve Command, Kostrad, while the Joint Team to Eradicate Corruption named three judges as suspects in bribery cases (*JP* 22 September 2000: 1; 29 December 2000: 1).

Big Chinese business tycoons were also targeted. Revelations about the misuse of Rp.144 trillion in BLBI funds led to investigations into forty-eight commercial banks by the Attorney General's office and recommendations for legal proceedings against several leading business figures, including Prajogo Pangestu, Marimutu Sinivasan and Sjamsul Nursalim. Secretary General of Forestry, Soeripto, submitted documents to the Attorney General's office alleging illegal use of Rp.346.8 billion in reforestation funds by Prajogo and Siti Hardijanti Rukmana (*Tempo* 25 March 2001: 28; *JP* 22 March 2001: 1). Forestry NGOs urged the arrest of Prajogo, Sukanto Tanoto (of the Raja Garuda Mas group) and Sjamsul Nursalim for violations of forestry regulations (*Antara* 26 April 2001).

Slowly, the wheels of justice appeared to grind away at former officals and business figures. The former head of Bulog, Rahardi Ramelan, was jailed in March 2002 and no less a figure than Akbar Tandjung, the Speaker of the DPR and senior Golkar figure, was detained as part of an investigation on charges of misuse of Rp.40 billion of Bulog funds alleged to have found their way into Golkar coffers to fund the 1999 election campaign.[42] Bank Indonesia Governor, Sjahril Sabirin, was sentenced to three years in jail for his role in the infamous Bank Bali case, although he continued to pursue his duties as Governor of Bank Indonesia. Most spectacularly, Indonesians were soon witnesses to what appeared to be the final humiliation of the Soeharto family and the beginnings of genuine rule of law as Soeharto himself was put under house arrest and repeatedly interrogated.

After being acquitted on earlier charges, two of the most powerful figures of the Soeharto era, Tommy Soeharto and Bob Hasan, faced new charges and were eventually convicted and sentenced. Tommy's case descended into farce as he escaped from custody, and spent a prolonged period on the run in the country where he had previously been untouchable. Although such figures as Tommy Soeharto no longer command their former immunity from prosecution, the prosecutors were to face difficulties in enforcing sentences. Ironically Tommy was eventually to be imprisoned, not for corruption, but for his role in the conspiracy to murder the judge who had provided the original sentence. Perhaps the biggest scalps were those of Bob Hasan, sentenced to six years in jail for his part in the diversion of forestry funds for aerial mapping, and several other Chinese business figures convicted or placed on trial for misappropriation of BLBI.

Yet the glacial progress of the Attorney General's office in assembling prosecutions was raising questions about its real commitment. The very largest Chinese business tycoons involved in BLBI scandals remained at large, notably Sjamsul Nursalim, who had received Rp.37 trillion in BLBI but continued to resist surrendering assets. Usman Atmadjaja, the head of Bank Danamon, another big BLBI recipient whose assets surrendered to IBRA were also allegedly given as guarantees to other debtors, continued to evade prosecution, allegedly living in Singapore (*Tempo Online* 19–25 March 2002). Despite the dramatic arrests and investigations, the government's success rate in obtaining convictions has been disappointing.

When Soeharto's son-in-law, Hashim Djojohadikusumo, was arrested and detained on charges of breaching legal lending limits as commissioner of Bank Industri, it was claimed by the Attorney General's Office (AGO) as a first step in serious moves to deal with banking fraud. Indeed, the arrest followed a prolonged reform within the AGO that involved large-scale rotation and training of prosecutors and replacement of assistant AGOs. At the same time, observers had become so cynical that Hashim's detention was widely suggested to be a show trial that had more to do with Indonesia's attempts to ensure rescheduling of its Paris Club debts (*Tempo* 19–25 March 2002).

In the end, the paltry inventory of successes by the Attorney General seemed to sustain the sceptics. After a protracted series of investigations, an initial acquittal and his re-arrest, all charges against Soeharto were dismissed in September 2000 on the grounds of his ill health (*JP* 4 August 2000: 1; 29 September: 1). No charges were brought against Siti Hardiyanti Rukmana, Probosutedjo or Sudwikatmono. Ginandjar walked free, as did the Bank Bali defendant, Djoko S. Chandra. Nothing further was heard of any attempts to investigate other well-connected suspects in the Bank Bali case, including prominent Golkar figures, A.A. Baramuli, Tanri Abeng and Setya Novanta. Three Supreme Court judges charged with corruption were acquitted in the District Courts or found their charges dropped (*Jakarta Post.Com* 29 December 2000).

Why then, was it so difficult to prosecute and why did corruption appear to remain so institutionally entrenched? While various cultural and social explanations of Indonesia's apparently endemic corruption had existed for decades, the general institutionalist position was that 'corruption and collusion are a response to incentives created by a particular set of institutions, not permanent features of Indonesian government and society' (Hamilton-Hart 2001: 79). Thus, as the World Bank had long argued, corruption could be alleviated by clear rules, disclosure of processes and enforcement of regulations as well as reliance on market forces to remove the very opportunities for rents (World Bank 1995: 29, 30). As we have seen, the IMF now demanded and received in its agreements with the Indonesian government regular public scrutiny and audit of state ministries and corporations by large independent bodies, providing public transparency and information. These were long-term projects. In the meantime, attention focused on bringing reforms to a profoundly corrupt judiciary still colonised by officials embedded in the old power relationships.

Ironically, prosecuting Soeharto might have appeared to have been the easiest task. Bringing the old dictator to account would ride a wave of popular support. In reality it was always going to be difficult. Habibie advisor, Dewi Fortuna Anwar, noted in 1999 that, 'One should not take it for granted that Soeharto has no power at all. I don't think that one should be so naïve in thinking that once you chop off the top, the roots will be all finished' (*Business Times Online* 1 June 1999). Indeed, to a large extent, reformers recognised that they remained limited in their ability to mobilise the power to root out the residual power of the Soeharto era embedded in the state apparatus. Old loyalties and ideological attachments were to remain in the police and the courts and the rampant corruption of the court system also presented opportunities for individuals with huge financial resources to determine decisions. As the banking scandals unravelled, Finance Minister, Bambang Sudibyo, observed that the government was still assessing whether it had the real power to put behind bars the 'powerful businessmen responsible for creating the messy banking sector and burdening the taxpayer' (*JP* 27 January 2000: 1).

Nowhere were the residues of former power more entrenched than in the judiciary and the courts. Together with the police and security forces, they remained in the hands of officials appointed in the Soeharto era. Indonesia's judges constituted a discreet corps of apparatchiks that had been an integral and necessary element in the smooth working of the grand alliance of state power, political oligarchy and corporate wealth that defined the Soeharto regime. They operated, as Tim Lindsey has noted, within a 'black state', outside the rule of law, where the real business of power and politics took place (Lindsey 2000: 288). As a *Jakarta Post* editorial noted, 'subordination of the judiciary [under Soeharto] paved the way for total control by the state of every aspect of public life in Indonesia' (22 November 1999: 4).

The acquittal in March 2000 of Djoko S. Chandra on all charges relating to the US$56 million Bank Bali case was a verdict that outraged reformers and reinforced calls for judicial reform. Prominent lawyer, T. Mulya Lubis noted, 'The decision to free the main suspect in the … Bank Bali case was … a setback (for justice) and a deadly blow to the government's efforts to revamp the legal system. This shows our judiciary has not been touched by legal reform.' The Djoko Chandra case, he noted, carried on from similar acquittals – those of Tommy Soeharto, Rudy Ramli of Bank Bali and Beddu Amang of Bulog (*JP* 11 March 2000: 1). As the *Jakarta Post* editorial noted, the Djoko case, 'sends a warning to the government and police that corruption cases may be dismissed if they fall into the hands of the wrong judges' (13 March 2000: 4).

In reforming the court system, the government was to focus on replacing the corps of judges with *ad hoc* judges drawn from the general legal profession. Thus the contest to shape reform took the form of a prolonged struggle between the government, parliament and the judiciary itself over the appointment of judges to the Jakarta courts and the Supreme Court, including a long stand-off between Wahid and parliament over the appointment of a new Chief Justice.[43] But this emphasis on the court system has two serious faults. Observers such as Tim Lindsey have noted that changing the courts is a generational matter as old and new and enthusiastic reforming practitioners replace irretrievably corrupt incumbents (*JP* 21 December 2000). This is reminiscent of observations made about the military as long ago as the 1970s, when it was expected that a new and modern cohort of officers would transform the institution. Whether there were ever any genuine reformers in the wings, the new officer cohorts soon became transformed into the images of their seniors, shackled by the very institutional and power structures in which they necessarily operated. As in the case of the military, the institutionalisation of judicial reform takes place in a hostile environment.

Courts and judges are also only one part of a more pervasive system of predatory interest. Tommy Soeharto openly boasted that he had remained free of capture during his period on the run after being sentenced to eighteen months in jail for land fraud because of collusion with the police. At the same time, the Attorney General's Office proved so ineffective that critics became increasingly impatient with Marzuki Darusman, the only Golkar member who retained his Cabinet post under Wahid. The World Bank bluntly noted:

> Corruption in the Attorney General's office and the police is compounded by a lack of capacity to effect successful prosecutions. Poor co-ordination among the various arms of legal and judicial systems – the Ministry of Justice, the Attorney General's office, the judiciary and the police – has thwarted efforts by the National Law Commission to push the reform agenda. And the independence of the judiciary has had the unfortunate effect of giving judges with the least interest in judicial reform a large say in them.
>
> (World Bank 2000: 38)

As progress to unravel this apparently seamless web of corruption and collusion in the legal and policing apparatus continued to stall, reformers and the public within Indonesia turned in desperation to a hero in the appointment of Baharuddin Lopa as Justice Minister (*Tempo Interaktif* 19–25 June 2001). In his first month in office, he fired twenty-seven Ministry officials, including twelve judges (*JP* 2 April 2001: 1). In a series of breathtaking moves the Ministry moved to re-open ten major corruption cases previously abandoned by the Attorney General's Office and to call in for questioning prominent political and business figures. These included Akbar Tandjung, the DPR Speaker, Arifin Panigoro, a businessman and now a key figure in the PDI-P, Ginandjar Kartasasmita and Nurdin Halid, prominent Golkar figures, and the business tycoons, Sjamsul Nursalim, Sinivasan Marimutu and Prajogo Pangestu (*Tempo Interaktif* 19–25 June 2001). Surrounding himself with a dedicated and highly qualified legal team, Lopa attracted wide support, both within the bureaucracy and more broadly. Yet, this galvanising of will in the Ministry was soon to end when Lopa suddenly and strangely died on a visit to the Middle East just as his programme was taking shape.

Could Lopa have made the difference? If it was simply a case of enforcing institutional reform on the rump of an old regime this might have been the case. There is no doubt that prosecutions stalled where the capacity of the state, measured in its narrowest institutional terms, was inadequate. Weak rules, undermanned institutions and poorly trained staff inevitably had their effect. But the problem was much deeper. In attempting to prosecute major (mostly Chinese) business figures, reformers faced important structural constraints. We propose that reform in the judiciary – either by purging it of its corrupt officials, or reforming its rules – was contingent on wider and more fundamental shifts in social and state power. As we have seen, not only the Indonesian government but the World Bank and the IMF were nervous about frightening Chinese-Indonesian investors away from bringing their money back into Indonesia. Although foreign investors were unwilling to come into a situation where there was a lack of legal and institutional certainty, Chinese investors could operate in these circumstances. On the contrary, it was such certainty, including systematic assaults on corruption that would deter their return.

It was also widely claimed that Chinese business figures possessed the wealth and contacts to bribe judges and officials. Economics Co-ordinating Minister, Kwik Kian Gie, complained of being undermined by, '"dark forces" who buy favours' (*Kompas Cybermedia* 6 March 2000; *Tempo* 26 March 2000: 102, 103). Addressing the PDI-P Congress in Semarang in April 2000, he asked his fellow PDI-P members to support him. 'Don't let me be finished off by the evil businessmen who always bribe their way around' (*Kontan Online* 3 April 2000). That such business figures were able to mobilise the power to prevent their prosecution may be attributable to the wealth they were able to dispense, especially within the courts but also across the bureaucracy in general. But it also signals that the sorts of arrangements between officials and business that had defined economic life under Soeharto were being replicated in the post-Soeharto era. The move against corruption was undermined because it had also become the very *modus operandi* of the new regime, simply adapting to the new arena of parties and parliaments.

This problem became apparent from the very early days of Habibie's Presidency when Attorney General, Andi M. Ghalib, who had taken charge of the anti-corruption programme, departed after being accused of accepting Rp.13 billion (US$1.7 million) from various sources including Prajogo Pangestu. In an investigation driven by the NGO, Indonesian Corruption Watch, under Teten Masduki, the banks accounts of Ghalib and his wife were revealed to contain US$2.6 million (*Detik.Com* 8 June 1999; *JP* 5 June 1999: 1; 15 June 1999: 1). This was a specific reminder that the same sort of relations between officials and business were continuing at the most basic level. But the problem extended far beyond simple pay-offs for personal enrichment. It soon became clear that boundaries between public and private, between business and politics, were just as murky in the post-Soeharto era as they had been before. Corruption simply became less centralised as business groups gathered around new centres of power in the major parties. The very dynamics of the new democratic politics required the main players to extend their control over key departments within the state apparatus and to forge alliances with powerful business interests.

No less a person than Wahid was to be drawn into this seemingly inescapable logic of political life. He had been subject to long and bitter recriminations concerning his acceptance of a US$2 million personal gift from the Sultan of Brunei, allegedly for assistance to Aceh and his involvement in the Rp.35 billion 'Bulogate' scandal referred to earlier. More significant was the larger web of business relations that was to radiate around his family and leading members of the PKB. Wahid's brother, Hasyim, was appointed as advisor to IBRA. Hasyim and his brother Salahuddin, as well as various office holders in the NU and PKB, including Rozy Munir, who became Minister of State Enterprises, were to become involved in a range of fund-raising activities and business ventures and alliances that included contracts from Pertamina, BPIS and PLN (Aditjondro 2000). Wahid became embroiled in speculation about his relationships with several large conglomerates as a result of generous government intervention in debt restructuring arrangements outlined earlier and his request to prosecutors to delay legal proceedings against Sinivasan Marimutu, Prajogo Pangestu and Sjamsul Nursalim (*JP* 20 October 2000: 1).

But the same political–business relationships were to be found in all the political parties. We have seen that they operated within Golkar before the 1999 elections in the Bank Bali case and in the revelations about Golkar's use of Bulog funds following the arrest of DPR Speaker and Golkar Chief, Akbar Tandjung. After Megawati became President, Taufik Kiemas (Megawati's husband) and Arifin Panigoro were to emerge within the PDI-P as key figures in this shadowy world. Just as individuals from within NU and PKB were placed in key administrative positions under Wahid, it was alleged that Taufik Kiemis had engineered the movement of close associates from his home province of Palembang into a range of key state agencies including parliamentary committees, the state workers' pension fund, Jamsostek, and IBRA. Taufik's reputation as a deal maker was reinforced, not only by his association with prominent Chinese business figures such as Tony Winata and Sjamsul Nursalim, but also by continuing allegations about his influence over policy and his role in mobilising state guarantees, access and resources in support of business deals and interests.[44]

Trying to deal with the situation by introducing technical regulatory and organisational reforms, rooting out old interests in the bureaucracy or hoping for individual heroes to do something dramatic ignores the point that the fall of Soeharto simply severed the apex from a system of power relations that continued to survive. Authority over the allocation of resources, contracts and monopolies had been shifted from a highly centralised system of state power to a more diffuse and chaotic environment of political parties, parliaments and provincial governments. Old business tycoons were able to rebuild their links with new political patrons (Indonesia Corruption Watch 2001). In these circumstances, the whole issue of corruption became interwoven with larger power struggles as illustrated in the pursuit of Wahid and in the way Akbar Tandjung's prosecution took place in the larger context of a struggle for power within Golkar with resilient pro-Soeharto elements. Prosecutions for corruption were, thus, more likely to succeed where targets had become political liabilities, as may have been the case with Tommy Soeharto, or where they were weakened in political struggles and unable to mobilise sufficient political protection.

The problems of reform

Although neo-liberals had been casual in their approach to the power of predatory coalitions in the Soeharto era, believing that deregulation and good macro-economic fundamentals would ultimately solve this problem, they changed their attitude as reform in the post-crisis period began to flounder (Hill, H. 1999b: 68–80). Clearly shaken by revelations that up to 30 per cent of its disbursements to Indonesia had been expropriated for corrupt purposes, the World Bank now warned explicitly about the need to halt corrupt projects and introduce accountability and transparency into the bureaucracy. In early 2001, the Bank decided to cut its loans to Indonesia by half from the average of US$1.3 billion in the 1990s. Cancellation of a US$300 million loan tranche in April 2001 for poverty assistance resulted from a belief that such loans could not be guaranteed (*Kompas Cybermedia* 20 June 2000; *JP* 29 January 2000: 9). The Bank's assessment of the first loan tranche noted that 'the slow pace of meaningful change in bureaucratic culture had rendered modest the impact of aid on the poor' (*JP* 31 August 2001: 1).

In effect, the World Bank was throwing its hands up in despair. It had poured millions into programmes of capacity building with disappointing results. Amongst its many critical observations, it noted that:

> The budget process is inadequate, reflecting the lack of an integrated macro-framework, ... the existence of considerable off-budget funds and the lack of interest in spending results. Budget execution and monitoring remain weak, with considerable funds wasted due to corruption, collusion and nepotism. Various unreconciled budget reports and annual financial reports to parliament are two years behind schedule.
>
> (World Bank 2000: 43)

As we have seen, a favourite theme for neo-liberal reformers during President Wahid's period in office was his own idiosyncratic and erratic behaviour.

Indeed, Wahid had presided over a rapidly revolving cast of ministers and senior officials. Investment and Public Enterprises Minister, Laksamana Sukardi, and Trade Minister, Jusuf Kalla, were sacked in a flurry of unsubstantiated accusations of corruption. Among other key ministers to depart were Finance Minister, Bambang Sudibyo, Economics Co-ordinating Minister, Kwik Kian Gie, and Forestry Minister, Nurmahmudi Ismail. Wahid became embroiled in an ongoing campaign to dismiss Bank Indonesia Governor, Sjahril Sabirin, and battles with parliament over the appointment of a Chief Judge of the Supreme Court. From its inception in January 1998 under Soeharto to the end of the Wahid period, IBRA had no less than six chief executives.[45] Economic policy drifted with no clear centre of decision-making. Wahid increasingly bypassed Economic Co-ordinating Minister Kwik for economic advice and policy direction, establishing a confusing range of advisory boards. These, however, were to be ineffectual and without influence, even though membership was eagerly sought by the parties and factions in parliament.[46]

It was a situation that led, in this view, to the continuing influence of 'vested interests' and the polemicists of nationalism and populism who undermined technocrats and prevented the implementation of sensible policy (Hill, H. 2000a; Pangestu 2000; Soesastro 2000). The World Bank lamented the often lonely and powerless political position of reforming technocrats in the state apparatus and their inability to resist vested interests (World Bank 2000: 43). It was no surprise when neo-liberals welcomed the appointment of Widjojo Nitisastro and two other economists to yet another of Wahid's economic advisory teams. Yet expectations that these appointments would somehow break the log-jam of reform were naïve, as noted by an editorial in the *Jakarta Post* of 22 March 2000.

> One wonders what Widjojo has up his sleeve to speed up the economic recovery while the major hurdles hindering the recovery are crystal clear: the inimical condition of the political apparatus, the security organisations and the legal system; the slow pace of decentralisation; and too much outside interference in the debt, corporate and banking restructuring. These difficulties did not confront Widjojo under the Soeharto administration.
>
> (*JP* 22 March 2000: 4)

Indeed, the difficulties of Wahid were not solely the consequence of a bizarre leadership style. Rather, this bizarre style might be considered a response to the overwhelming political obstacles to reform faced by Wahid and the meagre political resources he was able to mobilise to drive a reformist agenda. The circumstances of technocrats had changed. Under Soeharto they had been quarantined from serious involvement in institutional reform, limited to providing the macro-environment within which the state officials and the politico-business oligarchy might flourish. The sorts of structural reforms now on the agenda threw them into direct confrontation with these powerful interests. The *Jakarta Post* perceptively identifies this aspect of the problem: 'the bureaucracy inherited by the present government was designed, built and honed by former President Soeharto to serve his autocratic administration and is still mounting formidable opposition to Abdurrachman's reform programs' (*JP* 22 March 2000: 4).

What was conceived by neo-liberals as a problem of insulating rational technocrats from the very irrationality of politics was in reality a problem of constructing an effective political alliance that would domesticate the state apparatus and drive the neo-liberal agenda in the government and parliament. This never happened. None of the three post-Soeharto Presidents ever established their authority over the military and the civil bureaucracy, or controlled a disciplined majority in parliament. The fate of Presidents was to be, as Megawati has discovered, that of achieving accommodation with these forces. Instead, it was the conglomerates, the politico-business families and the entrenched core of state officials who set about effectively reorganising their interests despite the collapse of the highly centralised system of state power that had previously sustained them.

They did this in economic circumstances assumed by neo-liberals to functionally demand policy and institutional reform. Yet, even though corporate debt problems remained unresolved and the banking system continued to lie in ruins, a shattered but unreformed system of predatory capitalism was to prove economically workable. Business seemed to crank up profits even in the middle of massive institutional chaos, corporate insolvency, unresolved debt and paralysis in the banking system, and where corruption had wormed its way into the furthest corners of the economy. Perhaps their greatest achievement was in the reorganisation of their political power. A return to the old system of predatory capitalism under the umbrella of a highly centralised authoritarian state power was no longer possible. Things had gone too far for this. The critical achievement of the oligarchy was its metamorphosis within a new political democracy and within the framework of new political alliances with political and business interests, local officials, fixers and even criminals formerly operating on the fringes of the Soeharto regime as these now flooded into the new political arena. It is this remarkable metamorphosis that is the focus of the following chapter.

Notes

1 In brief, a sum of approximately US$80 million originating from IBRA and BI was paid by Bank Bali to a private firm, PT Era Giat Prima, allegedly to assist Bank Bali in the recovery of some US$120 million in inter-bank loans owed by defunct banks. A parliamentary report indicated that some of these funds found their way into Golkar campaign funds and cited several senior government officials for investigation. These included Habibie's chief advisor, Arnold Baramuli, former Finance Minister, Bambang Subianto, and former State Enterprise Minister, Tanri Abeng (*JP* 5 November 1999: 12; *Jakarta Post.Com* 16 November 1999; 23 November 1999).

2 The most active and influential of these organisations was Indonesia Corruption Watch, headed by Teten Masduki and associated with former Finance Minister Mar'ie Muhamad. For an extensive list see Hamilton-Hart 2001: 69.

3 The Japanese were most intransigent in early negotiations. Holding 38 per cent of Indonesia's private debt, they were not in a position to easily write it off, especially given that their own banking system was in desperate straits. They were also well aware that most of the big Indonesian conglomerates had substantial funds and healthy assets offshore, sometimes in Japanese banks (*AWSJ* 12 October 1998: 11; 4 November 1998: 1, 7).

4 A restructuring deal with Hashim Djojohadikusumo's Semen Cinibong, for example, collapsed when it was revealed suddenly that US$250 million had disappeared from its books (*AWSJ* 16 August 1999: 3). Few negotiations were as bitter as those between Bakrie and his Korean creditors. In a letter cited in *FEER* (28 October 1999: 42), they attacked Bakrie for lacking 'any sense of fairness and transparency'. They called the company's debt restructuring proposal 'worthless' and described the prospects of its operating units as 'dismal' (*JP* 21 March 2000: 8). Nevertheless, Bakrie finally agreed to a deal that would see foreign creditors own 95 per cent of the Bakrie family's stake in Bakrie Brothers and five other firms in return for US$1.05 billion in debt (*JP* 29 August 2000: 1; *Republika Online* 29 April 1999). Although apparently signed, sealed and delivered, it is still unclear whether this agreement holds.

5 The long and complex saga of recapitalisation is covered in a wide range of literature, specifically the surveys of recent developments in the *Bulletin of Indonesian Economic Studies* (Johnson 1998; Cameron 1999; Pardede 1999). It is also covered in World Bank Reports (World Bank 1999: 2.1) The Jakarta press also followed events closely (*JP* 22 August 1998: 1; 21 October 1998: 1; 4 March 1999: 1). Another source is the international financial press (*AWSJ* 4 March 1999: 1).

6 See the comments of Bank Indonesia Deputy Governor, Anwar Nasution (*JP* 16 October 2000: 10). For IBRA, the prospect that recapitalised banks might need ongoing injections of funds was heightened when the Sinar Mas group was left with US$11 billion in debt as pulp and falling paper prices led to a collapse of the junk bonds that underpinned the forestry group, APP. IBRA was now forced to undertake a further recapitalisation of BII, the former Sinar Mas bank taken over by the government in a previous rescue (Dick 2001: 23).

7 The list of debtors to IBRA was topped by Marimutu Sinivasan with Rp.16.966 trillion, followed by a range of leading business figures, including Prajogo Pangestu, Liem Sioe Liong, Bob Hasan and Eka Tjipta Widjaja. Also prominent were Tommy Soeharto and other Soeharto children and associates as well as several *pribumi* business figures such as Aburizal Bakrie. The same figures also dominated lists of major debtors to state banks and Bank Indonesia (*FK* 12 March 2000: 16; *Asiaweek* 7 May 1999: 60).

8 Standard and Poor's estimated that the cost would rise to Rp.600 trillion (US$87 billion) or around 82 per cent of GDP by the end of 1999. This compares to Thailand's 35 per cent, South Korea's 29 per cent and Malaysia's 22 per cent (*AWSJ* 11–12 June 1999: 3).

9 These were Liem Sioe Liong's Bank Central Asia, Sjamsul Nursalim's Bank Dagang Negara Indonesia, Atmadjaja's Bank Danamon and Bob Hasan's Bank Umum Nasional and Bank Modern.

10 There were no shortages of examples. According to a McKinsay audit, Aburizal Bakrie's Bank Nusa Nasional received assistance even though 99 per cent of its loans were doubtful and it had a CAR of minus 210 per cent (*FEER* 19 August 1999: 12). A Parliamentary Commission on Banking and Finance (Komisi IX) investigation of BLBI to the Soeharto Bank, Utama, revealed a Rp.531 billion loan despite a CAR of less than 2 per cent (*FK* 12 March 2000: 12– 23; *Kompas Cybermedia* 25 February 2000).

11 In June 2000, an audit of BLBI by the Supreme Audit Agency (BPK) and the Government Finance and Development Comptroller (BPKP) reported that they could account for only Rp.63.6 trillion of Rp.75 trillion in BLBI they were investigating. They also found that bank owners had diverted Rp.62.6 trillion for purposes other than recapitalisation, a figure subsequently revised to over Rp.138 trillion – almost 100 per cent (*Kontan Online* 8 February 1999; 3 July 2000; *JP* 8 September 1998: 1; 9 February 2000: 8; 9 August 2000: 1, 12; *AWSJ* 16 August 1999: 1, 8).

12 For example, BUN loaned Rp.3.3 trillion and Rp.118 billion to two of its owners, Kaharudin Ongko and Bob Hasan. Bank Surya allocated over Rp.1.6 trillion, or 94 per cent of its total credits to its owner, Sudwikatmono. The Hokindo and Centris groups received loans amounting to 850 per cent and 1,000 per cent of the capital of Bank Hokindo and Bank Centris respectively. This was eighty-five and one hundred times the allowed maximum limit. BDNI loaned Rp.16 trillion, or 65 per cent of its total credits, to its owner, Sjamsul Nursalim, of which 95 per cent became non-performing loans (*Kontan Online* 18 January 1999).

13 Some groups were even able to offload their corporate debt onto the banks. For example, a report from the Centre for Banking Crisis, a private watchdog organisation, claimed that the Widjaja family offloaded foreign exchange losses onto the ailing Bank Indonesia International (*AWSJ* 16 August 1999: 1, 8).

14 The state-owned Bank Rakyat Indonesia, for example, suffered a loss of Rp.3.7 trillion, of which Rp.2.1 trillion were losses from credits allocated to various big conglomerates. Over 53 per cent of the Bank's Rp.8.4 trillion loans to large enterprises were categorised as bad credit (*Suara Pembaruan* [internet edition] 18 January 1999). In the private sector, the 100 largest borrowers from the ten private banks frozen by the government were revealed have received Rp.31 trillion or 65 per cent of the total credits (*Kontan Online* 18 January 1999). It was no surprise that the Soeharto family and the big Chinese conglomerates dominated the lists of unco-operative debtors published by IBRA in 1998 (*FK* 21 September 1998: 10–18).

15 Among them, economist Rizal Ramli (later to be Co-ordinating Minister for Economics in the Wahid government) and former Bank Indonesia Governor, Soedradjat Djiwandono, urged more decisive action in seizing assets of recalcitrant debtors (*JP* 24 August 1988). Habibie's Co-ordinating Minister for Economics, Finance and Industry, Ginandjar Kartasasmita, named several potential targets for seizure, including forestry tycoon Bob Hasan's paper manufacturer, PT Kiani Kertas and the Liem group's profitable companies, Indomobil, Indofood and Indocement (*JP* 8 September 1998: 8).

16 MPR Chairman, Amien Rais, for example, publicly asked whether IBRA was laundering the debt of Soeharto (*Kontan Online* 26 April 1999; *Jawa Pos Online* 14 April 1999).

17 These included Liem Sioe Liong, Sjamsul Nursalim, the Bank Modern group and Sudwikatmono.

18 IBRA official, interview 23 March 2001

19 The Salim group's Indofood operations, for example, made Rp.209 billion in foreign exchange gains in 1999 compared to a loss of Rp.1.2 trillion in the previous year, enabling it to repay $400 million in debt without having to sell the lucrative Bogasari and its flour mills as earlier planned (*AWSJ* 25–26 February 2000: 1).

20 Among others, Sjamsul Nursalim, Edward Soerjadjaja, Eka Tjipta Widjaja and Liem Sioe Liong were all investing heavily in Malaysia, Singapore, the Philippines and China (*Tempo* 16 July 2000: 103,104).

21 Many of these inflated assessments were made by international accounting firms (Lande 2000)

22 In December 1999, IBRA seized fourteen hectares of land belonging to Siti Hardiyanti Rukmana to cover over US$100 million in debts to the state bank, Bapindo. It also announced progress in scheduling repayment of Rp.2.98 trillion owed by the Napan group, owned by Henry Pribadi, Ibrahim Risjad and Sudwikatmono (*JP* 22 December 1999: 8). In January 2000 it filed a bankruptcy suit against PT Tirtamas Comexindo, a company controlled by Hashim Djojohadikusumo, brother of General Prabowo and brother-in-law of Soeharto's second daughter, Siti Hedijati Prabowo (*AWSJ* 4 January 2000).

23 At the request of the IMF, IBRA's Oversight Committee (KPPT) undertook reviews of several debt restructuring agreements. In the case of Chandra Asri, it concluded that the principle of equitable treatment had not been followed. 'Looking at it from a purely commercial point of view, the restructuring is clearly disadvantageous to IBRA and advantageous to Marubeni.' It added that the quality of the collateral could not be quantified and that Pangestu's company, PT Zillion, provided no financial projections and hence no comment on its capacity to repay was possible. In the view of KPPT, IBRA stood to recover more of its debt if the company were sold or liquidated (KPPT 2001; *JP* 29 August 2001: 1). KPPT also argued that no financial due diligence or independent review were perfomed on the restructuring of the debt of PT Permadani Khatulistiwa, a company owned by prominent *pribumi* business interests.

24 It was revealed that Texmaco had received US$754.1 million plus Rp.1.9 trillion in trade credits from BNI in late 1997 and early 1998 following the intervention of Soeharto (*JP* 30 November 1999: 1; 6 December 1999: 8). In March 2000, the government once again came to the rescue of Texmaco, with IBRA taking over Rp.19 trillion of its debts (14 per cent of its budget) and the state bank, BNI, extending a US$96 million credit facility (*Kontan Online* 20 March 2000; *JP* 18 March 2000: 8).

25 Of this amount, non-performing loans accounted for Rp.289 trillion, equity investments in nationalised and recapitalised banks amounted to Rp.130 trillion, and claims on former bank owners' assets were Rp.127 trillion (World Bank 2000: 13).

26 While IBRA's asset disposal record languished at 2.5 per cent, comparable agencies elsewhere had better rates, including South Korea's KAMCO (38 per cent), Malaysia's Danaharta (40 per cent), and Thailand's FRA (78 per cent) (*JP* 21 June 2000: 8).

27 The former Director of CIDES, Umar Juoro, argued that the old conglomerates were now under pressure as growing debt forced them to surrender core corporate assets such as Astra, Bank Central Asia, Bank Danamon and Indofood. Economic recovery would require the entry of foreign investors now demanding transparency and accountability in governance. He suggested that democracy had removed the grip that oligarchies formerly held over the state. For example, conglomerates were now on the back foot in the face of community demands for the return of land formerly seized – he cited the case of the Sinar Mas shrimp farm closed by local protesters. If they were to survive it would be under systems of global governance defined by the market. However, he added, if the old conglomerates managed to hang onto these assets or buy them back from the state, however, the old system might survive (interview 10 March 2000). Prominent businessman, Sofjan Wanandi, felt that the major conglomerates and those closest to Soeharto would not survive. He saw new business groups emerging free of debt and based on export industries (interview 21 March 2001).

28 The Bank reported that the recovery rate on retail and corporate loans disposed to September 2000 was 57 per cent, while the sale of BCA Bank shares yielded only 13 per cent of book value (World Bank 2000: 13). In reality, the real problem for IBRA

was that the sale of Bank Niaga and BCA depleted most of the assets in its control. Expected cash recovery of Rp.27 trillion for 2001 was considered ambitious (Pangestu 2001: 144).

29 The failed bid by Standard Chartered Bank for BCA had included as its local partner the Berca Group, headed by Moerdaya Po, which had long been associated with Liem. While the new Farallon bid included the cash rich cigarette group, Djarum, as its local partner, this did not stop speculation that Liem was once again back in BCA (*Tempo* 19–25 March 2002).

30 The corporate losses and debts carried by the big conglomerates were huge. Liem, for example, had lost Bank Central Asia and the Soeharto family had lost most of its lucrative state monopolies. When Hashim Djojohadikusumo was arrested in 2002 he had already been forced to surrender his holdings in PT Semen Cinibong with debts of US$1.2 billion, and had lost his five private banks. He carried debts of Rp.3.2 trillion in BLBI (*Tempo* 19–25 March 2002).

31 For example, Bambang Trihatmodjo's sugar trading alone is estimated to have earned him as much as US$70 million per year. Most recently, Liem Sioe Liong had secured a US$657 million contract to import rice in 1997 and over US$90 million of this was reportedly handed on to Soeharto's daughter Siti Hutami Endang Adiningsi (*Time.Com* 24 May 1999).

32 Lucrative contracts for aircraft leasing and purchase, insurance and cargo handling for the state airline, Garuda, estimated to be worth US$18.27 million and Rp.27.1 billion per year, were cancelled or placed under review while other contracts for port management, water supply and purification, and the construction of power stations and a giant transport terminal in Jakarta, were revoked and transferred (*FK* 21 September 1998: 67; *JP* 1 June 1998: 1; 4 August 1999: 8; *FEER* 13 May 1999: 10–14).

33 Reports by Price Waterhouse Cooper and state auditors confirmed excessive charges to Pertamina as a result of mark ups and overpricing in shipping contracts and refinery and pipeline construction. In particular, US$116 million in losses was attributed to the unusually high insurance premiums paid to its insurer, PT Tugu Pratama, a company 35 per cent owned by Bob Hasan's Nusamba Group. Another US$90 million was due to excessive shipping costs (*JP* 12 October 1999: 1).

34 But the Bulog experience also reveals the limits of reform. In the case of Liem's flour-milling monopoly, Bogasari, the long period of monopoly had been used to build a position of market dominance in processing. It was well placed to compete with new entrants. In place since 1967 and still receiving 80 per cent of Bulog's orders, Bogasari's profits are estimated at around Rp.200 billion per annum (*FK* 22 September 1999: 88–89).

35 See the comments of Arif Arryman, a Director of the private think-tank, Econit, and a close associate of Rizal Ramli (*JP* 1 April 1999: 8), and those of former Pertamina Director, Ibnu Sutowo, and former Finance Minister and PAN party figure, Fuad Bawazier (*JP* 26 February 1999: 1; 3 May 1999: 1).

36 An Arthur Andersen audit of Bulog revealed leakages of Rp.3 trillion between 1996/1997 and 1998/1999. The Government Finance and Development Comptroller (BPKP), was also reported to be tracking a huge figure of Rp.100 trillion, creamed off Bulog accounts in the period since 1982 and parked in the accounts of 116 officials (*Kontan Online* 28 February 2000). A BPKP report found that expenditure from Bulog's non-budgetary funds in the period 1998–1999 was about Rp.2 trillion with disbursements going to businesses controlled by the Soeharto family and others associated with former Bulog chief, Bustanil Arifin. At the time, Bulog's non-budget funds stood at Rp.395 billion (*JP* 13 June 2000: 9; 20 July 2000: 4).

37 These were spread throughout the state apparatus, although concentrated in the Department of Forestry, Bulog, the State Secretariat and the Finance Department. Continuing practices of over-invoicing and collusion in false documentation in deals between private companies and government agencies were identified (*Tempo* 25 June 2000:

26–31; *Jawa Pos Online* 7 July 1999; *Kontan Online* 29 May 2000; 10 July 2000; 17 July 2000; *Kompas Cybermedia* 18 July 2000). In the financial year 1999/2000, the Supreme Audit Agency, BPK, reported irregularities of Rp.209 trillion in the management of the state budget and state companies, mainly Rp.204 trillion of BLBI issued in contravention of regulations and without proper security. In the eighteen months to June 2001, the Government Finance and Development Comptroller, BPKP, reported that Rp.103 trillion of state funds had gone missing, including Rp.4.4 trillion from Bulog, Rp.3 trillion from Pertamina and Rp.1.3 trillion from the Finance Ministry (Siregar 2001: 298).

38 Among these were the bizzare 'Bulogate' episode in 1999 involving the appropriation of Rp.35 billion from the Bulog Employees Foundation, Yantera, by President Wahid's masseur, Suwondo, and Bulog Deputy Chairman, Saupan (*Kompas* 16 October 2000). Tanjung was accused of diverting Rp.40 billion of Bulog funds from poverty alleviation programmes to bankroll Golkar's campaign in the 1999 elections (*Jakarta Post.Com* 13 March 2002).

39 Others arrested included IBRA Deputy Chairman, Pande Nasorahona. Bank Indonesia Governor, Sjahril Sabirin, was later to be convicted in connection with the case.

40 Tutut was called as a suspect in the investigation of an illegal compensation claim against Pertamina made by her company, PT Triharsa Bimanusa Tunggal, involving the government in a loss of over US$17.5 million (*Tempo* 20–26 February 2001: 49). Probosutedjo was questioned over his receipt of Rp.144 billion from the Reforestation Fund (*Indonesian Observer* 21 December 2000).

41 Former technocrat, Ali Wardhana, was declared a suspect in a case involving the embezzlement of Rp.1 trillion involved in the recovery of debt from the timber company, PT Barito Pacific, by the state-owned investment company, PT Bahana Pembinaan Usaha Indonesia (*JP* 1 May 2001; *Tempo* 26 February 2001: 49).

42 *Tempo Online* 9–15 April 2002. Akbar Tandjung also remains under the cloud of an earlier scandal involving the diversion of funds from a dredging project – the Taperum scandal – while he was State Secretary under Habibie (*Tempo* 26 February 2001: 18; *FK* 12 March 2000: 72).

43 The need to replace 70 per cent of the judges sitting in the Jakarta courts in July 2000, including a number of Commercial Court judges, and to make new appointments to the Supreme Court, gave the Wahid government an opportunity to make personnel changes. Yet, of the eighty-four candidates for appointment to the Supreme Court, 75 per cent were from the corps of career judges (*Suara Pembaruan* 14 June 2000). Nevertheless, after parliament completed 'fit and proper' tests for candidates to the Supreme Court (only eight out of forty-six candidates passed (*Straits Times* 23 August 2000: 23)), seventeen new nominees were appointed, including a number of *ad hoc* judges.

44 That Taufik was involved in an attempt to influence Freeport Mining to return Bob Hasan's 4.7 per cent share in PT Freeport Indonesia to Indonesian hands illustrates the central role he plays at this level of negotiation. Among allegations of his role as a deal maker and fixer are claims that he was at the centre of arrangements to enable the Bakrie group to raise US$150 million for the purchase of a coal mine from the state pension fund, Jamsostek, and Bank Mandiri, despite the continuing poor performance of the Bakrie group (*Time* 15 July 2002).

45 These were: Bambang Subianto (appointed by Soeharto); Iwan R. Prawiranata and Glenn Yusuf, who served in the Habibie period; Cacuk Sudarijanto, appointed by Wahid to drive the process of restructuring and assets sales; Edward Gerungan; and finally I Putu Gede Ary Suta, appointed in June 2001 by Wahid.

46 These were to include a National Economic Council (DEN – Dewan Ekonomi Nasional), a National Business Development Council (DPUN – Dewan Pembangunan Usaha Nasional), and a Technical Assistance Team (TAT) (*Tempo* 26 March 2000: 102, 103).

9

REORGANISING POLITICAL POWER

Constructing a new regime

As we have seen, the fall of Soeharto and the dismantling of his highly centralised authoritarian regime did not mean a swift and frictionless process of change towards liberal modes of governance and markets. Indeed, as Soeharto's grip began to falter, it was evident that the elements that had been nurtured under his system of rule could survive within the framework of a new, more democratic, regime, albeit through new alliances and vehicles. Thus, it is the reorganisation of the old predatory power relations within a new system of parties, parliaments and elections, and within new alliances, which is the central dynamic of politics in the post-Soeharto era. But the construction of a new regime creates new sources of tensions and contradictions. It is also a process that is not easily controlled as the door is opened to a range of new contending forces. Some seek to entrench their ascendancy within essentially the same predatory arrangements – new business interests and political entrepreneurs aiming to displace the old – but now they appeal frequently to nationalist and populist sentiment and imagery. Others seek more wide-ranging changes: typically small groups of liberal reformers and social radicals. But it is far from certain that forces with an abiding interest in a more profound transformation of the social order can triumph. The contest thus far indicates instead the salience of those that seek to maintain or capture the old system; even as they necessarily modify it through greater decentralisation and diffusion of powers, and through more loose and fluid new alliances.

In the process, politics in Indonesia has clearly acquired a vibrancy and unpredictability absent during Soeharto's heyday. Given Soeharto's seeming inability or unwillingness to institutionalise the process of political succession, it was perhaps inevitable that such trauma would take place in any case, whenever his long period of rule ended. However, Indonesia's descent into economic crisis clearly precipitated the process and added a new dimension of volatility.

Most immediately important in the transformation of Indonesian politics has been the establishment of free and open elections. Indeed, the parliamentary elections of June 1999 – presided over by the Habibie government – were the first meaningful ones since 1955 to the extent that they signalled that a change in government was possible. Likewise, institutions such as parliament and political parties have become far more significant. A mere rubberstamp for decades, the DPR and MPR have become genuine arenas of political contestation, while parties which previously were tightly controlled and orchestrated have developed as the main vehicles to promote the interests of contending social and political forces. Meanwhile, regional elites and institutions, including provincial parliaments, are emerging as important players as central state authority erodes. Contests over power now permeate down to the local level.

A series of reforms were to reduce the powers of the Presidency. This was best reflected – given the long rule of both Soekarno and Soeharto – in the new limit of just two five-year terms for any President. Moreover, an amendment to Article 20 of the 1945 Constitution – which would not have been possible under Soeharto – paved the way for an MPR decree that stated that bills passed by the legislature would have to be made law within one month regardless of whether the President had given approval.[1] Demonstrating the legislature's new found stature, Abdurrahman Wahid – elected President through an intricate process of wheeling and dealing among political elites to be discussed below – was to be summoned by the DPR to explain the sacking of two Cabinet Ministers barely nine months into his term (*Tempo* 9 July 2000: 28–29, *FK* 9 July 2000: 84–85). Later, he was to suffer the ignominy of a parliamentary investigation into corruption, eventually leading to his own ousting from power.[2] Such events would have been unthinkable during Soeharto's long period of rule.

More importantly, the position of the MPR – the state institution theoretically wielding the greatest formal authority – has been enhanced in practical terms. Presidents would now have to deliver 'progress reports' to annual sessions of the MPR – at which time their position could be under threat (*Kompas* 17 April 2000; *Republika* 18 April 2000; 24 May 2000; *JP* 18 April 2000), because these yearly meetings could, conceivably, develop into impeachment proceedings. Previously, by contrast, Soeharto had turned the five-yearly MPR meetings stipulated by the Constitution into mere rituals that legitimised his rule.

In spite of these momentous changes, the end of the New Order has not necessarily signalled a complete break with the past. Indeed, as we argue, contemporary Indonesian politics displays some remarkable continuities with the Soeharto era, especially in terms of the contending interests and forces reconstructing the political framework. Most striking is the salience of old and new forces with an interest in maintaining a system of arbitrary power and predatory markets formerly guaranteed under his rule. The problem for reformers is that the re-emergence of new oligarchic alliances may easily be accommodated within a form of democracy characterised by money politics and extra-legal appropriation of state power. To paraphrase Lenin, radical reform would necessitate not only the refusal of new forces 'to live in the old way', but also the inability of entrenched ones to 'live in the old way' (Lenin, quoted in Skocpol 1979: 47).

There can be little doubt that the New Order brand of corporatist authoritarian rule has unravelled, in spite of some lingering fears of a reactionary military coup, as unrest continues in post-Soeharto Indonesia. But a return to old-style corporatist authoritarianism is not a serious prospect, and even the military would have to deal with political actors in fundamentally different ways than those used under the New Order. Instead, the complex and fluid coalitions of oligarchy nurtured under Soeharto's rule are seeking to reconstitute the relations of power and the institutions of predatory capitalism within new political arrangements. These must now operate within a system of elections, parties and parliaments, rather than through the overt use of state repression.

There are now essentially two broad scenarios with regard to oligarchic power. First, significant sections of the New Order oligarchy could continue to survive on the basis of new alliances and money politics and thus reconstitute within a new, more open, and decentralised political format. Second, although oligarchy and predatory forms of power survive, old forces may be swept aside by new coalitions of political entrepreneurs and business interests, many of them regional and local. In reality, both the old and new intermingle and overlap in everyday politics in newly powerful political parties and parliaments, at both the national and the local level. A complicating factor, however, are the demands of external forces such as the IMF and the World Bank, whose influence over the process of change in Indonesia cannot be underestimated. It is largely due to the presence of such institutions as virtual domestic political actors – given their authority over economic policy making, especially since the economic crisis – that that cause of liberal economic reform has remained alive today in spite of the chronic weakness of those espousing it within Indonesia.

As these scenarios unfold, new political coalitions are acquiring guises that are more prominently nationalist and populist than under Soeharto, and are characterised by shifts to money politics as well as the rise of more diffuse power centres. The formation of these coalitions involves the adoption of an ideological armoury that includes the egalitarian symbolism of Islamic struggle, as well as more secular expressions of populism and nationalism. It also involves the deployment of new strategies of selective mass mobilisation and political thuggery through the instrument of paramilitaries and political gangsters. Currently, paramilitary or civilian militia forces linked, directly or indirectly, to political parties are playing an increasingly prominent and controversial role in Indonesia, giving rise to widespread discussion about the 'militarisation' of civil society (*Gatra* 21 August 1999; *Kompas* 10 May 2000; 11 May 2000; *JP* 11 May 2000; *Detik.Com* 16 June 2000).

At the same time, the position of the military itself is ambiguous. Some have suggested that the rise of paramilitary forces is directly related to the growing general distrust of the military-proper. As one prominent general once commented: 'If I were rich and did not trust the police and military, I would hire my own security guards. If people or groups do not trust the authorities, they tend to protect themselves through their own means.'[3] Nevertheless, sections of the military itself intend to ensure that it remains a major force in politics, notwithstanding often-stated intentions to revamp *dwifungsi*. It is instructive that just before the MPR convened to elect a President in October 1999, the military tried to force through legislation to transfer sweeping powers from the civilian administration in the event of a national emergency situation (see Government of Indonesia 1999a, 1999b). In spite of the fact that the new piece of legislation was in many respects less draconian than the legislation it would have replaced, it was widely regarded as a grab for power (*FK* 9 July 2000: 18–19). At the same time that talk of military withdrawal from politics had been commonplace, some senior officers continued to defend a military role in the MPR (*Kompas* 27 May 2000), although they were ultimately unsuccessful.

Nevertheless, there are signs that the old pacts of dominance are fracturing. Business in general can no longer just look to powerful centres of politico-bureaucratic power within the predatory departments of state to provide protection and monopoly. They must now form coalitions with parties and parliaments that compete to assert control over crucial gate-keeping institutions. As money politics and parties and parliaments consolidate as the primary vehicles of alliances that vie for control over state institutions and resources, the role of the military as bodyguard has become somewhat redundant. Chinese businesses, in particular, have, in any case, learnt that far from guaranteeing security and stability, the military may be a major incendiary for xenophobic and racist violence, as seen in May 1998.

Therefore, the military's continuing involvement in political life no longer focuses on protecting oligarchic power, as it did in the late Soeharto years. Nevertheless, like the military in post-Soviet Russia, they have an interest in the institutions of predatory capitalism. For instance, under present circumstances, the state budget is estimated to cover only 25 per cent of military operational costs. Thus, the military must maintain control over businesses that in the past have provided considerable extra-budgetary, non-transparent, funding (*JP* 14 June 2000). It is notable that the attempts of one crusading general, the late Agus Wirahadikusumah, to clean up the business practices of his elite Kostrad unit were thwarted as he was relieved from command (*Tempo* 7–13 August 2000).

Reforming political institutions

The promise of more free and fair elections in Indonesia in 1999 gave rise to hopes of a fairly smooth transition from authoritarianism to a liberal form of democracy that would signify a complete break with the past. Indeed, the relatively trouble free parliamentary elections of June, followed by the MPR session which elected Wahid in October 1999, appeared to lend credence to such hopes. But the advent of a new period more prominently emphasising electoralism, parties and parliaments must be understood from the vantage point of the inheritors of a crumbling regime struggling to ensure their survival. It must also be understood in relation to the strategies employed by a variety of interests to assure their ascendance in a changed social and political environment created by the fall of Soeharto. It is now, therefore, to the contest over reforming political institutions, and the interests involved, both old and new, that we must turn our attention.

A necessary starting point is an analysis of the reforms undertaken by the Habibie government. As already mentioned in Chapter 7, the formulation and passage of the set of new laws on politics were central to Habibie's plan for an 'orderly' and 'constitutional' political transition. Clearly, Habibie and his allies did not enjoy the option of resorting to authoritarian repression. Thus, a gamble had to be taken by opening up the electoral process and recasting Soeharto's successor as an ardent political reformer. Elections could have provided the Habibie government with the claim to legitimacy that it deperately lacked. But this required a careful crafting of the rules.

A key feature of the first phase of Habibie's plan of controlled, gradual reform was the freeing up of previous constraints on forming new political parties. Within months of Soeharto's fall, dozens of new political parties had been established, although in obvious contravention of the principle of a three-party system dominated by Golkar established under then-still-existing laws. By some counts, 180 or so political parties had been declared by early 1999, although out of the nearly 150 that registered, eventually only forty-eight fulfilled requirements to contest elections in June that year (*Kompas* 23 February 1999; 6 March 1999).

The paradox was that Habibie was also in no position to abandon entrenched forces. He had no choice but to rely on Golkar – one of the main pillars of the *status quo* – as an institutional power base, notwithstanding ultimately serious challenges to his authority within the party itself. Control over Golkar – and the ICMI – remained useful to Habibie, for it provided the opportunity to further cement a potentially useful alliance of the wealthy and powerful nurtured by the New Order with a disparate range of forces attracted to the ideological appeal of Islamic populism. In essence, the political machinery of Golkar, and the wealth with which it was endowed, was an invaluable resource for any bid to provide electoral legitimacy to Habibie's rule. Thus, Golkar remained a major political prize, in spite of its association with the depredations of the Soeharto era.

It would be a mistake, however, to draw the reformist/anti-reformist divide in terms of competition between Golkar and other parties. In fact, newer parties were also well populated by a variety of elements – political entrepreneurs and fixers, business and bureaucratic interests, both central and local – that were all a part of the vast network of political patronage that was the New Order, albeit sometimes ensconced only in the second or third layers of that network. For such interests, parties and parliaments were now the main avenue to political power and control over state institutions. Thus, among the major parties were dispersed different concentrations of old oligarchic forces, along with an array of newly ascendant secular nationalist or Islamic populist groups variously emphasising statism or social justice appeals, as well as small bands of reformist liberals.

Social radicalism, not surprisingly, remained largely the domain of the small, newly resurrected PRD, which was ill equipped in any case to be a serious contender for power.

In other words, most of these parties are not 'natural' political entities, carrying out 'aggregating' and 'articulating' functions, but constitute tactical alliances that variously draw on the same pool of predatory interests. Notwithstanding certain ideological schisms within and between parties, their function has primarily been to act as a vehicle to contest access to the spoils of state power.

Among the most important of the parties to emerge were the PDI-P (PDI-*Perjuangan* (Struggle)), the renamed, much larger, section of the PDI that continued to recognise the leadership of Megawati; the National Awakening Party (*Partai Kebangkitan Bangsa* – PKB) sponsored by Abdurrahman Wahid; and the National Mandate Party (*Partai Amanat Nasional* – PAN), led by Amien Rais. While all three figures – Megawati, Wahid and Rais – were widely hailed as reformers, the political vehicles they developed, as we shall see, would host a range of interests nurtured under the New Order, while providing room for the rise of new political fixers and entrepreneurs, nationally as well as locally.

It is also important to note that Pancasila's privileged status was downgraded during the initial phase of the Habibie plan of gradual reform. Although Soeharto had regarded the establishment of the *azas tunggal* as one of his most important political achievements (see Soeharto 1988), the November 1998 MPR session annulled a previous decree on its propagation. Political parties were also effectively allowed to proclaim adherence to any ideological stream – with the vague caveat that it did not contradict Pancasila.[4] This condition should be read primarily as an affirmation of the outlawed status of communism. Though the change provided an opportunity for new parties to develop distinctive political and economic platforms, it was instructive that few, aside from some emphasising their adherence to Islam, were to do so.

The loosening of previous requirements on adherence to Pancasila particularly heartened Muslim political activists, who in the past had been angered that their organisations had to replace Islam as founding principle with a mere secular creed.[5] This bolstered Habibie's standing within sections of organised Islam, the members of which were among the first to rally in support of his presidency.[6] Significantly, Habibie's support from some of the more militant Muslim groups, such as KISDI,[7] led to the widely perceived gulf developing between Islamic political forces and more secular nationalist opposition groups mainly rallying behind Megawati Soekarnoputri.

Furthermore, although the 1985 law on mass organisations – which stipulated that each group in civil society be 'represented' by a single corporatist organisation – had not been removed, new mass associations such as that of labour also began to be formed and recognised (Hadiz 1998; Ford 2000). Clearly such a new penchant for liberal reform constituted a bid to distance the Habibie government from Soeharto's. The liberalisation of state policy regarding press publication permits (*JP* 14 September 1999) – a hugely popular move – was also intended to achieve the same effect.

The second, and perhaps more critical, phase in the Habibie plan involved formulation and passage of new political laws that would govern the holding of fresh parliamentary elections. Indeed, by late 1998 public debate on political change was to centre mostly on electoral reform, especially as the idea of a transitional presidium disappeared from the demands of most activists except for the more militant student groups. As discussed earlier, leaders such as Wahid, Rais and Megawati were by then concentrating on developing political parties to contest elections scheduled for mid-1999. Given the heavily controlled electoral contests of the past, debates over electoral reform threw up a host of highly contentious issues not confronted before. A key issue in the debate concerned the number of appointed or indirectly elected members of the MPR – the body that would eventually elect a President. It will be recalled that Soeharto had previously handpicked more than half of the MPR's 1,000 members, which then comprised 425 elected DPR members, seventy-five military appointees, and 500 appointed regional and functional group representatives. With such a background, there was a natural concern among many that as many MPR seats as possible should be contested through elections.

Originally, under proposed new legislation formulated by a team supervised by the Ministry of the Interior, fifty-five appointed seats would have been reserved in the DPR for the military. In spite of representing a downsizing of the seats the military occupied in the previous parliament, that number of seats was widely considered disproportionate to the actual size of the military forces. Moreover eighty-one 'regional representatives' were to be elected by the regional parliaments formed by the 1997 elections dominated by Golkar, on top of sixty-nine 'functional group' representatives to be appointed by an array of still existing corporatist organisations, most of which previously had strong links to the former state party. Thus, out of the new 700-member MPR envisaged, up to 205 members would have notionally been under the control of potential supporters of Habibie, regardless of the results of general elections. This would have placed Habibie in a good position to win Presidency, if he could just maintain control of Golkar as well as assure the support of the military (Hadiz 1999).[8] As we shall see, however, Habibie struggled to do either. The laws as they eventually were passed, moreover, were quite different from the drafts initially proposed (see Government of Indonesia 1999c, 1999d, 1999e). Even Golkar legislators, instructively, were to pick serious quarrels with draft legislation produced by Habibie's Team Seven (*Kompas* 24 November 1998).

For example, in contrast to the proposal of this team –which favoured a single member constituency system – legislators eventually agreed upon a rather unique variant of the proportional representation system.[9] This meant that politicians did not have to develop independent bases of local support and would continue to rely on the national party apparatus, an arrangement that apparently suited the interests of a wide range of party bosses. For one thing, this set-up potentially allowed central party bureaucracies to exert considerable authority over local branches, in spite of the latter's large formal role in nominating candidates. The criteria having to be met in order to enter the elections were also made more lenient than originally proposed: a party required nine provincial branches and sub-branches in half of the *kabupaten* of these provinces, in contrast to the fourteen provincial branches initially stipulated. This allowed for a much greater number of parties to enter the elections than envisaged by Team Seven.

Another contentious issue in the debate on electoral reform specifically concerned the role of the military in politics. Under the new laws as passed in January 1999, the military were still given thirty-eight automatic seats in the 500-member national parliament (DPR), rather than the proposed 55, and 10 per cent of seats in the provincial and sub-provincial DPRD. While accommodating some of the public criticism over the continuing political role of the military, such numbers were still sufficient to conceivably allow it a great deal of bargaining power, especially as no single party was likely to garner an absolute majority in the national parliament (see Government of Indonesia 1999c, 1999d, 1999e).

The composition of the all-important MPR was also revamped. Under the new legislation, the President was still to be elected by a 700-member assembly, but by one consisting of 500 DPR members plus 135 Regional Representatives (five per province), and sixty-five Functional Group Representatives to be appointed by an Electoral Commission comprising government and political party representatives. Although the thirty-eight military appointees in the DPR meant that only 462 seats would effectively be contested in the June parliamentary elections, the mechanism for appointing regional and functional group representatives was made more open. For example, provincial level parliamentary elections were to be held simultaneously with national elections – unlike the proposal of Team Seven – so that regional representatives to the MPR could be appointed by fresh provincial legislatures (Government of Indonesia 1999c, 1999e). However, even this process would not go untainted by allegations of blatant money politics.[10]

Yet another thorny problem was the role of the bureaucracy in general elections, given its overt support of Golkar in the past. A new regulation prohibited civil servants from holding political party positions. This signalled an important change (*Kompas* 19 February 1999), for it meant that government officials were made to choose – at least temporarily – between holding to their bureaucratic office or their position in Golkar or other parties.[11] The change, of course, mostly affected Golkar, given its previous stranglehold over the bureaucracy. Interestingly, had the extrication of Golkar from the bureaucracy been undertaken earlier, this would have indicated a significant further development in the transformation of Golkar from the party of the state apparatus to the party of rich and powerful interests nurtured by state power. But the context within which the change was made reflected more clearly mounting popular pressure to eradicate the obvious sources of Golkar's past electoral supremacy.

The developments discussed above were all an integral part of the shaping of a new political format under which party and parliamentary elites occupy an increasingly strategic role in putting together coalitions of interests. But the rise of political parties did not necessarily reflect the greater capacity of society for self-organisation. A report by the revived newsmagazine, *Tempo*, suggested that up to ninety of the new parties initially established were bankrolled by the Soeharto family or its associates. The aim was apparently to influence the course of elections, and the MPR session later in 1999, that would elect a President and Vice-President (*Tempo* 15 February 1999: 22–23), or to help secure a deal for the Soeharto family, in the face of intense pressure for prosecution now confronting the Habibie government. One such party was the Republic Party, which boasted a platform that included *not* prosecuting Soeharto for alleged ill-gotten wealth, and included Pemuda Pancasila personnel and associates of 'Tutut' Soeharto among its members (*FEER* 27 May 1999: 20–22).

Within this new political format, political parties were set to become the main instrument through which power coalitions were built and alliances forged over the allocation of political and economic resources among economic and political elites. Why then, did this format prove so successful in maintaining the insulation of elites from reformist pressure, but most strikingly from workers and radical organisations? First of all, groups such as labour and the peasantry, among the most marginalised under Soeharto, proved unable to establish themselves in the political process. No significant social reformist party was to emerge. This was partly due to their own weakness and the legacy of their disorganisation during the New Order, but also a reflection of how old elites have been better placed to capture state institutions and gain access to state resources. Thus, not all social forces and interests could equally exploit all the new political opportunities. Workers and peasants were involved only as support bases for parties such as the PDI-P and the PKB during elections, drawn by populist appeals that had little reformist substance in real policy terms. It is this absence of a significant role for labour and workers' organisations in Indonesia that has distinguished it from some earlier democratisation experiences elsewhere (see Rueschemeyer *et al.* 1992; Therborn 1977).

It is in this sense as well that the contemporary Indonesian experience can be compared to that of Thailand and the Philippines. In both cases, as in Indonesia, democratic reforms have clearly resulted in greater public scrutiny being placed on the exercise of power. New, less easily manipulated forces are also unleashed in the process of reform. But as Hewison has pointed out, the democratic reforms in Thailand at the end of the 1980s can be seen less as a mechanism of popular sovereignty than as tools of powerful new political entrepreneurs to appropriate state power and capture networks of pork barrelling. Amongst the middle classes supportive of reform, there was, in Hewison's view, a fear of the 'dark forces' unleashed by formal democracy. Similarly, in the Philippines, Anderson argued that *cacique* democracy emerging in the wake of Marcos's fall has served primarily to reinstate the rule of powerful families and entrench oligarchy, patronage and money politics (Hewison 1993: 159–190; Anderson 1990: 33–47).

The new vehicles

The next phase in the Habibie plan of transition was the actual holding of parliamentary elections in the lead up to an MPR session to finally elect a President. On the surface the major contestants offered a choice between 'reform' and the *status quo*, though quite naturally, all were to adopt the rhetoric of the former and to distance themselves as much as possible from Soeharto's New Order. Even Golkar was to do this. As Golkar parliamentarian Ade Komaruddin claimed, the 'past did not matter', and moreover, the 'new parties were not identical with [the values of] reform'.[12]

In reality, elements that had been a part of the vast New Order system of patronage, albeit only in the fringes, were well represented in all the major parties. As mentioned earlier, the major parties are constituted by different concentrations of old politico-bureaucratic, military and business forces, secular nationalist or Islamic populist interests, as well as smaller numbers of reformist liberals. A prominent member of PAN, for example – and an alleged source of the party's financing – has been Fuad Bawazier, an old Soeharto ally and his last Minister of Finance.[13] The PDI-P has provided room for Arifin Panigoro, businessman and ally of Ginandjar Kartasasmita. Moreover, Chinese business groups, as noted later in this chapter, appear to have entered into new, though fluid, alliances with some of the major parties.

At the same time, for political 'fixers' and entrepreneurs, local notables and smaller business people, who had formerly attached themselves to Soeharto's juggernaut, party and parliamentary politics have become avenues to more direct political power. Even so-called labour parties were patronised by New Order stalwarts: two were established by individuals with links to the former state-backed official union, the FSPSI, while another was co-founded by the businessman Ibnu Hartono, a member of the Soeharto family. Not surprisingly, virtually all the parties expressed a commitment to end corruption and to democratic opening, but none ever really developed a concrete policy agenda on such critical issues as market reform, labour policy or the revamping of the legal framework. This is not surprising. Political parties have been less a vehicle to advocate contending policy agendas than machines for the capture of the terminals of patronage.

As was widely expected by many observers, the big electoral winners were the PDI-P, Golkar, PKB, PAN and the PPP, the New Order's 'Muslim' party. Only twelve other parties, out of forty-eight overall, gained any representation in the DPR. Among these were the Islamic-oriented PBB (Crescent and Star Party) and PK (Justice Party), and the pro-Megawati PKP (*Partai Keadilan dan Persatuan* – Justice and Unity Party), made up of disenchanted former military officers, bureaucrats and Golkar politicians. Amongst the electoral failures were the four ostensibly labour parties, including the National Labour Party formed by SBSI leader Muchtar Pakpahan, none of which won seats in the DPR, but also the People's Sovereign Party (PDR), formed by then Minister of Co-operatives Adi Sasono (who was also general secretary of ICMI), and appealing to Muslim support through populist slogans.

As was also widely expected, the PDI-P won the largest share of the vote, taking advantage of its image as a past 'victim' of New Order repression – although it was eventually to fail in installing Megawati as President. Elected as Vice-President instead, she was only to ascend to the Presidency in July 2001, as Wahid became increasingly estranged from members of the loose coalition that had preferred him. In spite of being almost universally vilified for representing the *status quo*, Golkar still managed to exploit its organisational and financial strength to come a distant second to the PDI-P, a relatively good result considering it was never going to win the 74.5 per cent attained in 1997. Meanwhile the PKB won the third highest number of votes but only emerged with the fourth largest number of seats (after the PPP) in the DPR. Its Java-centric support base was clearly disadvantaged by the national proportional representation system utilised, though its main sponsor, Abdurrahman Wahid, was later to win the Presidency. The other big winner was the PPP which, like Golkar, also benefited from a pre-existing political machinery. Among the major parties, PAN was to do quite poorly, despite its potential support base among the 28 million-member Muhammadiyah, which was once led by Amien Rais.

In spite of the representation of similar forces and interests in several of the major political parties, an important source of schism to emerge was along the lines of secular nationalism and Islamic populism. This was perhaps most clearly embodied in Golkar, which was characterised by parliamentarian Ade Komaruddin as containing competing factions, including 'reformers' and 'compromisers'.[14]

But in reality the most essential internal conflict has been between the ICMI elements brought into the party in the early 1990s, which had amassed considerable political influence since then, and the more secular, bureaucratic elements that had been its more traditional mainstay. The secular nationalist tradition in

Table 9.1 Performance of the big five political parties in the June 1999 elections

Party	*Percentage of votes*	*Number of DPR seats*
PDI-P	34	154
Golkar	21	120
PPP	11	58
PKB	13	51
PAN	7	35

Source: *Jakarta Post.Com* 27 July 1999 (http://www.thejakartapost.com:8890/elec99.htm)

Golkar and elsewhere has generally emphasised centralised bureaucratic rule and national consensus on policy, while Islamic populism tends to emphasise more local and small-scale economic and political interests with frequently strong xenophobic elements. The schism within Golkar was best reflected in the eventual abandonment of Habibie's Presidential election bid by the more secular nationalist allies of Akbar Tandjung, the party chairman.

In spite of these ideological issues, the contest over Golkar was important primarily because it signified that the party remained crucial to the oligarchy cemented under the New Order. It had a proven role in the allocation of patronage and in the forging of political, bureaucratic and military alliances down to the local level. Moreover, the party's largely intact machinery down to the villages, as well as its financial resources, made it a force which opponents could not ignore in the lead up to parliamentary elections.

In spite of its eventual electoral setback, Golkar was organisationally and financially the most prepared party to face the June 1999 parliamentary ballot. Its war chest reportedly stood at US$70 million in 1998 (McBeth 1998: 31), which compared favourably to most of the new parties that were dependent on a mere Rp.1 billion (US$125,000) in government subsidy. By mid-1999, the former state party was 'set to spend an average of 1 billion rupiah per district, or more than 300 billion rupiah in total' (*FEER* 6 May 1999: 26). Significantly, the electoral laws did put a cap on individual and corporate donations to parties, but failed to stipulate limits on party expenditure, thus favouring the party that already had the most resources at its disposal (see Government of Indonesia 1999c). In any case, as it turned out, the auditing process also stipulated in the legislation was never to be carried out with a great deal of seriousness, thus allowing for all kinds of financial irregularities during the elections to go unpunished.

Moreover, Golkar's resources were supplemented by its ability to make use of government facilities and funds, at least as alleged by its critics. Indeed, Golkar along with the pro-Habibie party, the PDR, was widely accused of misusing foreign aid earmarked to alleviate poverty (*JP* 27 May 1999; *Kompas* 3 June 1999).[15] Notably, Akbar Tandjung was later to be embroiled in a scandal involving the siphoning of Bulog funds to finance Golkar's campaign, even to the point of being convicted of graft (*Jakarta Post.Com* 5 September 2002). Nevertheless, it was significant that the party could no longer rely on direct financial support from Soeharto's Yayasan Dakab. Like the other *yayasan*, Dakab was transferred from the hands of Soeharto and his family into the hands of the government-proper due to intensified public scrutiny of their financial arrangements (*Kompas* 26 November 1998; 30 December 1998).

The search for yet more funds to finance campaigning was to entangle Golkar in more scandal, but demonstrated the many options the party had at its disposal to fill up the war chest. During investigations into the Bank Bali case, it transpired that the party had received Rp.15 billion in illegal donations from Marimutu Sinivasan, a garments industrialist who was also the party's deputy treasurer (*Kompas* 8 November 1999). Indeed, although big businesses were by this time hedging their bets by supporting different parties, Golkar's past links with established entrepreneurs were widely seen as ensuring it a continuing source of financial support (*Kompas* 29 March 1999).[16]

Moreover, in spite of mass defections,[17] the party still appeared to be in a position to use the bureaucracy for its own purposes – though less blatantly than before – especially in remoter areas where many of the newer parties did not have a strong foothold (*Kompas* 11 February 1999). A good Golkar result would undoubtedly have been helpful to most local bureaucratic elites that had been nurtured through the patronage networks that Soeharto built. Many of these local elites were rightly worried that *reformasi* would strike them down, as scores of village heads across Indonesia quickly found as angry mobs reportedly forced them to resign (Wagstaff 1999).

Golkar, however, was not without serious problems. It had been the ultimate symbol of the *status quo* under the New Order, and hence invited derision from a large section of the voting population.[18] As mentioned earlier, it was internally rocked by conflict between secular, bureaucratic elements led by Akbar Tandjung, and pro-Habibie elements, mostly entrenched in ICMI. Although Tanjung's own elevation in Golkar was partly owed to Habibie, he subsequently emerged as a strong rival, making full use of his experience as a seasoned party and state apparatchik.

Significantly, some Golkar stalwarts had rejected Habibie and found a new home in the PDI-P and the PKP, a minor party formed by former generals and bureaucrats. Former Golkar functionary, Jakob Tobing, for example, represented the PDI-P in the Indonesian Election Committee, and Golkar politicians such as Sarwono Kusumaatmadja and Sutradara Gintings had joined the PKP, formed by retired General Edi Sudrajat upon failing to wrest control of Golkar from Habibie and his allies.

The internal contradictions within Golkar were demonstrated during extended wrangling over the naming of a Presidential candidate before and after the June 1999 elections. Habibie supporters gained the clear upper hand when they were able to assure his nomination in a National Leadership Meeting in May 1999 – primarily over Akbar Tandjung – during which money politics was allegedly pervasive. The respected newsmagazine *Tempo* (24 May 1999: 23), for example, reported the occurrence of large-scale bribery of party regional delegates taking place on Habibie's behalf. But this ascendancy was not long lasting. The MPR session of October 1999 saw Habibie losing control, with about forty Golkar parliamentarians under Akbar Tandjung believed to have crossed over to the camp which rejected the President's accountability speech, effectively helping to bury his chances of re-election (*Kompas* 1 November 1999; *Tempo* 25–31 October 1999: 18–19).

With its electoral defeat in June and developments in the MPR session in October 1999 – which elected Wahid and Megawati as, respectively, President and Vice-President – Golkar's future has become uncertain. The party's ICMI wing was quashed with the sidelining of its champion, Habibie, and the marginalisation of allies such as activist Marwah Daud Ibrahim (*Kompas* 8 April 2000), while secular nationalist elements at least temporarily supported the Wahid–Megawati government in return for Cabinet seats. What is clear, however, is that Golkar's former role as the most effective vehicle to protect, nurture and incubate oligarchic interests has all but ended, and that it now has to compete with other parties in offering patronage and access to state resources.

Golkar was not the only party within which tensions and contradictions were embodied. PAN, for example, has been characterised by a serious rift between traditional Muhammadiyah followers and more secular, liberal intellectual reformers that had embraced Amien Rais. They did so initially because of his willingness to set up a party that was non-exclusively Muslim and because of his support for a relatively unambiguous political platform, which included exploring the option of a federalist Indonesia. This division was exemplified in differences between chairman Rais and one-time secretary general Faisal Basri, the liberal economist who, without Rais' strong ICMI connections, tended to eschew alliances with conservative Muslim forces. Other major figures in PAN have included such diverse individuals as A.M. Fatwa, the Muslim activist firebrand once gaoled by Soeharto, liberal academic Arief Arryman and the late Christian theologian/NGO activist Sumarthana (*Kompas* 1999: 486–487). Significantly, Faisal Basri and his allies were eventually to leave the party in frustration.

That such a diverse group could coalesce in PAN initially seemed to signal a new phase in Indonesian politics with a major party being established free of the primordial attachments which were the basis of many of the major parties in the 1950s. It should be noted, however, that in some ways PAN is an expression of the 1950s alliance between the Masyumi, the old party of the 'modernist' wing of Islam, and the intelligentsia-based Indonesian Socialist Party (PSI). Moreover, in spite of PAN's diversity, at least at the level of central leadership, it was from the Muhammadiyah community that overwhelming support was expected to derive. Indeed, support for PAN emanated from the urban Muslim middle class, small merchants and traders in Java, as well as large portions of the predominantly Muslim islands such as Sumatra – in other words old Masyumi bases of support. For such elements PAN was a vehicle through which they would contest access to state power and resources, especially given the long decline of the Muslim petty bourgeoisie during much of the New Order.

Nevertheless, PAN's 'intellectual' wing appeared quite influential at first. Compared to the other major political parties, PAN offered a fairly radical and concrete political platform, which would not have been possible without the presence of its more liberal intellectuals and activists. Besides consideration of federalism, the platform included amending the constitution as well as eliminating the military's political and economic role (*Kompas* 1999: 488–494), ideas which subsequently became more commonly expressed by others.

But it was unreasonable to expect that the influence of the liberal intellectuals would last. Rais presides over a traditionally conservative Muhammadiyah constituency guided by the vision of a kind of capitalist populism – geared to redress wealth imbalances in favour of *pribumi* (indigenous) Indonesians. As mentioned earlier, some of these, especially prior to the establishment of ICMI, had long resented their virtual exclusion from contests over state power and resources. Moreover, Rais' ICMI connections, in spite of his reputation as a Soeharto opponent, do link him closely with some of the most salient elements of the New Order political elite. These include the crafty, aforementioned Fuad Bawazier, whom parliamentarian Alvin Lie suggests had proved 'useful to the party', especially as an erstwhile 'Middle Axis operator'.[19]

Tensions and contradictions already within PAN came to the fore in the lead-up to the October MPR session, with Rais plotting a political revival after disappointing election results by turning to his natural allies among Muslim-based parties. He was to take the initiative in establishing the so-called 'Middle Axis' of Muslim parties to bolster his bargaining position *vis-à-vis* Megawati and Habibie, the two Presidential front-runners at the time, going even so far as to embrace erstwhile rival, Abdurrahman Wahid (*Kompas* 21 July 1999; 9 August 1999). Such a move constituted an abandonment of the more liberal and secular wing of PAN that had been attracted to it because of a non-exclusively Muslim platform.

Rais' partners in this 'Middle Axis' included such parties as the PPP, the PBB (Crescent and Star Party) and the PK (Justice Party). All these parties had leaders who had once been nurtured by Soeharto and Habibie through such vehicles as ICMI. Legal scholar Yusril Mahendra, for example, whose main claim to fame was that he was speechwriter to former President Soeharto, is chairman of the PBB. By Mahendra's own admission, the PBB's campaign fund was bankrolled by a Habibie donation of Rp.1 billion (*Kompas* 29 April 1999).[20] The party also included among its ranks Fadli Zon, a former student activist allied to Prabowo Subianto, and Eggi Sudjana, leader of a newly formed Islamic workers grouping, the PPMI, and KISDI leader Ahmad Sumargono, all of whom have strong ICMI associations.[21] Moreover, the PK was founded by employees of Habibie's BPPT, although its grassroots support derives from middle class Islamic study groups and student organisations such as KAMMI (*Kompas* 1999: 399–400; Mietzner 1999: 91).

But it was not altogether surprising that Rais should favour an alliance with Muslim forces rather than the liberal intelligentsia. Besides ICMI connections, they shared an interest in halting the rise of secular nationalist forces arrayed behind Megawati Soekarnoputri.[22] It is important to emphasise that the issue was not only one of political ideology, though this was stressed in much of the rhetoric of the time. There would have been genuine fears that a coalition of forces focused on the PDI-P could threaten the increasingly privileged position (in relation to state and bureaucratic power) enjoyed by those nurtured within the ICMI patronage network in the late Soeharto era.

The PDI-P also displayed deep divisions that suggested that the party was largely a tactical alliance of a range of otherwise disparate forces. Its own major divide was between the nationalist politicians who had been nurtured by the old PDI or PNI, military and Golkar officials who had lately joined the party and had a vested interest in a strong, centralised bureaucracy, and a small group of liberal intellectuals. In contrast to the latter, the first two groups were suspicious of free markets, with old PDI/PNI stalwarts particularly happy to deploy the populist and nationalist imagery and jargon of Soekarnoism. But tensions between these two have also been brewing, as the old PDI/PNI loyalists have had to accept with some frustration what they view as the disproportionate influence of come-lately Golkar/military recruits.[23] Indeed, PDI-P parliamentarian M. Yamin suggests that the old PNI/PDI elements – which are the 'heart of the party' – harbour resentment over the quick rise of 'newcomers' to positions of power.[24]

But it was essentially the aura of Soekarnoism that guaranteed the PDI-P a huge following among the urban poor, workers and sections of the peasantry – support which, together with that from much of Indonesia's religious minorities – secured the PDI-P's electoral victory. Interestingly, however, in spite of such support the PDI-P has had no organic links to any lower class social movement. Its position on labour, for example, has not been clear at all, confined to vague comments about 'protecting workers as a special and humane [*sic*] factor of production' and the intent to develop a 'social security system without the excessiveness occurring in Western Europe' (PDI-P 1999: 15).

Also, in spite of Megawati Soekarnoputri's popularity among the poor, labour and peasant organisation representatives have remained absent from the PDI-P leadership. This representation was perhaps deemed unnecessary, given the assessment by parliamentarian M. Yamin that the party 'did not see mass support for [labour leader] Muchtar Pakpahan'.[25] Instead, PDI *satgas* – among the most high profile of the current array of civilian militias – have reportedly been hired by industrialists to quell labour unrest from time to time (*Tempo Online* 3 May 2000).[26]

Rather than being based in labour movements or other grassroots organisations as were the European labour parties of the nineteenth century, the PDI-P leadership at the end of the Soeharto era was centred firmly on politicians of the old PDI, such as Sabam Sirait, Aberson Marle Sihaloho, Alex Litaay, Subagyo Anam and Taufik Kiemas – Megawati's husband. They were supplemented by military officers led by retired General Theo Sjafei (regarded as an old Benny Moerdani ally) and a few middle ranking Golkar refugees such as Jakob Tobing, Potsdam Hutasoit and Arifin Panigoro, a businessman with close links to former Soeharto economic minister, Ginandjar Kartasasmita (*Tempo* 3 October 1999: 25).

Some of these leaders already represented the old PDI in Soeharto's parliament while a few – like Anam and Kiemas – are middle level entrepreneurs, respectively in the forestry and petrol businesses. Liberal intellectuals such as economist Kwik Kian Gie and banker Laksamana Sukardi, together with academics such as Mochtar Buchori and Dimyati Hartono, and regular *Kompas* columnist Theo Toemion, co-existed somewhat uneasily within the party and provided it with professional and intellectual credentials. However, like the intellectuals in PAN, some were eventually to be marginalised as well. The experience of PAN and the PDI-P demonstrates that the liberal urban intelligentsia, as well as the working class, have no effective party vehicle of their own.

Significantly, the internal make-up of the PDI-P clearly has much in common with the secular nationalist wing of Golkar represented by Akbar Tandjung and Marzuki Darusman. In spite of Tandjung's background in the Muslim student association, the HMI, he was a chief opponent within Golkar of control by Habibie's ICMI wing. Not surprisingly, top PDI-P official Arifin Panigoro assessed an alliance with Golkar as the 'easiest' to establish (*Detik.Com* 17 July 2000)

Moreover, in spite of its origins as a pro-reform, anti-Soeharto party, the PDI-P is, ironically, in many ways a natural successor to Golkar as incubator and protector of the oligarchic project. As already observed, the internal composition and political base of the two parties are strikingly similar but for the Islamic populism embraced by the ICMI wing of the party which had thrown its support to Habibie. That the PDI-P comes replete with a small band of liberal intellectuals seemingly out of place in a bastion of statist-nationalism is itself reminiscent of the position of Soeharto's old economic technocrats within Golkar.

It should be recalled that before the New Order, the Soekarno-linked Indonesian Nationalist Party (PNI) was the party of the *pamong praja*, the civilian bureaucracy. Thus, the idea of benign bureaucratic rule is firmly ingrained in the PDI-P, its effective successor. The PDI-P has also inherited the organicist idea of unity between ruler and ruled which was central to Soekarnoist romanticism, and later adopted and propagated in a more bureaucratic form by Golkar ideologues from the days of Ali Moertopo. Once installed as Vice-President, Megawati herself was to make remarks about the need for the unity and consensus that are so important to this organicist vision (*Kompas* 1 November 1999).

Equally importantly, the party's stated interest in establishing 'an autonomous industrial structure' (PDI-P 1999: 13) ensures a continuing emphasis on a dominant state role in the economy which, in the context, opens up possibilities for the emergence of new political, economic and bureaucratic alliances aspiring toward oligarchic status. Not surprisingly, the party has been opposed to the replacement of Pancasila as state ideology, and has been very reluctant to support major changes to the 1945 Constitution, which stresses such a dominant state role (PDI-P 1999: 9–10).

Given its similarities to Golkar, it is not inconceivable that the PDI-P is attractive to some sections of the New Order oligarchy seeking new allies, and indeed some major business support was quickly demonstrated for the PDI-P. Some press reports suggest that Indonesian–Chinese businesses were counted among the PDI-P's strongest supporters (*FEER* 6 May 1999: 26). It may also be significant that allegations of illegal donations by the Lippo conglomerate to the party were widespread before the October 1999 Presidential elections (*Republika* 30 August 1999). The picture that emerges is of a Chinese bourgeoisie no longer trusting the political protection and patronage that Golkar is able to offer and looking for other partners, including the PDI-P.

However, some of the internal contradictions within the PDI-P were to come to the surface as well after it had to resign itself to its failure to install Megawati as President in 1999. The party congress in March 2000, for example, saw the ousting of many of its liberal intellectuals from key positions and the growing stranglehold over the party of the ambitious politician-businessman Taufik Kiemas. Filmmaker and journalist Eros Djarot, a long-time confidante of Megawati, was one victim of the Congress (*Tempo Online* 8 March 2000), along with Buchori and Hartono, who lost his position to Panigoro as head of the PDI-P faction in the DPR. Laksamana Sukardi, in the meantime, was to be unceremoniously dropped from the Wahid Cabinet (*Kompas* 25 April 2000), with the apparent consent of Megawati, as was Kwik Kian Gie (although these two were to re-emerge in Cabinet positions as Megawati claimed the Presidency in 2001). Such developments clearly demonstrated the growing dominance of the mainstream populist and bureaucratic stream within the party over that of the liberal intellectuals.

Conflict within the PDI-P has, moreover, also surfaced at the local level. Controversy raged within the party, for example, when it failed to win the mayor's office in Medan, in spite of being the largest faction in the city's legislative body. As it transpired, PDI-P members in the legislature had been bribed to vote for another candidate (*Kompas* 22 March 2000). Greater turmoil within the party was only averted due to the unifying factor of Megawati herself.[27] The power of her influence was demonstrated time and time again, including when her personal intervention curbed potential violent infighting between supporters of different officials within the party's East Java branch in July 2000 (*Kompas* 11 July 2000). But in time even her aura of unassailability was to suffer badly – many within her party broke ranks when, as President, she initially supported unpopular increases in the prices of fuel, electricity and telephone services in early 2003, as Indonesia continued to struggle out of the quagmire of economic crisis under her administration.

Of the major parties, the PKB had initially appeared to be the most solid and relatively homogenous, and much of this was due to the overwhelming personal authority of its sponsor and unofficial leader, the charismatic Abdurrahman Wahid.[28] This was to change however, as key lieutenant Matori Abdul Djalil abandoned him in July 2001 in support of parliamentarians that were then seeking his impeachment. Though not a Muslim party, the PKB relies on the NU's impressive network of patronage that extends down to rural villages, especially in Java, underpinned by local notables, *pesantren* (religious schools) and *kyai* (religious leaders). The NU's membership is even larger than its more geographically dispersed rival, the Muhammadiyah, commonly estimated at between 30 and 35 million. The PKB's platform is nationalistic – with much emphasis on maintaining unity and integrity – but quite vague, apart from conveying a strong appeal to populist and egalitarian sentiment. Significantly, it also firmly takes the position that 'The Unitary State of the Republic of Indonesia based on Pancasila is the final form of the Indonesian nation' (*Kompas* 1999: 420).

However, the PKB was not immediately supported by all sections of the NU. Indeed, rival parties emerged from the NU such as the PKU (*Partai Kebangkitan Ummat*) and PNU (*Partai Nahdlatul Ulama*), which espouse a more strictly defined Islamic base. Leaders of these fringe parties – who typically had ICMI connections – were considered sympathetic to Habibie's election bid (*Tempo Online* 26 May 1999), and criticised the alliance with Megawati to which Wahid and the PKB appeared committed prior to the holding of the MPR session in October 1999.

But all of this apparently ended with the ascension of Wahid to the Presidency, which has placed those associated with the PKB and the wider NU community in a good position to involve themselves in the jostling over control of state institutions. A sign of the NU's influence was the instalment of Rozy Munir, a close Wahid confidante, in the position of Minister of Investment and State Enterprises (*Kompas* 29 April 2000), replacing the PDI-P's Laksamana Sukardi – though a subsequent Cabinet reshuffle later dissolved the post.

Significantly, amongst the civilian militia forces today, the PKB and the NU-linked *Banser* have the highest profiles. Involved intimately in the 1960s in the mass slaying of communists and alleged communist sympathisers, the *Banser* repeatedly flexed its muscles in order to protect the Wahid Presidency and the interests of the PKB/NU. (*JP* 10 May 2000). Like the PDI-P's paramilitary wing, the *Banser* is also allegedly involved in the quelling of labour unrest on behalf of industrialists.[29] The combination of access to state power and resources, and well-developed civilian militia capable of effective political thuggery, made the PKB a formidable force among the array of coalitions involved in the contests to construct a new political regime in Indonesia, especially during the brief Wahid Presidency.

The limits of reform

Pact of mutual protection: the Wahid presidency

As mentioned earlier, Wahid's ascension to the Presidency in October 1999 was the result of an intricate process of bargaining among political elites. The spectre of violent conflict between sections of organised Islam and supporters of Megawati Soekarnoputri – and therefore the threat of wider social disturbances – provided the impetus for a compromise among political elites. Indeed, unrest was widely believed to be the potential outcome of either a Habibie or a Megawati victory. Wahid was acceptable to a degree to both camps, and so, as he openly admits, his election was designed no less to avert 'a civil war'.[30] As Golkar leader Marzuki Darusman put it in September 1999, the 'parties have to be aware of the possibility of a new radical movement, or people power. So we, the parties, have to sit down and talk together, negotiate with each other' (quoted in Bourchier 2000: 30). However, what glued together the divergent interests that supported the late Wahid bid for the Presidency was not just an abiding interest in maintaining stability, but more fundamentally, one in safeguarding a new political regime within which new alliances primarily expressed as political parties would be dominant.

Thus, Wahid's surprise election in many ways represented a pact of mutual protection. More than just social disturbance, a Habibie or Megawati victory was feared, for example, to be the catalyst for military intervention in the name of restoring order, a development that would have seen party elites subordinated to military interests.

The drive to form some kind of pact of mutual protection among elites was perhaps best indicated by the more or less equal division of the political prizes among most of the big winners of the parliamentary elections. These prizes were divided among the so-called 'Middle Axis' parties led by Amien Rais, and Golkar, the PDI-P and the PKB, (*Kompas* 1 November 1999; *Tempo* 25–31 October 1999: 18–19; *Asiaweek* 29 October 1999: 25–36). Rais emerged as Speaker of the MPR, and Tandjung as Speaker of the DPR, while Wahid and Megawati were respectively elected to the Presidency and Vice-Presidency (the latter by defeating the PPP's Hamzah Haz).

The process was not without its casualties. The main losers were obviously Habibie – whose ambitious election plans were dashed – and his ICMI allies within Golkar. In turn this defeat was caused, as mentioned earlier, by the defection of about forty Golkar MPR members, who joined ranks with those rejecting his accountability speech – rebuffed by a slim margin of 355 to 322 votes (*Asiaweek* 29 October 1999: 28).

In spite of the persistence of his supporters, Habibie's position was in reality already seriously weakened prior to the MPR session by various developments. First was the Bank Bali scandal, which implicated close associates of Habibie, such as Supreme Advisory Council chair, Baramuli. Second was the independence vote in East Timor in August 1999, which Habibie had allowed, and which cast him as the villain who made possible Indonesia's 'loss' of East Timor (*Tempo* 3 October 1999: 22–23).[31] More importantly, the Tandjung camp within Golkar had begun to sense that its political survival could only be ensured by participation in a wider pact of mutual protection that would necessarily include Wahid, Megawati and Rais. Thus, like Soeharto before him, Habibie had become a liability.

The process that culminated in the Wahid victory, however, was also not particularly smooth. Megawati's failed Presidential bid, for example, momentarily pushed Jakarta to the brink of renewed mass rioting (*Agence France Presse* 20 October 1999: 40). It also involved intermittent last minute Presidential challenges by Akbar Tandjung and Yusril Mahendra, as well as vacillating by Wiranto on a Vice-Presidential bid,[32] demonstrating the intensity of the wheeling and dealing behind the scenes.

The embodiment of the eventual political compromise among elites was of course the so-called National Unity Cabinet that was announced soon after Wahid's victory. Constructed on the basis of negotiations between Wahid, Rais, Wiranto, Tandjung and Megawati, with each nominating members (*Kompas* 30 October 1999), it demonstrated another division of the political prizes among salient elites.

Amien Rais – whose Middle Axis alliance was so instrumental in stitching together a front to elect Wahid – was a big winner in this process. Amongst his associates represented in the Cabinet was the relatively unknown Bambang Sudibyo, an academic who was catapulted to head the Ministry of Finance. Another academic with links to PAN was Yahya Muhaimin, who became Minister of Education. Moreover, the wider Middle Axis was represented in the Cabinet by the appointment of the PBB's Yusril Mahendra as Minister of Laws, the PK's Nurmahmudi Ismail as Minister of Forestry, and the PPP's Hamzah Haz as Social Welfare Minister – though he would be among the first victims of Wahid's constant Cabinet reshuffling (*Kompas* 1 December 1999).[33]

The PKB – like PAN – a party that attained relatively modest results in the June parliamentary elections, emerged as another winner in the horse-trading process, clearly due to its hold on the Presidency. Wahid's associates in the new Cabinet were to include Alwi Shihab, who took over the Foreign Ministry, and political scientist Muhammad Hikam, who became Minister of Research and Technology. Though defeated in the Presidential poll, the PDI-P was embraced, not only through the ascendance of Megawati to the Vice-Presidency, but also through the appointment of Kwik Kian Gie as Chief Economic Minister and Laksamana Sukardi as Minister of Investment and State Enterprises.

Two pillars of the New Order, Golkar and the military, were also not to be left out in the cold. The Akbar Tandjung wing of Golkar was rewarded with such appointments as Marzuki Darusman as Attorney General and former FSPSI chairman Bomer Pasaribu as head of Manpower. Interestingly, a carrot was handed to Habibie's supporters within Golkar by the appointment of Jusuf Kalla as Minister of Trade and Industry, though he would be another early reshuffling victim. Generals Susilo Bambang Yudhoyono and Agum Gumelar, both of whom had been cultivating a reformist image by appearing in a host of democracy-related public meetings and television programmes, were respectively handed two strategic portfolios – the Departments of Mines and Energy and of Communications. The former was to be given a broader brief subsequently as co-ordinating minister in charge of politics and security.

The end product of this pact of mutual protection was to sustain one important tradition of the New Order – the absence of an effective political opposition. Essentially, the opportunities presented by access to state institutions and their resources remained too enticing for the major political forces to take an 'outsider' position. Thus, the basis for the lack of a formal political opposition was not cultural, but decidedly material. In spite of this, the National Unity Cabinet proved to be tenuous and contradiction-ridden. Although most evident in the form of Wahid's constant reshuffling of his Cabinet, the actual problem was the often-feverish contest for control over strategic state institutions.

Thus, yet another MPR session the following year – in which Wahid's 'progress' report was roundly criticised by the parliamentary representatives of most of the major political parties – spelled the beginning of the end. It eventually led to a parliamentary investigation into the President's role in the so-called 'Bulogate' and 'Bruneigate' scandals. These allegedly involved Wahid's mobilisation of political funds by siphoning off money from Bulog's employees' fund (Baswir 2000), and the receipt of an illegal donation from the Sultan of Brunei (*Adil* 15 June 2000).

Dogged by the threat of impeachment, Wahid faced another equally intractable problem: his inability to exert control over the executive arm of government, its ministries and bodies, especially given a hostile legislature and a military jealously guarding its own turf. Wahid, for example, is regarded to have failed in promoting reformers to key positions within the military, but not before causing acrimony with top generals. In addition, he confronted problems with a corruption-ridden judiciary, which led him to reshuffle the Supreme Court (*Kompas* 21 March 2000).[34] The difficulties encountered in prosecuting Soeharto and his family were to highlight the unwieldliness of the courts, the bureaucracy and the security forces, in which interests nurtured under the New Order continued to predominate, as well as Wahid's own lack of authority.[35] But it must be said that Wahid had the uncanny knack of compounding his own problems: he did his reputation and credibility no favour by attempting to shield such New Order tycoons as Prajogo Pangestu, Sjamsul Nursalim and Marimutu Sinivisan from prosecution (*Kompas* 20 October 2002).

The persistence of predatory politics: Megawati and beyond

The end of Soeharto's long rule over Indonesia has not seen an end to one of its defining themes: the appropriation of state power by its officials to further private interests. This clearly continues to be the main theme of Indonesian political economy, albeit in an environment which is more politically open and fluid. Such fluidity was demonstrated in the quickly shifting political alliances that brought Wahid to power and later ousted him in July 2001; and which stalled Megawati's rise to the Presidency, only to facilitate it subsequently.

What the end of Soeharto's rule has shown, furthermore, is the emergence of competing new coalitions of interests partly expressed in the form of new political parties. As discussed earlier, these reflect alliances of both old interests that aim to maintain their position in a new political environment, as well as newly ascendant ones intent on entrenching themselves within essentially the same predatory arrangements. Relative newcomer Alvin Lie, for example, a PAN parliamentarian with a medium-scale business background in Central Java, admits that he has little in common with Muslim firebrands in the party such as A.M. Fatwa, and that parties are mere 'tactical alliances'.[36] The contest is thus not primarily about agendas or even ideologies, but about furthering dominance over the institutions of the state, and their resources and coercive power, and about developing networks of patronage and protection.

In this context, political survival increasingly depends on the ability to play the game of money politics, which necessitates the constant mobilisation of political funds. Wahid's involvement in money scandals was one of the clearest demonstrations of this, as were his tentative attempts to protect a number of New Order tycoons. In a sad reminder of the limits to reform, Wahid – who had built a personal reputation as a reformer – was drawn irreversibly into the logic of such a system. It is no surprise that his successor, Megawati, another politician with reformist credentials, has since been embroiled in controversy over the use of non-budgetary and non-transparent 'Presidential Aid' funds (*Tempo Online* 22–28 April 2002), as well as the alleged business and political designs of her husband.

It is important to note that the contest over state power, and for control over state institutions and resources, has not been confined to coalitions of interests operating in the capital city of Jakarta. Developing their own systems of patronage down to the local level, coalitions expressed as political parties have also competed intensely over control of local machineries of power. Thus, as we shall see, the election processes of some mayors and *bupati* (regents) have been particularly controversial, filled with allegations of widespread bribery (*Kompas* 22 March 2000; 17 April 2000; *Tempo Online* 29 February 2000; *Detik.Com* 17 July 2000) and political intimidation.

There are reasons to strongly suspect that local interests, once part of the New Order's extensive network of patronage, are now also reconstituting (Hadiz 2003). For example, one preliminary study concluded that local political elites are now largely comprised of entrepreneurs who 'matured' under the New Order (IPCOS 2000). Schemes to bolster regional autonomy, including the introduction of legislation to meet rising regional demands for control over resources, demonstrate the enhanced bargaining position of local elites. It is important to remember that 'mini-revolutions' in the regions only rarely accompanied the fall of Soeharto, in spite of the reports about the sudden demise of numerous local village chiefs noted earlier. It is to this dimension of the reorganisation of power in post-Soeharto Indonesia that we now turn our attention.

The reconstitution of local power

It is notable that local political dynamics after the fall of Soeharto have mirrored those at the national level, especially in terms of the essential predatory logic. Contemporary developments in two Indonesian provinces – Yogyakarta, a major cultural and political centre, and North Sumatra, a major hub of both manufacturing and agricultural-based industries – provide ample reason to make such an observation.

In Yogyakarta, the PDI-P emerged victorious in the 1999 parliamentary elections. Of the six national parliamentary seats that represent the Special Region of Yogyakarta, two were PDI-P, while the remainder were equally divided amongst PAN, PKB, Golkar, and the PPP. More importantly, PDI-P became the dominant force in Yogyakarta's provincial parliament, controlling eighteen of the fifty-four seats. The same pattern was largely replicated in the various sub-provincial parliaments. Likewise in North Sumatra, the PDI-P emerged as the dominant force. It won ten of the twenty-four national parliamentary seats there, as well as thirty of the eighty-five seats in the provincial parliament, thereby emerging as the strongest faction. It also controlled no less than 228 of the 690 seats in the various sub-provincial legislatures, leaving Golkar trailing a distant second with merely 145 seats.[37]

As in Jakarta, it is useful to understand political parties in Yogyakarta and North Sumatra as the vehicles of emerging coalitions of interests, older and newer, forged in battles to secure control over state power and its resources. Again, the demarcation lines at the local level are rarely between 'reformist' and pro-*status quo* forces, for these will intermingle within individual party vehicles. As in other places in Indonesia, the authority and power of local state institutions have been significantly enhanced with the erosion of central state authority – formally 'acknowledged' by the implementation of laws in January 2001 geared to decentralise administrative and fiscal powers to sub-provincial governments.

But the prospect of formal decentralisation immediately gave rise to questions about local corruption and the emergence of petty official fiefdoms. The fear of the emergence of local bossism in particular led one provincial parliamentarian in Yogyakarta to suggest that 'opportunists' will be especially interested in controlling sub-provincial legislatures and governments, where unbridled corruption can now grow.[38] In North Sumatra, some local legislators admitted that corruption became an increasingly serious problem in local state institutions soon after local autonomy became imminent.[39]

Furthermore, under the present Indonesian electoral system, local legislatures became crucial sites of political battles during elections for *bupati* and for mayor, thus providing fertile ground for the proliferation of money politics. In Yogyakarta this was witnessed in the election of the *bupati* of Sleman, which involved contending forces unabashedly deploying both bribery and intimidation in Parliament House. Indeed, allegations of beatings, kidnappings, the use of paramilitary organisations, and even bomb threats were pervasive.[40] In North Sumatra, the election of the *bupati* of Karo was a particularly ugly process, which involved the mysterious burning of the local parliament building.[41] Another notable case involved the aforementioned debacle for the PDI-P in the city of Medan, the capital of North Sumatra. Controversy raged within the party when its official candidate – a long-time bureaucrat named Ridwan

Batubara – failed to win the mayoralty, in spite of the party's strong position in the city's legislative body, because party members reportedly accepted bribes to vote for a local businessman, Abdillah (*Kompas* 22 March 2000).

What these examples show is a situation in which local state institutions are emerging as a site for the auctioning of powerful positions and the distribution of political largesse. Such an observation is important to decipher much of what is happening in the immediate post-Soeharto period. Like Anderson's study of the significance of political murders in Thailand in the 1980s in relation to the rise of parliaments in that country (Anderson 1990), the fact that so much effort is now being invested to gain control over local offices in Indonesia is clearly indicative of their growing value.

It is important to recognise that naked force has an important role too in the new political format. In Yogyakarta, 'Islamic' paramilitary groups have been at least as ubiquitous as that of the *satgas*, or paramilitary wing, of the politically ascendant PDI-P. Groups such as *Gerakan Pemuda Ka'bah* (GPK), loosely linked to the PPP, are prominent, as are *Front Pembela Islam* (FPI), which allegedly involves co-operation between several Islamic-oriented parties, including the nominally secular PAN.[42] In general, party-linked paramilitary organisations or civilian militia frequently function as goons when these parties need to flex their muscles – especially during election time. It is significant also that there have been allegations about the links of some civilian militia to underworld activities.

In North Sumatra, for example, protection rackets, illegal gambling, prostitution and the like still appear to be the domain of surviving New Order 'youth' organisations such as Pemuda Pancasila, and the powerful Ikatan Pemuda Karya. In the past, these have been widely feared state-backed organised crime outfits (see Ryter 1998). It is significant that a number of these organisations' members currently occupy local political offices, and that some have migrated from Golkar to other parties, including PAN, the PDI-P, and the PKP.

Indeed, the dynamics in North Sumatra are illuminating. Of the twenty-two *bupati* and mayors winning elections there since *reformasi*, at least six have backgrounds as local entrepreneurs, thereby showing the growing attractiveness of direct bureaucratic power to individuals engaged in business.[43] Not surprisingly, a number of these businessmen/politicians simultaneously have links with old New Order youth/gangster organisations. Most of the other local electoral victors are long-time bureaucrats, indicating a strong degree of continuity with the New Order. Given the wider context, there now appear to be great opportunities for the emergence of local oligarchies fusing local business, bureaucratic and business interests in diffuse, predatory, networks of patronage.

The military and the balkanisation thesis

As constantly shifting coalitions of predatory interests continue to compete over the spoils of state power and the capacity to develop networks of patronage, both nationally and locally, the Megawati government confronts, like Wahid's before it, the problem of growing separatist sentiment, especially in resource-rich Aceh and Papua (Irian Jaya). With Acehnese and Papuans clamouring for East Timor-style independence referendums (supported by at least some local officials and notables), and the East Kalimantan legislature calling for a federal state structure (*Kompas* 11 November 1999), addressing longstanding grievances remains a priority. But Aceh and Papua are not East Timor: their secession would spark real fears of the balkanisation of Indonesia, and thus no less than all-out war broke out in Aceh in 2003. Disconcertingly for the nationalist Megawati, separatist and federalist sentiments have also been expressed in oil-rich Riau (*Kompas* 16 November 1999). Moreover, in few other places has the government faced greater challenges than in Maluku, where Christians and Muslims have been involved in murderous sectarian violence since 1999 (*Tempo* 23 January 2000: 19–27; *FK* 9 July 2000: 12–17).

It is in this context that debate continues about the military's political role. The military has alternately been portrayed in the public debate as either the guardian of national integrity or the cynical instigator of local unrest, with some of its personnel being accused of actively fanning the flames in Maluku (Aditjondro 2000); for example, by covertly supporting militant Lasykar Jihad fighters transported from Java. In addition, the military is accused of holding on to control over lucrative, non-transparent business ventures, either to benefit particular individuals or to make up for budgetary shortfalls (*Kompas Cybermedia* 2 October 2002).

In spite of such allegations, the military was to capitalise on the United States' war on terror in the aftermath of the World Trade Center and Pentagon attacks of 11 September 2001, and also the bomb blasts at two nightspots popular with foreign tourists on the island of Bali on 12 October 2002. The latter caused nearly 200 deaths, including those of a large number of Australians. A result has been the push for a resumption of contacts between the Indonesian military and that of the United States (*Kompas Cybermedia* 19 April 2002), which had earlier been cut off because of scrutiny over Indonesia's human rights record. A similar resumption of controversial contacts between the Australian military and the infamous Kopassus soon also became a possibility. Indeed, Singaporean[44] and Australian[45] politicians were to express the view that the Indonesian military was the main bastion against anarchy and chaos – and a bulwark against Islamic extremism – in spite of widespread allegations of continuing military complicity in outbreaks of violence (*JP* 15 August 2002) and past tutelage of radical Islamic fringe groups (International Crisis Group 2002). Moreover, Megawati soon went out of her way to repair relations between the Presidency and the military leadership that were damaged during Gus Dur's tenure (*JP* 19 August 2002). A Presidential decree on anti-terrorism, pushed through just after the Bali bombing, reignited fears among human rights activists about the potentially sweeping powers that it allotted to the security apparatus – and invited disturbing comparisons with conditions under the New Order (*Jakarta Post.Com* 30 October 2002). It has been suggested that the military was shrewdly seizing the opportunity presented by the international anti-terrorism drive to enhance its own domestic political position.

The critics have a point. While constitutional amendments in 2002 (*Tempo Online* 11 August 2002), as well as proposed new political legislation, have effectively abrogated the formerly sacred principle of military representation in the DPR and MPR, it would be incorrect to assume that military political influence is exclusively exercised through the seats in national (and regional) legislative bodies, or through retired officers holding executive positions in the civilian bureaucracy. A major foundation of military capacity to pursue its own institutional interests is the territorial command system that extends from the capital down to the regions, districts and towns and villages. This command structure essentially establishes a military counterpart to the official civilian bureaucracy at each level, and enables officers to influence the daily affairs of government (Crouch 1999: 145–146) and even enter into political and business alliances. While such edicts as the one which compels active officers to choose between remaining in the military or concentrating on civilian bureaucratic careers (*Kompas* 31 March 1999) are important developments, the scrapping of the territorial command structure is more crucial.

It is because of the military's need to secure its institutional and material interests that the spectre of a 'Pakistan-style' solution – to endemic instability and the perceived threat of national disintegration – is a scenario taken seriously in some political commentaries.[46] Even the reform-minded late General Wirahadikusumah thought that the Indonesian middle class would support a military takeover should civilians fail to safeguard stability.[47] In some ways, such an assessment is supported by a poll that suggests wide public support for overriding military powers during national emergencies.[48] Apparently growing nostalgia in some quarters for New Order orderliness (*JP* 15 May 2002) might now cause some anxiety for coalitions of interests ascendant within political parties and parliaments, nationally as well as locally. Indeed, when outbreaks of violence take place, the military's political stock has tended to rise, domestically and, as mentioned earlier, internationally. This has naturally led to conspiracy theories about the military's role in fermenting unrest.

In spite of such real and possible military political manoeuvrings, reverting to a military-led, centralised authoritarianism would be tantamount to pushing the clock back too far. It is bound to be resisted, not least by those who have done so well in the post-Soeharto era so far by capturing and appropriating the institutions of Indonesia's new democracy. The more immediate challenge for the military is to safeguard its institutional and material interests in the context of a post-authoritarian Indonesia.

Notes

1 See Majelis Permusyawaratan Rakyat Republik Indonesia 1999.

2 Which finally arose ostensibly from his dismissal of the national police chief without parliamentary approval.

3 Kostrad Chief Lt. General Agus Wirahadikusumah, quoted in *JP* 26 April 2000.
4 See Majelis Permusyawaratan Rakyat Republik Indonesia 1998; Government of Indonesia 1999c.
5 Eggy Sudjana, chairman, *Persatuan Pekerja Muslim Indonesia* (Indonesian Muslim Workers Union), interview 20 November 1998, in which he declared that the new policy underscored the democratic credentials of the Habibie government.
6 The day after Habibie was sworn in as President, students occupying Parliament House were attacked by mobs, partly recruited from among toughs and the unemployed, that identified themselves as 'Muslim' by their garb and slogans. They were believed to have been organised by Habibie allies or groups that had enjoyed the patronage of General Prabowo Subianto, then commander of Kostrad. (See 'Eggi Sudjana dan Fadli Zon Pimpin Pemuda Mesjid dan Preman Serbu MPR/DPR', *SiaR* 22 May 1998.) For a less critical account of these groups see Ecip 1998: 149–152, where it is suggested that students who chanted 'hang Habibie' provoked the confrontation with the pro-Habibie group.
7 The Indonesian Committee for Islamic World Solidarity, KISDI, was initially formed to display solidarity with the plight of Palestinians, and later, Bosnian Muslims. It soon became a grouping of some of the most militant Muslim political activists.
8 This is on the basis of various drafts of the bill circulated in public for discussion.
9 Elections for (DPR) members were to be organised on a proportional representation basis by province. However, the law specifies that actual winning candidates are to be determined according to how parties perform at the district, rather than provincial, level.
10 For example, Herman Abdul Rachman, PPP member of the Yogyakarta Special Region parliament, interview 14 December 2000. He suggests that those who sought to be appointed Yogyakarta's delegate to the MPR offered large bribes to the legislature.
11 There were stipulations, however, about the possibility of later rejoining the civil service.
12 Ade Komaruddin, interview 19 December 2000.
13 Alvin Lie, PAN national parliamentarian, interview 21 December 2000, and Imawan Wahyudi, PAN legislator in the Yogyakarta Special Region, interview 11 December 2000. This is an apparently thorny issue within the party, as Bawazier's presence clearly compromises PAN's reformist credentials.
14 Ade Komaruddin, Golkar parliamentarian, interview 19 December 2000.
15 Led by Adi Sasono, it was comprised of former student, NGO and Muslim mass organisation activists. Sasono's patronage allowed the PDR access to bureaucratic privilege, specifically to the resources and networks down to the village and community levels directly under the control of his department. A latecomer even among new political parties – it was only established in January 1999 – the party was geared to bolster the position of Habibie, especially *vis-à-vis* his rivals within Golkar itself. The tabloid *Indikator* (17–23 December 1998: 3) reported on widespread allegations that the PDR had misused large amounts of funds under the jurisdiction of Sasono's Department.
16 One member of the Chinese business community in Jakarta suggested that a division had emerged between medium and small merchants who felt betrayed by Golkar (because they were not saved from the looting of May 1998), and inclined to support opposition political parties, and big conglomerates who still tended to support Golkar (personal communication 20 May 1999). The conglomerates of course had a vested interest in securing a government that would go easy on their bad debts.
17 Elements of the old Golkar had formed or joined a number of small parties, such as the Partai MKGR, made up of the old Golkar sub-organisation.

18 As witnessed in the destruction of Golkar floats in Jakarta during the official opening of the 1999 campaign period in 19 May 1998.
19 Alvin Lie, interview 21 December 2000.
20 Habibie himself denied this claim.
21 Yusril Mahendra was to find it difficult to maintain control over the party, which fractured into two competing groups in 2000/2001.
22 Thus many Muslim parties argued that Islam does not accept a woman as President (see *Jawa Pos* 27 April 1999).
23 For example, it has been suggested that new recruits such as (retired) General Theo Sjafei are particularly influential on party policies affecting the military. Subagyo Anam, PDI-P parliamentarian, interview 27 September 1999.
24 M. Yamin, interview 18 December 2000. He is regarded as a close ally of the influential Taufik Kiemas – Megawati's husband.
25 M. Yamin, interview 18 December 2000.
26 Some members of the New Order-sponsored organisation of thugs, Pemuda Pancasila (see Ryter 1998), have apparently also joined up with the PDI-P, perhaps even entering the ranks of its *satgas* (literally, task force). By June 1999, there were Pemuda Pancasila rallies in support of Megawati's presidential bid (*JP* 30 June 1999). This was ironic, given the alleged role of the organisation in the storming of Megawati's headquarters on 27 July 1996 (*FK* 9 July 2000: 40).
27 Indeed it has been suggested that that the contending factions within the PDI-P are mostly united by their awareness of a reliance on Megawati's popular appeal. Subagyo Anam, PDI-P parliamentarian, interview 27 September 1999
28 Ex-PPP figure, Matori Abdul Djalil, a somewhat bland politician, was the first official PKB Chairman.
29 Leaders of Solidaritas Buruh, Yogyakarta, interview 15 December 2000.
30 Remarks made at *Forum Rembug Nasional*, Bali, 1 July 2000.
31 A referendum was held in East Timor on 31 August 1999, in which its people voted overwhelmingly for independence. Habibie, who may have underestimated the popularity of the independence movement, had made the referendum possible.
32 Wiranto had earlier rejected an offer from Habibie to be his running mate, putting a decisive nail into Habibie's political coffin. Later, upon Wahid's victory, he announced that he was standing for the Vice-Presidency, only to withdraw once again.
33 Yusril Mahendra also only lasted until early 2001, before re-emerging in the Megawati government that succeeded Wahid's.
34 The DPR, interestingly, was to nominate Muladi as Chief Justice during this reshuffling. Muladi, Habibie's Minister of Justice, was rejected by Wahid, leading to another tense showdown between the President and parliament.
35 Soeharto, charged with misappropriating *yayasan* funds, was spared from prosecution when a Jakarta court dropped the case, on the grounds of ill health. Youngest son Tommy was convicted for corruption, but sparked a national manhunt when he eluded gaol.
36 Alvin Lie, interview 20 December 2000.
37 Data on Yogyakarta was provided by Ridaya Laode, while Elfenda Ananda tabulated the data on North Sumatra.
38 Interview with Herman Abdul Rahman, member of DPRD-I Yogyakarta for the PPP, interview 14 December 2000.
39 Victor Simamora, member of the North Sumatra provincial parliament for the small Partai Bhineka Tunggal Ika, interview 3 July 2001. He made headlines in local newspapers when he suggested that some of his colleagues had offered themselves for bribes in the tendering of projects. Also, O.K. Azhari, PDI-P member of the Medan municipal parliament, interview 5 July 2001.

40 Hafidh Asrom, businessman and defeated candidate for the *bupati*-ship of Sleman, interview 9 December 2000.
41 John Andreas Purba, PDI-P member of Karo sub-provincial parliament, interview 6 July 2001.
42 Syukri Fadholi, then head of the PPP faction in the Yogya DPRD, and now Deputy Mayor of Yogyakarta, interview 15 December 2000, and Herman Abdul Rahman, member of DPRD-I Yogyakarta for the PPP, interview 14 December 2000.
43 Amir Purba, Dean, Faculty of Social and Political Sciences, Islamic University of North Sumatra (UISU), interview 5 July 2001; and data kindly complied and supplied to me by Elfenda Ananda.
44 See the transcript of a speech by Senior Minister Lee Kuan Yew on the occasion of the 1st International Institute for Strategic Studies Asia Security Conference, Singapore, 31 May 2002, Singapore Government Press Release, Media Division, Ministry of Information, Communications and the Arts, at http://sg.news.yahoo.com /020530/57/2qiur.html
45 See the transcript of a joint press conference by Senator Robert Hill, Australian Minister of Defence, and Paul Wolfowitz, US Deputy Secretary of Defence, Shangri-la Hotel, Singapore 1 June 2002 at http://www.minister.defence.gov.au/Hilltpl.cfm? CurrentId=1558
46 See comments by political scientists Kusnanto Anggoro and Cornelis Lay in *Kompas* 12 April 2000. See also Anggoro 2000.
47 Interview 21 May 1999.
48 The poll appeared in the 9 July 2000 edition of *Tempo*, a news magazine widely-read among the urban middle classes. Though admittedly 'unscientific', the poll showed 67 per cent approval of the law among visitors of the magazine's website.

10

CAN OLIGARCHY SURVIVE?

The apparent descent into money politics and frequently brutal social conflict that have accompanied Indonesia's move away from authoritarian rule, and from the economic shadow cast by a highly centralised system of state power, may be seen simply as the pathologies of a volatile transition to a market economy and to liberal democracy. Just as capitalism in Europe and North America in the eighteenth and nineteenth centuries was characterised by chaos, corruption, vote buying, environmental destruction and the untrammelled power of private interests, the fall of authoritarian regimes, it might be suggested, requires a similar period in which institutions might be built to tame 'savage capitalism'. However, what is taking place in Indonesia, we have argued, is the reorganisation of the power relations incubated within the Soeharto regime, rather than their fundamental transformation. The disorganisation of civil society, the co-option of Indonesia's capitalist and middle classes into a system of predatory politics, and the violent destruction of working class politics, have prohibited the emergence of coherent liberal and social democratic coalitions and forces from the ashes of centralised authoritarian rule. But some disagree with these rather gloomy predictions. They argue that democracy itself, however flawed, will provide the institutional opportunities for progressive politics, and that the inexorable discipline of global markets will ultimately force economic life in Indonesia into the constraints of regulatory capitalism.

Can democratic constitutionalism transform Indonesia?

As we have seen, the post-Soeharto era was welcomed by a broad range of academic and political observers as the beginning of a new, benign era of markets and democracy in Indonesia. Among the more optimistic observers was the sociologist, Arief Budiman, who argued in an essay on Indonesia's democratisation process that 'Democracy, in spite of backlashes, is inevitable for all countries' (1999: 41). While acknowledging possible obstructions, he suggested that both the internal as well as the external conditions for a democratic triumph were favourable in Indonesia – these respectively included the rise of public awareness of democracy, the rise of professional organisations, the tarnished image of the military and the 'growing global capitalism that rejects authoritarian political systems' (Budiman 1999: 56–57). In a similar vein, Van Klinken wrote that Indonesia was not only experiencing the 'decay of authoritarianism but the early stages of democratic transition', a process 'difficult to reverse'. It was a process, he noted, that confirmed, 'the now standard view in the literature on democratisation is that it proceeds from decay of an authoritarian system, through transition, to consolidation, and finally maturation' (Van Klinken 1999: 59–60).

These authors were correct in their observations that the apparatus of centralised authoritarian rule could no longer be sustained. They were also right to the extent that the disintegration of authoritarian rule presented opportunities for a wide range of interests formerly operating at the margins of the New Order to force their way into the political system. But less clear were the larger implications of market and democratic reforms for the entrenched systems of power, and whether they really represented a more deep-seated social and political transition. As it became obvious that many of the old politico-business families and conglomerates were surviving the protracted and contested struggles to restructure debt and deal with corruption, important questions emerged about the nature of the markets and regulatory systems that were being constructed. At the same time, Indonesia's new democracy failed to eradicate many of the basic features that had defined politics in the Soeharto era. It seemed that the new entrants were less the tax-paying middle classes demanding accountability and representation long-anticipated (Sjahrir 1992) than new predators seeking a share of the action. Nor did democratic politics bring with it the unambiguous transition to rule of law, human rights and accountability that many had hoped for.

In particular, the performance of the government of President Wahid, who had solid credentials as a reformer, dealt a blow to the hopes of the more optimistic observers of Indonesia's 'transition to democracy'. But it was not only Wahid who was drawn into the mire of money politics and into accommodation with those old interests entrenched in the state apparatus and the world of business. We have now also seen that predatory politics is being reassembled under President Megawati at the same time that she has been forced into a deepening accommodation with the military. Not surprisingly, even the most enthusiastic analysts began to reassess many of their assumptions about the economic and political regimes that might follow the demise of authoritarianism.

Significantly, many observers of Indonesia's new political framework were influenced heavily by variations of the still growing 'democratic transitions' literature. Often cited was the revisionist modernisation theorist Samuel Huntington who, in one of his permutations, announced that a 'Third Wave of Democratisation' (1991) was enveloping the globe as economic growth and the consolidation of middle classes forced authoritarian regimes to give way in the wake of the Cold War. This was in itself a major irony, given that a slightly later version of Huntington essentially argued that the cultural values that underpinned democracy belonged to Western civilisation only (1993). Also at least implicitly adopted was the position laid out most clearly in the literature epitomised by O'Donnell and Schmitter's famous contribution to the seminal *Transitions from Authoritarian Rule* (1986) project. In the very beginning of this contribution, the flagship of a broader, ambitious project involving numerous collaborators, the authors stated that they dealt with:

> transitions from certain authoritarian regimes toward an uncertain 'something else'. That 'something' can be the insaturation of political democracy or the restoration of a new, and possibly more severe form of authoritarian rule. The outcome can also be simply confusion, that is, the rotation in power of successive governments which fail to provide any enduring or predictable solution to the problem of institutionalising political power. Transitions can also develop into widespread violent confrontations.
>
> (O'Donnell and Schmitter 1986: 3)

Thus, following O'Donnell and Schmitter (and their collaborators), many of the analyses of post-Soeharto Indonesia highlighted the 'elements of accident and unpredictability, of crucial decisions taken in a hurry with very inadequate information' (1986: 3). These relied mostly on detailing the 'dilemmas', 'choices and 'processes' that present themselves in indeterminate situations when authoritarianism unravels (1986: 5).

The conclusion was that almost anything seemed equally possible, but that a desired liberal, democratic outcome would be more certain if the right actors would only make the right choices and enter into the right negotiations and agreements. Crouch (2000), as mentioned earlier, thus largely dismissed analyses of the structural dimension of political change in favour of an approach that favoured accidental events and the choices made by actors in very fluid situations. Though he was cautious about predicting future outcomes, positive or otherwise, Crouch's preferred approach is, in effect, similar to that of the authors of the 'transitions' literature.

Indeed, although O'Donnell and Schmitter sometimes invoked 'structural' factors, their analysis was heavily weighted toward the calculations and immediate reactions of political elites to events. Like another theorist sometimes referred to in the Indonesian debates – Di Palma (1990) – much emphasis was laid in their work on elite 'pacts' that emerge in the process of negotiating and 'crafting' political transitions. In spite of a chapter on 'resurrecting civil society', it was obvious that the approach they were advocating was extremely actor- and elite-based, one in which social structures and social forces were of secondary significance.

Many of the analyses of Indonesia immediately prior to and after the fall of Soeharto have equally emphasised the reaction of personalities to events as they unfold. Some are so overwhelmed by the task of chronicling the rapid, often complex sequences of events, and actions taken by actors, that these overtake and subsume any overarching analysis of their broader significance (e.g. Van Dijk 2001). Moreover, the idea that there could be a transition to 'something' other than a liberal form of democracy was somewhat lost in the discussion. For many observers of post-Soeharto Indonesia, the conditions for benign, democratic 'good governance', and the triumph of the rule of law, were basically technical in nature: legislation guaranteeing freedoms for political parties, 'good' election laws, and a number of other legal and institutional reforms (Lindsey 1999; also see Hadiz 1999).

This aspect of the problem is undoubtedly crucial. Nevertheless, the position taken in this book has been that institutions and the ways in which they actually work are contingent on the outcome of contests between social forces. These contests can be bitter, protracted and violent. Though elite negotiation is invariably part of the process, the crafting of new institutional frameworks governing politics and markets should be understood primarily in the context of broader constellations of social power and interests and the way they are able to organise politically (Bellin 2000). In the case of post-Soeharto Indonesia, it is important to recognise that the shaping of new institutional frameworks has largely been the purview of those who were at least partly nurtured and incubated under the New Order. As the book has shown, these are interests that have been able to secure their position via new and shifting alliances; they have been able to essentially reinvent themselves within Indonesia's new democracy, and indeed, to appropriate it. New political players who flooded into the system as it opened had little choice but to operate within the power relations and predatory processes already in place.

Unlike others, we have also not characterised Indonesia as being in the middle of a 'transition' period (e.g. Manning and Van Diermen 2000; Kingsbury and Budiman 2001). We reject the view, in fact, that Indonesia is at some intermediary point between predatory rule and the ultimate triumph of liberal forms of markets and democracy. We have argued, on the contrary, that the essential new patterns and dynamics of social, economic and political power have now been established. From this point of view, the violence, money politics and alleged political kidnappings that routinely take place in Indonesia today are not the symptoms of the growing pains of an infant liberal democracy, but fundamental to the logic of 'something else' that is increasingly well entrenched. This 'something else' is a form of democracy driven increasingly by the logic of money politics and political intimidation; elements laid down in the Soeharto years, although within the context of a highly centralised system of authoritarian rule.

The approach taken in this book also implies that narrower issues relating to the question of leadership are of secondary importance. As Liddle conceded, a leader-centred approach could simply descend into 'idiosyncratic description' of individuals in which there is no way of telling what outcomes are possible under any set of circumstances (Liddle 1991: 242). By the same token it is inadequate to primarily allocate 'blame' for failure to extend *reformasi* on the quirks in the personalities of major political figures. The Wahid Presidency did not falter only because he was an eccentric, as some may suggest, but because of the logic of an emerging system in which the illicit mobilisation of political funds was essential to political survival. The same logic has come to be imposed on the Megawati Presidency, as criticism emerges about the ambitions of her businessman husband, Taufik Kiemas, and about her use of non-accountable special presidential aid funds (*Tempo Online* 22–28 April 2002). Moreover, the intractability of Indonesia's economic reform problems is also not due simply to the absence of the right economic technocrats or economic teams. It results instead from the continuing salience of predatory interests and their capture of the institutions of state power. Such an explanation for failure also applies to success. It is equally inadequate to explain the opening up of the Indonesian political process as simply the product of the benign enlightenment of particular individuals. As shown earlier (Chapters 7 and 9), democratising was essential for the survival of the interests nurtured under Soeharto's authoritarian rule, as the institutional structures of the New Order became unviable.

What is significant in Indonesia is that any leader has to survive within a particular framework of power relations established over the previous decades and in the absence of genuine effective vehicles of reform. As suggested earlier, this was demonstrated most clearly by the failed experience of the Wahid government, during which many of the hopes for reform were dashed. In particular, Wahid's attempt to challenge the military by promoting allies appeared to deliver the lesson that no President could survive without its support. It was no surprise, therefore, that under Megawati the military regained control over appointment of its high command and re-established its influence over policies towards Aceh. Significantly, it also was able to negotiate an apparent immunity from prosecution for war crimes, especially with regard to East Timor (*JP* 16 August 2002). As an ultimate irony, President Megawati even backed the re-election of incumbent Jakarta Governor, Sutiyoso, the former army commander in the capital city widely considered responsible for the devastating assault on PDI headquarters in 1996 that had created such bitterness and resentment for Megawati and her supporters just a few years earlier (*Kompas Cybermedia* 10 June 2002).

The fact that democratic and market outcomes may preserve the authority of former elites might be explained simply in terms of the dominance of hardliners over moderates within the elite pacts that conduct the negotiations (McFaul 2002). For transitions theorists, this is not necessarily a problem for the long term. They argue that the very fact that new institutions are in place means that possibilities for important shifts in the structure of power and opportunities to consolidate liberal democracies are created. Illiberal democracies, it can be speculated, will ultimately become liberal ones as progressive forces and constitutional liberalism are nurtured within the cradle of stable regimes (Zakaria 1997). While transitions theorists recognised that democratic transitions were conducted within constraints and required bargains that often left intact the interests of the entrenched elites, the mere existence of democratic institutions nevertheless opens the door for more thoroughgoing social reform. Linz (1997: 408) has argued that 'even bad democracies are better than authoritarian rule or chaos since we can assume that they may undergo processes of re-equilibration, and with improved conditions and leadership may become fully consolidated'. The assumption here is that institutions have a life of their own, forcing political life into specific channels and, in the case of democracy, opening new opportunities for formerly marginal forces through constitutional arrangements. Thus, there is scope for the crafting of democracy; for agency to triumph over structure.

However, democracies where various oligarchies, cartels and rapacious predatory interests control power have shown a great capacity for survival. This has been demonstrated over the past half-century in the Philippines and Pakistan. The point is that while agency is undoubtedly important, not all outcomes are equally possible under any given set of circumstances and within any constellation of power. As Marx famously remarked: 'Men make their own history, but they do not make it just as they please; they do not make it under circumstances chosen by themselves, but under circumstances directly encountered, given and transmitted from the past' (Marx 1963: 15).

At the same time, there is nothing immutable about these circumstances. They are not culturally pre-determined on the basis of some timeless set of value systems or entrenched institutional pathways. Nor do they persist merely because of a lack of social capital. The essential lesson then is that grafting (or crafting) institutions does not necessarily bring change; whether these take root or are simply expropriated is more fundamentally the result of shifts in the structure of the state and of social power that take place through processes of conflict and struggle. It is from this standpoint that we may also understand the real significance of the Asian economic crisis of 1997/1998 for Indonesia. By fracturing the old power coalitions and unravelling the highly centralised system of authoritarian rule that had sustained them, the crisis produced that significant potential for a real reordering of social and political power. Successive governments now faced a critical press, a parliament that has become a sea of shifting and undisciplined interests, and a vast array of professional organisations and non-government organisations dedicated to scrutinising human rights and corruption, as well as social welfare and local community groups. But is the mere existence of what may approximate a civil society with its increasingly autonomous groups enough to really shift the balance of social power?

It is not only the presence of a strong reformist impulse in Indonesian society that is critical, but also how such reformist interests are politically organised into a disciplined and coherent force able to capture state power. There is no doubt enthusiasm for an end to the corruption and arbitrary rule that pervade Indonesian society at many levels, not only within the urban middle classes. What is an essential starting point for reform in Indonesia, however, is the very thing that remains lacking – specifically, a disciplined and even ruthless political party driven and defined by any coherent ideological agenda for liberal reform and transparent governance. Yet, as we have shown, among the parties dominating the political landscape no liberal party has emerged committed to market reform or to rule of law – no surprise given the resilience of a civil and military state apparatus enmeshed in the old structures of political capitalism. No genuinely social democratic party has emerged with an agenda based on social justice – currently implausible given the weakness of organised labour.

What then can break the mould? Whether economic and political regimes are captured by interests entrenched over almost four decades of predatory authoritarian rule or new liberal or social democratic alliances are successfully established will be increasingly determined in the context of a deepening globalisation of neo-liberal power. Will this more intense engagement with neo-liberal globalism inexorably drive a liberal or social democratic transition or will it instead offer a lifeline for entrenched oligarchies and predatory officials?

Will neo-liberal globalisation enforce market reforms?

As we have seen, leading spokespersons in the IMF and the US government, among them IMF Chief Michel Camdessus and the Chairman of the US Federal Reserve Board, Alan Greenspan, saw the economic crisis as an opportunity to impose liberal market reforms. Programmes forced on the Indonesian government in the IMF Letters of Intent appeared to remove the very oxygen that had sustained the politico-business oligarchy. Just as Indonesian governments and businesses were expected to submit to the discipline of global capital markets, meeting the expectations of open policies and good governance, a surge of foreign direct investment was expected to replace insolvent local companies, delivering an immediate dividend in improved corporate governance and rule of law.

However, there are important divisions within the neo-liberal camp itself over the question of markets and democracy between the policy ideologues – the true believers – of the World Bank and the IMF, and the complex business and financial communities that operate on the ground. The latter are much more ambiguous about markets and democracy. Hence, the assumption that global capital might become the Trojan Horse for a liberal convergence towards markets, good governance and even democratic reform is difficult to sustain. After all, the incentive to globalise production and investment often exists precisely because different regions offer different opportunities to secure competitive advantages. In the case of Indonesia and other low wage economies, the uncertainties of rule of law and the absence of transparency and accountability are offset by the advantages of escape from high wages, organised labour, environmental regulation and the social welfare regimes in the West. As we have seen, foreign firms in Indonesia had often benefited by calling in the military to control labour within their own factories.

In other words, as two fee-market economists, Zingales and McCormack (2003), have recognised, 'there is a difference between being for free markets and being for the elite that has benefited from them … a large share of the incumbent super rich can be against free markets because it sees them only as competition and not as opportunity'. So long as high profits were likely and there was a perception that a powerful centralised state would provide political guarantees against chaos in an unregulated economy, investors were prepared to enter the dangerous world of investment in Indonesia. A survey of foreign investors in Indonesia conducted just before the crisis showed that:

> Many international investors were very enthusiastic. Bureaucratic strings, corruption, insider trading and the weak financial system did not deter investors … Almost all business players truly understood the weakness of the legal system, the lack of transparency in decision-making and the role of political forces … But there were still no signs of hesitation on the part of investors.
>
> (*Kompas Cybermedia* 22 July 1998)

Quite apart from the question of their real liberal credentials, the expected surge of direct foreign investment into post-crisis Indonesia did not happen. Domestic interests entrenched in business and the bureaucracy were able to hold up the juggernaut of global markets where their assets were threatened by the prospect of fire sales to foreign companies or their authority in state-owned companies was placed at risk. Formidable, and often unlikely, alliances between state officials, workers, investors and nationalists in parliament intervened to great effect in disputes over the sale, for example, of Bank Bali, Bank Central Asia and Bank Niaga, and in the privatisation of Semen Gresik. They have made it difficult for foreign companies trying to assume control of bankrupted local public utilities, including those in the lucrative water supply business (*AWSJ* 2–4 June 2000: 1, 6; 30 May 2000: 1, 10).

Foreign creditors also faced a capricious court system willing to back domestic investors in bankruptcy cases and to support the manoeuvres of insolvent local partners seeking to retain their assets against buy-outs by foreign partners.[1] Their ability to control events was further eroded as central authority over economic regulation and corruption was progressively devolved. Foreign mining companies now confronted claims from local communities over land deals done under the Soeharto regime, and faced regional governments demanding a share of taxes, royalties, profits and ownership in resource ventures under the terms of new decentralisation laws.[2] For these resource companies, decentralisation had greatly expanded the complexity and number of government demands for taxes and rents and given rise to unpredictable political alliances that now included regional governments, local business figures and politicians.[3]

But when the political guarantees failed to protect them, would Indonesia's foreign investors now press for market and regulatory reform? Some international companies, backed by their own governments, attempted to enforce legal contracts and to collect debts by engaging their Indonesian antagonists in the bankruptcy and civil courts. These often became part of broader demands for greater transparency (*AWSJ* 4 November 1998: 1, 7: *JP* 3 November 1998: 1; 4 November 1998: 8). Although a few companies, notably the World Bank-owned International Finance Corporation (IFC), led the charge in the courts, many other foreign investors simply decided to limit their Indonesian plans or walked away, citing the difficulties of corruption, the disintegration of power and problems of dealing with officials and the courts (*Reuters* 4 September 2002).

But many foreign business interests decided to defend the privileges gained under the old regime, even where this might have meant throwing a lifeline to cronies now under assault from reformers in the state. It must be remembered that foreign investors had been willing to enter huge public infrastructure contracts issued without public tender and to 'cut in' Soeharto family members where huge profits could be made from the state guarantees of markets and pricing. Even though these were often obtained through opaque processes involving bribery, foreign business proved willing to defend their property rights, along with the cronies and predatory interests involved. Thus, foreign energy companies, vigorously supported by governments that had provided billions of dollars through export banks and insurers to support their bids, moved against PLN when it defaulted on payments and sought to cancel contracts (e.g. *FEER* 21 October 1999: 63–64; *AWSJ* 9 March 2000: 1).

Rather than a demand for liberal reform we find an implicit and sometimes explicit nostalgia for the Soeharto regime among investors – as we do among sections of the middle class (*JP* 15 May 2002). Their interests lie not so much in whether the regime is democratic or whether transparent market institutions prevail as long as a degree of predictability and protection for property rights, whatever their origins, is guarantee.

In any case, private capital flows have continued to haemorrhage at a rate of around US$10 billion per annum since 1997, after a period of capital inflows that peaked at US$11.5 billion in 1996 (IMF 2002: 51–54). Where foreign investors continued to be active it was increasingly within Indonesia's capital markets as they began to boom in 2002 and 2003.[4] Yet, this was a sector where investment was most disarticulated from real reforms in the corporate and wider political world. In reality, the disciplines of global capital markets proved elastic in a situation where fund managers could not ignore the growth prospects of companies in an economy that was now surging back into the void created by the crisis. Even where few of the fundamental structural problems of banking, corporate debt restructuring, legal reform, corporate governance or corruption had been solved, fund managers could hardly cede an advantage to their competitors.[5] They were entering Indonesian capital markets more on the terms of Indonesia's new system of fragmented oligarchic capitalism than on their own and in danger of reliving the same financial stampedes that had taken place in 1997.

Will neo-liberal globalisation enforce democracy?

Neo-liberals have always claimed a connection between democracy and the free market. In one sense, democracy was embraced as that liberating mechanism within which civil coalitions might be formed to challenge state control and open the door to market reform (World Bank 1997a: 334). Yet, this essentially liberal view of democracy as a force that frees individuals from the state sits uneasily with the neo-liberal view of democracy as a mechanism within which the interests of private investors might be safeguarded and insulated from predatory forces. This more functional view of democracy was illustrated, for example, in an editorial in *The Australian*, which commented on the constitutional reforms in August 2002 that introduced direct presidential elections and brought an end to military representation in parliament. 'If properly implemented, the reforms could bolster economic stability and the return of the foreign capital that has fled amid the political turbulence and corruption scandals of recent years' (14 August 2002: 10).

It is a system of democracy in which markets must be protected from politics and government is unimpeded by the intervention of vested interests – which might include labour or other social democratic reformers as well as various business or welfare cartels. It must coincide with the principles of the regulatory state, as distinct from either the European social democratic state or the East Asian developmental state (Jayasuriya 2000). It must necessarily be insulated from the sort of social change that could challenge vast inequalities in power and wealth that are embedded in the free market formula and protect the rights of property from the tyranny of a social democratic majority (Dorn 1993). It might be characterised as a system of a system of 'low intensity democracy' (Smith 2000; Gills 2000), or liberal authoritarianism, as Jayasuriya puts it (2000).

Hence, we find increasing unease amongst Western, mainly US, business and government sources at the growing social instability and dissent that have been unleashed by democratic changes and decentralisation of state authority in Indonesia, in particular, the growing restiveness of labour and the militancy of human rights groups and their willingness to take legal action against companies. At the same time, while the US has discovered that its interests in the post-Cold War period might be best served by democracies rather than the dictators it formerly supported, the fear of democracy opening the door to anti-market forces was to be paralleled by a fear of the 'failed state' opening the door to social violence, global criminal activity and terrorism. Such uncertainties have been given another dimension by the events of 11 September 2001 and the Bali bombings of 12 October 2002. America's democratic and market agendas were now to be overtaken by other considerations about order and security.

Thus the new security apprehensions reinforce neo-liberal fears of Huntington's 'democratic distemper'. The rise of neo-conservative views in the US government signalled a willingness, not merely to support or bring down particular rulers, but actively to impose a new global order and export a system of democratic governments (Fidler and Baker 2003; Mallaby 2002).

This new policy was manifested in the neo-liberal retreat from enthusiastic embrace of administrative decentralisation and unconstrained liberal social and political agendas. It was a retreat that was to clash with other policies towards human rights and the position of formerly disgraced military establishments (Hoffman 2002).

These dilemmas are illustrated in the case of US State Department action concerning the activities of Exxon Mobil in Aceh where the task of securing US commercial advantage and the new priorities of waging the war against terror triumphed over unease about the relationship between Exxon and the military and the latter's role in human rights violations. It urged a United States court to dismiss a lawsuit brought by a human rights group over its operations in Aceh, arguing that the case would harm Washington's campaign against terrorism and that it might provide opportunities for Chinese oil companies. It was stated that, 'We would expect that foreign companies, such as from the People's Republic of China' would be 'far less concerned about human rights abuses, or about upholding best business practices' (Perlez 2002).

Not only did the nationalist rhetoric of President Megawati and her uncompromising position with regard to regional separatist movements sit more easily in the new circumstances, the American government also faced difficult choices over its attitude towards the Indonesian military. It had previously cut support and co-operation in response to international unease about human rights abuses committed in East Timor and during brutal campaigns in the dying days of the Soeharto regime. However, the prospect that terrorism might gain a grip within the failed state of post-Soeharto Indonesia made the military once again an attractive option. US Defence Secretary Paul Wolfowitz led the charge to reopen relations with the Indonesian military. He was, he claimed, under no illusions about the past record of the TNI, but Indonesia now desperately needed the discipline and effectiveness that only the military could provide (cited in *Agence France Presse* 16 October 2002). The US thus brought the TNI in from the cold with a US$50 million package for counter-terrorism training while watching the Indonesian Human Rights Court acquit several Indonesian military officers of human rights abuses in East Timor (*Tempo Online* 11 August 2002). Following Wolfowitz down this path were his counterparts within the Australian government, including Defence Minister Robert Hill and Foreign Minister Alexander Downer.

Where to now?

Neo-liberals had recognised that political obstacles would stand in the way of the inexorable drive towards liberal market capitalism and democracy. The World Bank noted that, 'Deep distributional conflicts and constraints embedded in state institutions are at the heart of explanations for so many countries' failure to reform.' But, the Bank continues, these problems are not immutable. It makes the enlightening observation that, 'Ultimately, change comes when the incentives to throw out the old policies and old institutional arrangements becomes stronger than the incentives to keep them.' But for whom are the incentives and how would this happen? The Bank suggests, 'An economic crisis or an external threat, or the arrival of a new government with fewer vested interests in the old system' (World Bank 1997a: 334). Although all these things appeared to happen in Indonesia, the reformist impulse proved to be politically weak. How do we explain this?

An important part of the explanation lies in the fact that entrenched interests remained relatively intact and aggressive in the economic sphere, despite the devastating shocks of the crisis that it was assumed would make reform impossible to resist. As the IMF reported, at the end of 2001, many of the major business groups simply refused to repay debt; over half of the corporate external debt, totaling US$31 billion, remained non-current in meeting payments falling due (IMF 2002: 57). Closing off lines of global credit held few fears for investors who could continue business without being forced into bankruptcy and who could, in any case, retain profitable business activities and access other sources of informal credit. Most of Indonesia's corporate moguls were able to retain many of their assets by refusing to surrender assets in settlement of bank debts and, together with nationalist allies in parliament, holding up the asset disposal process (IMF 2002: 28–50) Such action was logical where the costs could be transferred to the state and where government had proven unable to enforce bankruptcy in the courts.

In other words, the assumed link between structural shocks and reform of the economic regime has not been realised. Where reforms threatened the very social order in which political ascendancy was embedded, however painful the process, social interest took precedence over any notions of efficiency or rationality in the economy. Where debt-ridden private investors were not subjected to effective rules, they could survive by defaulting and by bypassing domestic banks, reorganising in flourishing export sectors or domestic consumer markets. Ultimately, structural shocks are significant in the downfall of regimes to the extent that they lead to the destruction or weakening of entrenched coalitions and enable new, reformist coalitions to organise and emerge.

The World Bank recognised the politics of the matter and that reform could be delayed, often for painfully long periods when 'those in power stick with outdated policies because it is in their interest to do so' (World Bank 1997a: 334). Nevertheless, it argued that reform-minded elites could speed the process by properly explaining the benefits of reform, spelling out a vision, assembling supportive coalitions and providing compensation for those who would be the losers from structural adjustment.[6] Such a view assumes that such 'rational' leaders already posses the power to build coalitions, make policy and spell out visions. Yet, in the case of Indonesia, the sort of leaders the World Bank talks about are scattered, isolated and powerless, through the state apparatus and within society. Even the Bank admits this:

> Progress on governance has been ... left largely to a few key reformers who have been moving forward in their respective spheres, garnering whatever support they can muster from senior leaders. These initiatives appear *ad hoc* and are floundering under resistance from well-entrenched vested interests.
>
> (World Bank 2000: 43)

What the World Bank does not tell us is how those reformers are expected to seize power from vested interests, and why political parties with little interest in cohesive liberal reformist policy agendas still constitute the only game in town.

Thus the primary question for Indonesia is not whether Indonesia's political elites will be rational in their policy choices but how reformist interests might organise and come to power. In reality, the possibility that cohesive reformist parties might emerge from the wreckage, driven by a coherent agenda of market liberalism rather than being swallowed in a system of power relations embedded in the pursuit of rents appears even more remote than ever. The fall of Soeharto's system of administrative oligarchy has not unlocked the door for the neo-liberal regulatory state. Rather, it has produced, in its initial stages, a frightening descent into the uncertainties of a 'wild' or 'savage' capitalism and a reorganisation of oligarchy within institutions where political opportunism is the central currency. But such a system produces its own limits and it is in the effort to contain ever-unravelling power and authority that a new regime will evolve. Thus we see the reassertion of those secular forces and interests forged around the old centralised state apparatus; not only the military and the central bureaucracy, but also key political organisations such as Golkar and President Megawati's PDI-P. Ironically, the US has little option but to turn to these to restore a suitable environment for investment and to forestall what is feared to be an epidemic of reactionary populism within a failed state. This does not mean a return to the centralised authoritarianism of Soeharto, but it does mean a system of democratic rule where the state apparatus will provide some form of order in which oligarchies rather than markets will prevail.

Notes

1 No case illustrated more clearly the traps for foreign investors than the drawn-out struggle between the Canadian firm, Manufacturers Life Insurance, and a British Virgin Islands company suddenly claiming ownership of the 40 per cent share of Manulife Indonesia formerly held by a bankrupted local partner – Suyanto Gondokusumo of the giant Dharmala group. It was a case marked by allegations of violence, threats and forgery (*AWSJ* 6 December 2000: 1, 10). The decision by the presiding judge in the Central Jakarta Commercial Court in June 2002 to declare the highly profitable Manulife bankrupt on a technicality – later overturned – was a clear reminder that local interests held the whip hand in this arena (*Tempo* 25 June–1 July 2002).

2 See, for example, the tug-of-war over taxes between the central government and local authorities that threatened to close the US mining company, Newmont (*Kompas Cybermedia* 13 April 2000; *JP* 1 November 2000: 12).

3 In the complex jostling to secure a share of the US$800 million BHP Kaltim mine in East Kalimantan, for example, we find not only the provincial government but various players from the past – Liem Sioe Liong, Prabowo Subianto and Aburizal Bakrie – prominent within the various contending consortia (*Australian Financial Review* 15 May 2002: 54).

4 The robustness of Jakarta's stock exchange is best illustrated in the government's IPO for the state-owned Bank Mandiri, which was subscribed at least 6.7 times and expected to raise US$328 million, of which two-thirds would be taken up by foreign investors. This follows the sale of 51 per cent of the recapitalised Bank Danamon to Temasek Holdings and Deutschebank for US$336 million. Nevertheless, as mentioned above, foreign investors continued to be wary of direct investment in the real sector.

5 For example, Rosser points out that the progress in accounting and financial reporting after the crisis has been uncertain and uneven (Rosser 2003).

6 The Bank stated:

> Reform-oriented political leaders and elites can speed reform by making decisions that widen the people's options, articulate the benefits clearly, and ensure that policies are more inclusive. In recent years farsighted political leaders have transformed the options for their people through decisive reform. They were successful because they made the benefits of change clear to all, and built coalitions that gave greater voice to often-silent beneficiaries. They also succeeded – and this is crucial – because they spelt out a longer term vision for their society, allowing people to see beyond the immediate pain of adjustment. Effective leaders give their people a sense of owning the reforms.
>
> (World Bank 1997a: 335)

BIBLIOGRAPHY

Abeng, Tanri (2001) *Indonesia Inc. Privatising State Enterprises*, Singapore: Times Academic Press.

Aditjondro, George (1995a) 'Bali's Jakarta Colony', Bali Modernity Workshop, University of Wollongong.

—— (1995b) 'The Emerging Oligarchy in Indonesia: Will They Survive the Post-Soeharto Era?', *Report for the Special Research Centre on Political and Social Change in Asia*, Perth: Murdoch University.

—— (1995c) 'The Habibie Family's Business Interests', and 'The Pos Kota Group', Perth: Asia Research Centre, Murdoch University

—— (2000) 'Chopping the Global Tentacles of the Soeharto Oligarchy', presented at the conference *Towards Democracy in Indonesia*, University of Auckland, 1 April.

Aidit, D.N. (1970) 'The Three Political Forces' (extracted from Aidit's General Report to the Sixth National Congress of the PKI in September 1959), in Herbert Feith and Lance Castles (eds) *Indonesian Political Thinking 1945–1965*, Ithaca: Cornell University Press, pp. 257–265.

Allison, J.M. (1969) 'Indonesia: Year of the Pragmatists', *Asian Survey* 9 (2).

American Embassy (1994) 'Labour Trends in Indonesia', unpublished report, Jakarta.

Amsden, Alice (1989) *Asia's Next Giant: South Korea and Late Industrialisation*, New York: Oxford University Press.

Anderson, Benedict (1983) 'Old State, New Society: Indonesia's New Order in Comparative Historical Perspective', *Journal of Asian Studies* 42 (3).

—— (1990) 'Murder and Progress in Modern Siam', *New Left Review* 181: 33–48.

Anderson, B. and Ruth McVey (1971) *A Preliminary Analysis of the October 1 1965 Coup in Indonesia*, Ithaca: Cornell University Modern Indonesia Project.

Anggoro, Kusnanto (2000) 'How to Sell TNI on Democratic Controls', *Jakarta Post* 7 March.

Anspach, R. (1969) 'Indonesia', in F. Golay *et al.* (eds) *Underdevelopment and Economic Nationalism in Southeast Asia*, Ithaca: Cornell University Press.

Antlov, Hans (1995) *Exemplary Centre, Administrative Periphery: Rural Leadership and the New Order in Java*, Richmond: Curzon.

Arief, S. and A. Sasono (1973) *Indonesia: Dependency and Underdevelopment*, Kuala Lumpur: META.

Arndt, Heinz (1967) 'Economic Disorder and the Task Ahead', in Tan, T.K. (ed.) *Soekarno's Guided Indonesia*, Brisbane: Jacaranda, pp. 129–142.

Asaat (1956) 'Speech to the All-Indonesian National Importers' Conference', in Herbert Feith and Lance Castles (eds) *Indonesian Political Thinking 1945–1965*, Ithaca: Cornell University Press.

Aspinall, Edward (1999), 'The Indonesian Student Uprising of 1998', in Arief Budiman, Barbara Hatley and Damien Kingsbury (eds) *Reformasi: Crisis and Change in Indonesia*, Melbourne: Monash Asia Institute.

Aspinall, Edward, Herb Feith and Gerry van Klinken (eds) (1999) *The Last Days of President Suharto*, Melbourne: Monash Asia Institute.

Bardhan, Pranab (1989) 'The New Institutional Economics and Development Theory: A Brief Critical Assessment', *World Development* 17 (9): 1389–1395.

Barro, Robert (1993) 'Pushing Democracy is No Key to Prosperity', *Asian Wall Street Journal* 27 December: 6.

Baswir, Revrisond (2000) 'Buloggate dan dan Manejemen Uang Publik', *Kompas* 5 June.

Bates, Robert (1981) *Markets and States in Tropical Africa*, Berkeley: University of California Press.

Beeson, Mark (1998) 'Indonesia, the East Asian Crisis and the Commodification of the Nation State', *New Political Economy* 3 (3): 357–374.

Beeson, Mark and Richard Robison (2000) 'Introduction: Interpreting The Crisis', in R. Robison, M. Beeson, K. Jayasuriya and H. Kim (eds) *Politics and Markets in the Wake of the Asian Crisis*, London: Routledge, pp. 3–24.

Bellin, Eva (2000) 'Contingent Democrats: Industrialists, Labour and Industrialisation in Late-Developing Countries', *World Politics* 52 (January): 175–205.

Bello, W. (1998) 'The End of the Asian Miracle', *Inside Indonesia* April–June: 7–10.

Benda, Harry J. (1964) 'Democracy in Indonesia', *The Journal of Asian Studies* May: 449–456.

Bhattacharya, A. and M. Pangestu, (1992) 'Indonesia: Development and Transformation Since 1965 and the Role of Public Policy', paper prepared for the World Bank Workshop on the Role of Government and East Asian Success, East West Centre, November.

Boileau, J. (1983) *Golkar: Functional Group Politics in Indonesia*, Jakarta: Centre for Strategic and International Studies.

Bourchier, David (1987) 'The Petition of Fifty', *Inside Indonesia* 10 (April): 7–10.

—— (1994) 'Setiakawan: The New Order's First Free Trade Union', in David Bourchier (ed.) *Indonesia's Emerging Proletariat: Workers and Their Struggles*, Clayton: Monash University Centre of Southeast Asian Studies.

—— (2000) 'Habibie's Interregnum: *Reformasi*, Elections, Regionalism and the Struggle for Power', in Chris Manning and Peter van Diermen (eds) *Indonesia in Transition: Social Aspects of Reformasi and Crisis*, Singapore: Institute of Southeast Asian Studies.

Bresnan, John (1993) *Managing Indonesia: the Modern Political Economy*, New York: Columbia University Press.

Buchanan, James M. and Gordon Tullock (1962) *The Calculus of Consent*, Anne Arbor: University of Michigan Press.

Budianta, Melani (2003) 'The Blessed Tragedy: The Making of Indonesian Women Activism During the *Reformasi* Years', in Ariel Heryanto and Sumit Mandal (eds) *Challenging Authoritarianism in Southeast Asia: Comparing Indonesia and Malaysia*, London: RoutledgeCurzon, pp.145–177.

Budiman, Arief (1998) 'Pemimpin, Di Mana Kau?', *Tempo* 17–23 November: 76.

—— (1999) 'The 1998 Crisis: Change and Continuity in Indonesia', in Arief Budiman, Barbara Hatley and Damien Kingsbury (eds) *Reformasi: Crisis and Change in Indonesia*, Melbourne: Monash Asia Institute.

Camdessus, M. (1997) 'Asia will Survive with Realistic Economic Policies', Parts I and II, *The Jakarta Post* 8 December: 5; 9 December: 5.

Cameron, Lisa (1999) 'Survey Of Recent Developments', *Bulletin of Indonesian Economic Studies* 35 (1): 3–40.

Castles, Lance (1965) 'Socialism and Private Business: the Latest Phase', *Bulletin of Indonesian Economic Studies* 1 (1).

—— (1967) 'The Fate of the Private Entrepreneur', in T.K. Tan (ed.) *Sukarno's Guided Indonesia*, Brisbane: Jacaranda.

Cawson, Alan (1986) *Corporatism and Political Theory*, Oxford: Basil Blackwell.

Chalmers, Ian (1997) 'The Rise of Statist Nationalism', in Ian Chalmers and Vedi R. Hadiz (eds) *The Politics of Economic Development in Indonesia: Contending Perspectives*, London: Routledge.

Chalmers, Ian and Vedi R. Hadiz (eds) (1997) *The Politics of Economic Development in Indonesia: Contending Perspectives*, London: Routledge.

Chaudhry, Kiren Aziz (1993) 'The Myths of the Market and the Common History of Late Developers', *Politics and Society* 21 (3): 245–274.

—— (1994) 'Economic Liberalisation and the Lineages of the Rentier State', *Comparative Politics* 27 (1): 1–25.

—— (1997) *The Price of Wealth: Economies and Institutions in the Middle East*, Ithaca and London: Cornell University Press.

Cohen, Margot (1998) 'Unguided Missiles', *Far Eastern Economic Review* 26 November.

Cole, David C. and Betty F. Slade (1996) *Building a Modern Financial System: The Indonesian Experience*, Cambridge: Cambridge University Press.

—— (1998) 'Why has Indonesia's Fiscal Crisis been so Bad?', *Bulletin of Indonesian Economic Studies* 34 (2): 61–66.

Coleman, James S. (1988) 'Social Capital in the Creation of Human Capital', *American Journal of Sociology* 94: S95–S120.

Cribb, Robert and Colin Brown (1995) *Modern Indonesia: A History Since 1945*, London and New York: Longmans.

Crouch, Harold (1978) *The Army and Politics in Indonesia*, Ithaca: Cornell University Press.

—— (1999) 'Wiranto and Habibie: Military–Civilian Relations Since May 1998', in Arief Budiman, Barbara Hatley and Damien Kingsbury (eds) *Reformasi: Crisis and Change in Indonesia*, Melbourne: Monash Asia Institute.

—— (2000) 'The Perils of Prediction: Understanding the Indonesian Transition 1998–1999', unpublished paper.

Cumings, Bruce (1989) 'The Abortive Abertura: South Korea in the Light of the Latin American Experience', *New Left Review* 173: 5–32.

—— (1999) 'The Asian Crisis, Democracy and the End of Late Development', in T.J. Pempel (ed.) *Politics of the Asian Economic Crisis*, Ithaca: Cornell University Press, pp. 17–44.

Dale, Reginald (1998) 'Asia Crisis will Bolster US Prestige', *International Herald Tribune* 20 January.

Department of Foreign Affairs (1993) 'Workers' Rights: Issues and Perspectives', Jakarta: Republic of Indonesia.

Di Palma, Giuseppe (1990) *To Craft Democracies*, Berkeley: University of California Press.

Dick, Howard (1985) 'Survey of Recent Developments', *Bulletin of Indonesian Economic Studies* 3 (1).

—— (2001) 'Survey of Recent Developments', *Bulletin of Indonesian Economic Studies* 37 (1): 7–42.

Dobb, Maurice (1947) *Studies in the Development of Capitalism*, New York: International.

Donor, Richard (1992) 'Limits of State Strength: Towards an Institutional View of Economic Development', *World Politics* 44 (3): 398–431.

Dorn, James A. (1993) 'Economic Liberty and Democracy in East Asia', *Orbis* 37 (4): 599–619.

Draper, Hal (1977) *Karl Marx's Theory of Revolution: State and Bureaucracy*, New York and London: Monthly Review Press.

Easton, David (1965) *A Systems Analysis of Political Life*, New York: Wiley.

Ecip, Sinansari S. (ed.) (1998) *Kronologi Situasi Penggulingan Soeharto*, Bandung: Mizan.

Econit (1996) *Dampak Kelangkaan Semu Pupuk Urea Terhadap Impor Beras dan Kesejahteraan Petani Tahun*, Jakarta.

Edwards, Richard C., Michael Reich and Thomas E. Weisskopf (1972) *The Capitalist System: a Radical Analysis of American Society*, New Jersey: Prentice-Hall.

Eichengreen, B. (1997) 'The Tyranny of the Financial Markets', *Current History* November: 377–382.

Eldridge, Philip J. (1995) *Non-Government Organizations and Democratic Participation in Indonesia*, Kuala Lumpur: Oxford University Press.

Emmerson, Donald (1976) *Indonesia's Elite: Political Culture and Cultural Politics*, Ithaca and London: Cornell University Press.

—— (1978) 'The Bureaucracy in Political Context: Weakness in Strength', in Karl D. Jackson and Lucian W. Pye (eds) *Political Power and Communications in Indonesia*, Berkeley: University of California Press, pp. 82–136.

Engels, Frederich (1968) 'The Origin of the Family, Private Property and the State', in Karl Marx and Frederich Engels *Selected Works*, London: Laurence and Wishart.

Evans, Peter (1989) 'Predatory, Developmental and other Apparatuses: A Comparative Political Economy Perspective on the Third World State', *Sociological Forum* 4 (4): 561–585.

—— (1995) *Embedded Autonomy: States and Industrial Transformation*, Princeton: Princeton University Press.

Evans, Peter, Dietrich Rueschemeyer and Theda Skocpol (eds) (1985) *Bringing the State Back In*, Cambridge: Cambridge University Press.

Fatton, Robert (1988) 'Bringing the Ruling Class Back In', *Comparative Politics* 20 (3): 253–264.

Fauzi, Noer (1997) 'Penghancuran Populisme dan Pembangunan Kapitalisme: Dinamika Politik Agraria Indonesia Pasca-Kolonial', *Reformasi Agraria: Perubahan Politik, Sengketa dan Agenda Pembaruan Agraria di Indonesia*, Jakarta: Konsorsium Pembaruan Agraria and Lembaga Penerbit Fakultas Ekonomi Universitas Indonesia.

—— (1998) 'Ratification of Customary Land Tenure Systems: An Agenda for Indonesian NGOs in Response to Land Administration Project – Study of Indigenous Lands', prepared for the Eleventh INFID Conference on Democratization in the Era of Globalization, Bonn, Germany, 4–6 May.

Feith, Herbert, (1962) *The Decline of Constitutional Democracy in Indonesia*, Ithaca: Cornell University Press.

—— (1963) 'Dynamics of Guided Democracy', in Ruth McVey (ed.) *Indonesia*, New Haven: Yale University Press.

Fidler, Stephen and Gerard Baker (2003) 'America's Democratic Imperialists: How the Neo-Conservatives Rose from Humility to Empire in Two Years', *Financial Times* 6 March: 11.

Fine, Ben (2001) *Social Capital Versus Social Theory*, London: Routledge.

Fischer S. (1988) 'The Asian Crisis: A View From the IMF', an address delivered at the Midwinter Conference of the Bankers' Association for Foreign Trade, 22 January (http://www.imf.org/external/np/speeches/1998/112298.htm).

Fitzpatrick, Daniel (1998) 'Corporate Governance, Economic Crisis and the Indonesian Banking Sector', *The Indonesian Banking Sector*, 9 (2): 178–193.

Ford, Michele (2000) 'Continuity and Change in Indonesian Labour Relations', *Southeast Asian Journal of Social Science* 28 (2): 59–88.

Forum Demokrasi (1998) 'Pernyataan' (signed by Bondan Gunawan and T. Mulya Lubis), 19 May.

Forum Kota (1998) 'Pernyataan Sikap Komunitas Mahasiswa se-Jabotabek', 28 May.

Forrester, Geoff (1998) 'Introduction', in Geoff Forrester and R.J. May (eds) *The Fall of Soeharto*, Bathurst: Crawford Publishing House.

Frankel, Jefrey (1998) 'The Asian Model, The Crisis and the Fund', paper delivered at the US International Trade Commission, April, (http://www.stern.nyu.edu~nroubini/asia/AsiaHomepage.html).

Friedman, Thomas L. (1997) 'Quit the Whining! Globalization Isn't a Choice', *International Herald Tribune* 30 September: 10.

Gadjah Mada University (1998) 'Pernyataan Universitas Gadjah Mada' (signed by Professor Ichlasul Amal), 19 May.

Garnaut, Ross (1998) 'The Financial Crisis: A Watershed in Economic Thought about East Asia', *Asian Pacific Economic Literature* May: 1–11.

Gill, S. and D. Law (1988) *The Global Political Economy: Perspectives, Problems and Policies*, London: Harvester Wheatsheaf.

Gills, Barry (2000) 'American Power, Neo-Liberal Economic Globalisation and Low Intensity Democracy: An Unstable Trinity', in Michael Cox, G. John Ikenberry and Takashi Inoguchi (eds) *American Democracy Promotion: Impulses, Strategies and Impacts*, Oxford: Oxford University Press, pp. 326–344.

Glassburner, B (1978), 'Political Economy and the Soeharto Regime', *Bulletin of Indonesian Economic Studies* 14 (3).

Gomez, Edmund (2002) 'Political Business in Malaysia: Party Factionalism, Corporate Development and Economic Crisis', in Edmund Gomez (ed.) *Politics and Business in East Asia*, London: Routledge, pp. 82–114.

Gomez, E.T. and K.S. Jomo (1997) *Malaysia's Political Economy: Politics, Patronage and Profits*, Cambridge: Cambridge University Press.

Gontha, Peter (1992) 'Dampak Positip dan Negatip: "Tight Money Policy" dan Pembatasan Offshore Loan', unpublished paper, 22 April.

Gourevitch, Peter A. (1993) 'Democracy and Economic Policy: Elective Affinities and Circumstantial Conjunctures', *World Development* 21 (8): 1271–1280.

Government of Indonesia (1985) 'Yayasan "Dana Abadi Karya Bhakti"', *Anggaran Dasar Yayasan Yayasan* 10: 1–7.

—— (1999a) *RUU Keselamatan dan Keamanan Negara 1999*.

Government of Indonesia (1999b) *RUU Penanggulangan Keadaan Bahaya 1999.*

—— (1999c) Undang-Undang Republik Indonesia No. 2 Tahun 1999 Tentang Partai Politik.

—— (1999d) Undang-Undang Republik Indonesia No. 3 Tahun 1999 Tentang Pemilihan Umum.

—— (1999e) Undang-Undang Republik Indonesia No. 4 Tahun 1999 Tentang Susunan dan Kedudukan Majelis Permusyawaratan Rakyat, Dewan Perwakilan Rakyat, dan Dewan Rakyat Daerah.

—— (2000) *APBN (Anggaran Belandja Negara)*, State Budget, 31 July.

—— (2001) *RAPBN (Rencana Anggara Belandja Negara)*, Planned State Budget.

Government of Indonesia *LOI* (1997–2001), *Letters of Intent and Memorandum of Economic and Financial Policies*, cited on 9 August 2003 (http://www.imf.org/external/NP/LOI/).

Gramsci, Antonio (1971) *Selections from the Prison Notebooks of Antonio Gramsci*, trans. Q. Hoare and G. Nowell Smith, New York: International Publishers.

Grindle, Merle S. (1991) 'The New Political Economy: Positive Economics and Negative Politics', in Gerald M. Meier (ed.) *Politics and Policy Making in Developing Countries*, San Francisco: International Centre for Economic Growth.

GSP Information Centre (1993) 'GSP Sub-committee of the Trade Policy Staff Committee 1992 GSP Annual Review: Workers Rights Review Summary, Case 007-CP-92, Indonesia', Washington: GSP Information Centre.

Habibie, B.J. (1983) 'Some Thoughts Concerning a Strategy for the Industrial Transformation of a Developing Country', address delivered to the Deutsche Gesellschaft fur Luft-und Raumfahrt in Bonn, Germany.

—— (1986) 'Industrialisasi Transformasi, Tecknologi dan Pembengiman Bangsa', *Prisma* 1: 42–53.

Hadiz, Vedi R. (1997) *Workers and the State in New Order Indonesia*, London: Routledge.

—— (1998) 'Reformasi Total? Labour After Soeharto', *Indonesia* 66 (October): 115–124.

—— (1999) 'Contesting Political Change After Soeharto', in Arief Budiman, Barbara Hatley and Damien Kingsbury (eds) *Reformasi: Crisis and Change in Indonesia*, Melbourne: Monash Asia Institute.

—— (2003) 'Power and Politics in North Sumatra: the Uncompleted *Reformasi*', in Ed Aspinall and Greg Fealy (eds) *Local Power and Politics in Indonesia: Democratisation and Decentralisation*, Singapore: Australian National University and Institute of Southeast Asian Studies, pp. 119–131.

Haggard, Stephan and Robert Kaufman (eds) (1992) *The Politics of Economic Adjustment*, Princeton: Princeton University Press.

—— (1995) *The Political Economy of Democratic Transitions*, Princeton: Princeton University Press.

Haggard, Stephan and Chung-in Moon (1990) 'Institutions and Economic Policy: Theory and a Korean Case Study', *World Politics* XLLI (2): 210–237.

Hamilton, Gary (1999) 'Asian Business Networks in Transition: or, What Alan Greenspan does not know about the Asian Business Crisis', in T.J. Pempel (ed.) *The Politics of the Asian Economic Crisis*, Ithaca: Cornell University Press.

Hamilton-Hart, Natasha (2000) 'Indonesia: Reforming the Institutions of Financial Governance', in Gregory Noble and John Ravenhil (eds) *The Asian Financial Crisis and the Architecture of Global Finance*, Cambridge: Cambridge University Press, pp. 108–131.

—— (2001) 'Anti-corruption Strategies in Indonesia', *Bulletin of Indonesian Economic Studies* 37 (1): 65–82.

—— (2002) *Asian States, Asian Bankers: Central Banking in Southeast Asia*, Ithaca: Cornell University Press.

Hanke, Steve (2003) 'Soeharto too, was a victim of regime change', *The Australian* 29 April: 13

Harberger, Arnold (1993) 'Secrets of Success: A Handful of Heroes', *American Economic Review* 83. Cited in Hector E. Schamis (1999) 'Distributional Coalitions and the Politics of Economic Reform in Latin America', *World Politics* 51 (January).

Harris, J., F. Schiantarelli and M. Siregar (1992) *The Effect of Financial Liberalisation on Firms' Capital Structure and Investment Decisions*, Jakarta: World Bank.

Harris, Nigel (1988) 'New Bourgeoisies', *The Journal of Development Studies* 24 (2): 237–249.

Harriss, John (2002) *Depoliticising Development: The World Bank and Social Capital*, London: Routledge.

Harsono, Andreas (1999), 'Megawati Stance May Hamper Timor Independence Move', *The American Reporter* 8 February.

Hartarto (1985) 'Menumbuhkan Pohon Industri dan Keterkaitannya', *Prisma* 5: 65–70.

Hartcher, Peter (1999) 'The Father of Reform', *Australian Financial Review Magazine* May: 54–62.

Hawkins, Everett D. (1963) 'Labour in Transition', in Ruth McVey (ed.) *Indonesia*, New Haven: Yale University Press.

Hefner, Robert W. (1993) 'Islam, State, and Civil Society: ICMI and the Struggle for the Indonesian Middle Class', *Indonesia* 56 (October): 1–35.

—— (2000) *Civil Islam: Muslims and Democratisation in Indonesia*, Princeton: Princeton University Press.

Hewison, Kevin (1993) 'Of Regimes, State and Pluralities: Thai Politics Enters the 1990s', in K. Hewison, R. Robison and G. Rodan (eds) *Southeast Asia in the 1990s: Authoritarianism, Capitalism and Democracy* Sydney: Allen and Unwin, pp. 159–190.

—— (2001) 'Thailand's Development through Boom and Bust', in Garry Rodan, Kevin Hewison and Richard Robison (eds) *The Political Economy of Southeast Asia: Conflicts, Crises and Change*, Melbourne: Routledge, pp. 71–103.

Hill, David T. (1994) *The Press in New Order Indonesia*, Asia Paper no. 4, Perth: University of Western Australia Press in association with Asia Research Centre,

Hill, Hal (1984) 'High Time for High Tech – Or Is It?', *Far Eastern Economic Review* 12 July: 24.

—— (1987) 'Is Public Loss Better than Private Gain', *Far Eastern Economic Review* 3 December.

—— (1990), 'Ownership in Indonesia', in H. Hill and T. Hull (eds), *Indonesia Assessment 1990*, Canberra: Australian National University.

—— (ed.) (1994) *Indonesia's New Order: The Dynamics of Socio-Economic Transformation*, Honolulu: University of Hawaii Press.

—— (1996) *The Indonesian Economy Since 1966: Asia's Emerging Giant*, Cambridge: Cambridge University Press.

—— (1997) 'Myths about Tigers: Indonesian Development Policy Debates', *The Pacific Review* 10 (2): 256–273.

—— (1999a) 'An Overview of the Issues', in Heinz Arndt and Hal Hill (eds) *Southeast Asia's Economic Crisis: Origins, Lessons and the Way Forward*, Sydney: Allen and Unwin, pp. 1–15.

Hill, Hal (1999b) *The Indonesian Economy in Crisis: Causes, Consequences and Crises*, Singapore: Institute of Southeast Asian Studies.
—— (1999c) 'What's Scaring Indonesia's Investors', *Asian Wall Street Journal* 5–6 March: 10.
—— (2000a) 'Indonesia: the Strange and Sudden Death of a Tiger Economy', *Oxford Development Studies* 28 (2): 117–139.
—— (2000b) 'Indonesia to Keep on Muddling Through in the Next Few Years', *Jakarta Post* 21 December.
HMI (Himpunan Mahasiswa Islam) (1984) *Pandangan Kritis terhadap RUU Keormasan*, Jakarta: HMI.
Hobohn, Sarwar (1995) 'Survey of Recent Development', *Bulletin of Indonesian Economic Studies* 31 (1).
Hobsbawm, Eric (1990) *Echoes of the Marseillaise*, London: Verso.
Hoffman, Stanley (2002) 'Clash of Globalisations', *Foreign Affairs* 81 (4).
Horsley, N. (1997) 'Asia Needs a New Model', *Asian Wall Street Journal* 9 December: 8.
Hughes H. (1998) 'IMF is Right on Indonesia', *Australian Financial Review* 12 March: 18.
Huntington, Samuel (1968) *Political Order in Changing Societies*, New Haven: Yale University Press.
—— (1991) *The Third Wave: Democratization in the Late Twentieth Century*, Norman, OK: University of Oklahoma Press.
—— (1993) 'The Clash of Civilizations', *Foreign Affairs* 72 (3): 22–49.
Hutchcroft, Paul D. (1998) *Booty Capitalism: The Politics of Banking in the Philippines*, Ithaca: Cornell University Press.
IMF (2000) 'Report of the IMF Executive Board Article IV Consultation with the Government of Indonesia' (reproduced in full in *Jakarta Post* 27 September 2000: 5).
—— (2002) *Indonesia: Selected Issues*, Washington, DC: IMF.
Indonesia Corruption Watch (2001) *Corruption Outlook 2001: Korupsi Dari Istana ke Multi-partai*, Jakarta: Indonesia Corruption Watch.
Indonesian Environmental Forum (WALHI) (1991) *Sustainability and Economic Rent in the Forestry Sector*, report of 10 October, Jakarta.
Indrawati, Sri Mulyani (2001) 'Indonesian Economy: Maintaining Recovery. Challenges under Current Political Uncertainty', paper presented to the Indonesia–Australia Bilateral Conference, Jakarta, 22–23 March.
Ingleson, John (1986) *In Search of Justice: Workers and Unions in Colonial Java, 1908–1926*, Singapore: Oxford University Press.
International Crisis Group (2002) 'Al-Qaeda in Southeast Asia: The case of the "Ngruki Network"', Jakarta and Brussels, 8 August.
International Labour Organization (1998) *Employment Challenges of the Indonesian Economic Crisis*, Jakarta: United Nations Development Programme.
IPCOS (Institute for Policy and Community Development Studies) (2000) 'Kepemimpinan Politik Pasca-Pemilu 1999: Arus Bawah Politik', preliminary study.
ISEI (Ikatan Sarjana Ekonomi Indonesia) (1990) 'Penjabaran Demokrasi Ekonomi', report issued by ISEI, Jakarta, 15 August.
Jayasuriya, Kanishka (1997) *NIASnytt* 4 (December): 19–27.
—— (ed.) (1999) *Law, Capitalism and Power in Asia: the Rule of Law and Legal Institutions*, London and New York: Routhledge.
—— (2000) 'Authoritarian Liberalism, Governance and the Emergence of the Regulatory State in Post-Crisis East Asia', in Richard Robison, Mark Beeson, Kanishka Jayasuriya

and Hyuk Rae Kim (eds) *Politics and Markets in the Wake of the Asian Crisis*, London: Routledge, pp. 315–330.

Jayasuriya, Kanishka and Andrew Rosser (1999) 'Economic Orthodoxy and the East Asian Crisis', paper presented at the Annual Conference of the Australian Political Science Association, 27–29 September.

—— (2001) 'Economic Crisis and the Political Economy of Liberalisation in Southeast Asia', in Garry Rodan, Kevin Hewison and Richard Robison (eds) *The Political Economy of Southeast Asia: An Introduction*, Melbourne: Oxford University Press.

Jenkins, David (1984) *Suharto and His Generals: Indonesian Military Politics, 1975–1983*, Ithaca: Cornell Modern Indonesia Project Monograph No. 64.

—— (1998) 'The Handover', *Sydney Morning Herald* 23 May 1998.

Jessop, Bob (1983) 'Capitalism and Democracy: the Best Possible Shell?', in David Held *et al.* (eds) *States and Societies*, Oxford: Basil Blackwell, pp. 272–289.

Johnson, Chalmers (1982) *MITI and the Japanese Miracle: the Growth of Industrial Policy 1925–1975*, Stanford, CA: Stanford University Press.

Johnson, Colin (1998) 'Survey of Recent Developments', *Bulletin of Indonesian Economic Studies* 34 (2): 3–60

Jones, Steven (1986) 'Soeharto's Kin Linked with Plastics Monopoly', *Asian Wall Street Journal* 25 November.

Jones, Steven and Raphael Pura (1986) 'Suharto-Linked Monopolies Hobble Economy', 'Soeharto's Kin Linked With Plastics Monopoly', 'Suharto Family Tied to Oil Trade', *Asian Wall Street Journal* 24–26 November.

Kammen, Douglas (1997) 'A Time to Strike: Industrial Strikes and Changing Class Relations in New Order Indonesia', unpublished PhD Thesis, Cornell University, Ithaca.

Kartasasmita, Ginandjar (1985) 'Daya Tahan Ekonomi dan Kekuatan Dalam Negeri', *Prisma* 6: 42–47.

King, Dwight Y. (1982) *Interest Groups and Political Linkage in Indonesia: 1800–1965*, Detroit: Cellar Book Shop.

—— (2000) 'The 1999 Electoral Reforms in Indonesia: Debate, Design and Implementation', *Southeast Asian Journal of Social Science* 28 (2): 89–110.

Kingdom, John (1992) *No Such Thing as Society? Individualism and Community*, Buckingham: Open University Press.

Kingsbury, Damien and Arief Budiman (eds) (2001) *Indonesia: the Uncertain Transition*, Adelaide: Crawford House Publishing.

KPPT (Komite Pemantau Pelaksanaan Tugas BPPN) (2001) 'PT Chandra Asri Petrochemical Centre' and 'PT Permadani Khatulistiwa Nusantara', reports issued by KPPT, Jakarta, 29 August.

Kompas (1999) *Partai-Partai Politik Indonesia: Ideologi, Strategi dan Program*, Jakarta: Kompas.

Krueger, Anne O. (1974) 'The Political Economy of the Rent-Seeking Society', *American Economic Review* 64: 291–303.

Krugman, P. (1998) 'What Happened to Asia?', cited 9 August 2003 (http://web.mit.edu/krugman/www/disinter.html).

Kurth, J. (1979) 'Industrial Change and Political Change: A European Perspective', in David Collier (ed.) *The New Authoritarianism in Latin America*, Princeton: Princeton University Press.

Kwik Kian Gie (1993) 'A Tale of a Conglomerate', *Economic and Business Review Indonesia* 5 June: 26–27; 12 June: 26–27.

Lal, Deepak (1983) *The Poverty of Development Economics*, London: Institute of Economic Affairs.

Lande, Laurie (2000) 'Cleaning up Indonesia's Financial Debris', cited 9 August 2003 (http://groups.yahoo.com/group/berita-bhinneka/message/29398?source=1).

Lane, Max (1989a) 'Students on the Move', *Inside Indonesia* 19: 10–15.

—— (1989b) *State, Society and Democratisation in Contemporary Indonesia: A Class Approach*, Canberra: unpublished mimeo.

Leftwitch, Adrian (1994) 'Governance, the State and Economic Development', *Development and Change* 25 (2): 363–386.

Lenin, V.I. (1963) *Selected Works*, vol. 1: 296. Cited in Bob Jessop, 'Capitalism and Democracy: the Best Possible Shell', in David Held *et al.* (ed.) (1983) *States and Societies*, Oxford: Basil Blackwell, p. 279.

Lev, D. (1966) *The Transition to Guided Democracy: Indonesian Politics 1957–59*, Ithaca: Cornell Modern Indonesia Project.

Levine, D. (1969) 'History and Social Structure in the Study of Contemporary Indonesia', *Indonesia* 7: 5–19

Leys, Colin (1996) *The Rise and Fall of Development Theory*, London: James Currey.

Liddle, R. William (1989) 'Development or Democracy', *Far Eastern Economic Review* 9 November: 22, 23.

—— (1991) 'The Relative Autonomy of the Third World Politician: Soeharto and Indonesian Economic Development in Comparative Perspective', *International Studies Quarterly* 35: 403–427.

—— (1992a) 'The Politics of Development Policy', *World Development* 20 (6): 793–807.

—— (1992b) 'Regime in Crisis? Presidential Succession, the East Timor Massacre and Prospects for Democratisation in Indonesia', paper presented at the 44th Annual Meeting of the Association for Asian Studies, Washington, DC, 2–5 April.

—— (1999) 'Indonesia's Unexpected Failure of Leadership', in Adam Schwarz and Jonathan Paris (eds) *The Politics of Post-Suharto Indonesia*, New York: Council on Foreign Relations Press, pp. 16–39.

—— (2001) 'Indonesia's Democratic Transition: Playing by the Rules', in A. Reynolds (ed.) *The Architecture of Democracy*, Oxford: Oxford University Press, pp. 373–399.

Lindsey, Timothy (1999) 'Indonesia's *Negara Hukum*: Walking the Tightrope to the Rule of Law', in Arief Budiman, Barbara Hatley and Damien Kingsbury (eds), *Reformasi: Crisis and Change in Indonesia*, Melbourne: Monash Asia Institute.

—— (2000) 'Black Letter, Black Market and Bad Faith: Corruption and the Failure of Law Reform', in Chris Manning and Peter van Diermen (eds) *Indonesia in Transition: Social Aspects of Reformasi and Crisis*, Singapore: Institute of Southeast Asian Studies, pp. 278–292.

Linnan, David K (1999) 'Insolvency Reform and the Indonesian Financial Crisis', *Bulletin of Indonesian Economic Studies* 35 (2): 107–137.

Linz, Juan J. (1997), 'Some Thoughts on the Victory and Future of Democracy', in Axel Hadenius (ed.) *Democracy's Victory and Crisis*, Cambridge: Cambridge University Press, pp. 404–426.

Lubis, Todung Mulya (1993) *In Search of Human Rights: Legal-Political Dilemmas of Indonesia's New Order, 1966–1990*, Jakarta: PT Gramedia Pustaka Utama in cooperation with SPES Foundation.

Lucas, Anton and Carol Warren (2000) 'Agrarian Reform in the Era of Reformasi', in Chris Manning and Peter van Dierman (eds) *Indonesia in Transition: Social Aspects of Reformasi and Crisis*, Singapore: Institute of Southeast Asian Studies, pp. 220–238.

McBeth, John (1998) 'Political Update', in Geoff Forrester (ed.) *Post-Soeharto Indonesia: Renewal or Chaos*, Bathurst: Crawford Publishing.

McBeth John and Michael Vatiokiotis (1998) 'The Endgame', *Far Eastern Economic Review* 28 May.

MacDougall, John (1993) 'The Golkar Congress – Suharto Gets His Way', *Inside Indonesia* 37 (December).

McFaul, Michael (2002) 'The Fourth Wave of Democracy and Dictatorship: Noncooperative Transitions in the Postcommunist World', *World Politics* 54 (January): 212–244.

MacIntyre, Andrew (1989) *Politics, Policy and Participation: Business–Government Relations in Indonesia*, PhD Thesis, Australian National University, Canberra.

—— (1991) *Business and Politics in Indonesia*, Sydney: Allen and Unwin.

MacIntyre, Andrew (1992) 'Politics and the Reorientation of Economic Policy in Indonesia', in A. MacIntyre and K. Jayasuriya (eds) *The Dynamics of Economic Policy Reform in South-east Asia and the South-west Pacific*, Singapore: Oxford University Press, pp. 138–157.

—— (1993) 'The Politics of Finance in Indonesia: Command, Confusion, and Competition', in Stephen Haggard, Chung H. Lee and Sylvia Maxfield (eds) *The Politics of Finance in Developing Countries*, Ithaca: Cornell University Press.

—— (1994) 'Business, Government and Development: Northeast and Southeast Asian Comparisons', in A. MacIntyre (ed.) *Business and Government in Industrialising Asia*, Sydney: Allen and Unwin, pp. 1–28.

—— (1999) 'Political Institutions and Economic Crisis in Thailand and Indonesia', in T.J. Pempel (ed.) *Politics of the Asian Economic Crisis*, Ithaca: Cornell University Press, pp. 143–162.

—— (2001) 'Institutions and Investors: The Politics of the Financial Crisis in Southeast Asia', *International Organization* 55 (1): 81–122.

McLeod, Ross (1994) 'Introduction', in Ross McLeod (ed.) *Indonesia Assessment 1994: Finance as a Key Sector in Indonesia's Development*, Singapore: Institute of Southeast Asian Studies; Canberra: Research School of Pacific Studies, Australian National University.

—— (1997) 'Postcript to the Survey of Recent Developments: On Courses and Cures For the Rupiah Crisis', *Bulletin of Indonesian Economic Studies* 33 (3): 35–42

—— (1998) 'Indonesia', in Ross McLeod and Ross Garnaut (eds) *East Asia in Crisis: From Being a Miracle to Needing One*, London: Routledge, pp. 31–48.

—— (1999) 'Crisis-Driven Changes to the Banking Laws and Regulations', *Bulletin of Indonesian Economic Studies* 35 (2): 147–154.

—— (2000a) 'Soeharto's Indonesia: A Better Class of Corruption', *Agenda* 7 (2): 99–112.

—— (2000b) "Survey of Recent Developments', *Bulletin of Indonesian Economic Studies* 36 (2): 5–40.

McLeod, Ross and R. Garnaut (eds) (1998) *East Asia in Crisis: From Being a Miracle to Needing One?*, London: Routledge.

McVey, R. (1992) 'The Materialization of the Southeast Asian Entrepreneurs', in R. McVey (ed.) *Southeast Asian Capitalists*, Ithaca: Cornell Southeast Asia Program.

Magenda, Burhan (1991) *East Kalimantan: The Decline of a Commercial Aristocracy*, Ithaca: Cornell Modern Indonesia Project Monograph Series no. 70.

Mahbubani, Kyshore (1993) 'The Danger of Decadence', *Foreign Affairs* 72 (4): 10–14.

Majelis Permusyawaratan Rakyat Republik Indonesia (MPR) (1998) 'Ketetapan-ketetapan MPR-RI Hasil Sidang Istimewa 1998', document issued by MPR.

—— (1999) 'Ketetapan-Ketetapan MPR RI Hasil Sidang Umum MPR RI Tahun 1999', document issued by MPR.

Mallaby, Sebastian (2002) 'The Reluctant Imperialist: Terrorism, Failed States and the Case for American Empire', *Foreign Affairs* 81 (2): 2–7.

Manning, Chris and Peter Van Dierman (eds) (2000) *Indonesia in Transition: Social Aspects of Reformasi and Crisis*, Singapore: Institute of Southeast Asian Studies.

Marx, Karl (1904) *A Contribution to the Critique of Political Economy*, Chicago: Charles H. Kerr and Company, pp. 10–15. Cited in Lewis C. Feuer (1969) *Marx and Engels: Basic Writings on Politics and Philosophy*, Glasgow: Fontana.

—— (1963) *The 18th Brumaire of Louis Bonaparte*, New York: International Publishers.

Marx, Karl and F. Engels (1956) *The Holy Family*, Moscow: Foreign Language Publishing House. Cited in Ralph Milliband (1989) 'Marx and the State', in Graham Duncan (ed.) *Democracy and the Capitalist State*, Cambridge: Cambridge University Press, pp. 80–85.

—— (1962) *Karl Marx and Frederick Engels: Selected Works*, Moscow: Foreign Languages Publishing House.

Mietzner, Marcus (1999) 'From Soeharto to Habibie: the Indonesian Armed Forces and Political Islam during the Transition', in Geoff Forrester (ed.) *Post-Soeharto Indonesia: Renewal or Chaos?*, Bathurst: Crawford Publishing, pp. 65–104.

Miliband, Ralph (1983) 'State Power and Class Interests', *New Left Review* 138: 57–68.

—— (1989) *Divided Societies: Class Struggle in Contemporary Capitalism*, Oxford: Oxford University Press.

Moertopo, Ali (1973) *The Acceleration and Modernization of 25 Years Development*, Jakarta: Yayasan Proklamasi and Center for Strategic and International Studies.

—— (1975) *Buruh dan Tani dalam Pembangunan*, Jakarta: Center for Strategic and International Studies.

—— (1981) *Strategi Pembangunan Nasional*, Jakarta: Yayasan Proklamasi and Center for Strategic and International Studies.

Moon, Chung-in and Sang Young Ryu (2000) 'The State, Structural Rigidity, and the End of Asian Capitalism: A Comparative Study of Japan and South Korea', in Richard Robison *et al.* (eds) *Politics and Markets in the Wake of the Asian Crisis*, London: Routledge, pp. 77–98.

Moore, Barrington (1966) *Social Origins of Dictatorship and Democracy*, London: Penguin.

Mortimer, R. (1969) 'Class, Social Cleavage and Indonesian Communism', *Indonesia* 8.

—— (1973) 'Indonesia: Growth or Development', in Rex Mortimer (ed.) *Showcase State: The Illusion of Indonesia's Accelerated Modernization*, Sydney: Angus and Robertson.

Muir, Ross, (1991) 'Survey of Recent Developments', *Bulletin of Indonesian Economic Studies* 27 (3).

Mussa, M. and G. Hacche (1998) 'Take the IMF Medicine and You Will Soon Mend', *International Herald Tribune* 17–18 January: 6.

Naipaul, V.S. (1998) 'Indonesia: The Man of the Moment', *New York Review of Books* 11 June.

Nasution, Adnan Buyung (1992) *The Aspiration for Constitutional Government in Indonesia: A Socio-legal Study of the Indonesian Konstituante 1956–1959*, Jakarta: Pustaka Sinar Harapan.

Nasution, Anwar (1992) 'The Years of Living Dangerously: The Impacts of Financial Sector Policy Reforms and Increasing Private Sector Indebtedness in Indonesia, 1983–1992', *Indonesian Quarterly* 20 (4): 405–435.

—— (1995) 'Survey of Recent Developments', *Bulletin of Indonesian Economic Studies* 31 (2).

Nelson, Joan (1990) *Economic Crisis and Policy Choice: The Politics of Adjustment in the Third World*, Princeton: Princeton University Press.

North, Douglass (1981) *Structure and Change in Economic History*, New York: Norton.

—— (1994) 'Economic Performance Through Time', *American Economic Review* 84 (3): 359–368.

—— (1995) 'The New Institutional Economics and Third World Development', in John Harris, Jane Hunter and Colin M. Lewis (eds) *The New Institutional Economics and Third World Development*, London: Routledge, pp. 17–26.

North, Douglass and R. Thomas (1970) 'An Economic Theory of the Growth of the Western World', *The Economic History Review* 23 (1): 1–17.

O'Donnell, Guillermo and Philippe C. Schmitter (1986) 'Tentative Conclusions about Uncertain Democracies', in Guillermo O'Donnell, Philippe C. Schmitter and Laurence Whitehead (eds) *Transitions from Authoritarian Rule: Prospects for Democracy*, Baltimore and London: Johns Hopkins University Press, pp. 3–72.

Olson, Mancur (1982) *The Rise and Decline of Nations: Economic Growth, Stagflation and Structural Rigidities*, New Haven: Yale University Press.

Pangaribuan, Robinson (1988) *Perkembangan Kekuasaan Sekretariat Negara Dalam Jajaran Politik Nasional Periode 1945–1987*, Jakarta: Universitas Indonesia, Fakultas Ilmu Sosial dan Ilmu Politik, Skripsi Sarjana.

—— (1995) *The Indonesian State Secretariat 1945–1993*, trans. and ed. V.R. Hadiz, Perth: Asia Research Centre, Murdoch University.

Pangestu, Mari (1989) 'Economic Policy Reforms in Indonesia', *Indonesian Quarterly* 17 (3): 218–233.

—— (1991) 'The Role of the Private Sector in Indonesia: Deregulation and Privatisation', *Indonesian Quarterly* 19 (1): 27–51.

—— (1995) 'The Role of the State and Economic Development in Indonesia', *Indonesian Quarterly* 21 (3).

—— (1996) *Economic Reform, Deregulation and Privatization: The Indonesian Experience*, Jakarta: Centre for Strategic and International Studies.

—— (1998) 'More Misery Ahead', *Far Eastern Economic Review* 19 February: 52–53.

—— (2000) 'Seize Momentum, Ensure Change in Right Direction', *Jakarta Post* 21 December: 4.

—— (2001) 'Survey of Recent Developments', *Bulletin of Indonesian Economic Studies* 37 (2): 141–71.

Pardede, Raden (1999) 'Survey Of Recent Developments', *Bulletin of Indonesian Economic Studies* 35 (2): 3–39.

Pauker, Guy (1968) 'Indonesia: The Age of Reason', *Asian Survey* 13 (2).

PDI-P (Partai Demokrasi Indonesia-Perjuangan) (1999) *PDI Perjuangan Menjawab*, Jakarta: PDIP.

Pempel, T.J. (1998) *Regime Shift: Comparative Dynamics of the Japanese Political Economy*, Ithaca and London: Cornell University Press.

Pereira, Luois, Carlos Bresser, and Yoshiaki Nakano (1998) 'The Missing Social Contract: Governability and Reform in Latin America', in Philip D. Oxhorn and Graciela

Ducatenzeiler (eds) *What Kind of Democracy? What Kind of Market? Latin America in the Age of Neo Liberalism*, Pennsylvania: Pennsylvania University Press.
Perlez, Jane (2002) 'US Backs Oil Giant on Lawsuit in Indonesia', *New York Times* 8 August: 11.
Perraton, J., D. Goldblat, D. Held and A. McGrew (1997) 'The Globalisation of Economic Activity', *New Political Economy* 2 (2): 257–278.
Petras, James (1989) 'State, Regime and the Democratic Muddle', *Journal of Contemporary Asia* 19 (1).
Pincus, Jonathan and Rizal Ramli (1998) 'Indonesia: From Showcase to Basket Case', *Cambridge Journal of Economics* 22: 728–732.
Polanyi, Karl (1957) *The Great Transformation*, Boston: Beacon.
Prawiranegara, Sjafruddin (1983) 'Pancasila Sebagai Azas Tunggal', in *Perihal: Pancasila Sebagai Azas Tunggal*, Jakarta: Dewan Dakwah Islamiah Indonesia, pp. 7–18.
Prawiro, Radius (1990) 'Kembali ke Kebijaksanaan Ekonomi Pasar', *Eksekutif* January: 48–53.
PT Data Consult Inc. (1991) *Anatomy of Indonesian Conglomerates*, Jakarta: PT Data Consult.
Pura, Raphael (1986) 'Suharto Family Tied to Indonesian Oil Trade', *Asian Wall Street Journal* 24 November.
—— (1995a) 'Bob Hasan Builds an Empire in the Forest', *Asian Wall Street Journal* 20–21 January: 1, 4.
—— (1995b) 'Indonesian Plywood Cartel under Fire as Sales Shrink', *Asian Wall Street Journal* 23 January: 1, 4.
Putnam, R. (1993) *Making Democracy Work: Civic Traditions in Modern Italy*, Princeton: Princeton University Press.
Radelet, S. (1995) 'Indonesian Foreign Debt: Headed for a Crisis or Financing Sustainable Growth?', *Bulletin of Indonesia Economic Studies* 31 (3): 39–72.
Radelet, S. and J. Sachs (1997) 'Asia's Re-emergence', *Foreign Affairs* 76 (6): 44–59.
—— (1998) 'The Onset of the East Asian Financial Crisis', Harvard Institute for International Development, Cambridge, MA. Draft paper cited on 9 August 2003 (http://www.stern.nyu.edu/globalmacro/).
Rais, Amien (1998) *Suara Amien Rais, Suara Rakyat*, Jakarta: Gema Insani Press.
Ramage, Douglas (1997) *Politics in Indonesia: Democracy, Islam and the Ideology of Tolerance*, London: Routledge.
Ramli, Rizal (1991) 'Hutang Luar Negeri Indonesia: Kontraksi and Beban Ekonomi', *Prisma* September: 3–11.
—— (1992) 'Kayu', *Tempo* 13 June: 77.
Reeve, David (1978) 'Sukarnoism and Indonesia's "Functional Group" State: Developing "Indonesian Democracy" ', *Review of Indonesian and Malay Affairs* 12 (2): 43–94.
—— (1979) 'Sukarnoism and Indonesia's "Functional Group" State: Implementing "Indonesian Democracy"', *Review of Indonesian and Malay Affairs* 13 (1): 53–115.
—— (1990) 'The Corporatist State: The Case of Golkar', in Arief Budiman (ed.) *State and Civil Society in Indonesia*, Clayton: Monash University Centre of Southeast Asian Studies, Monograph No. 22.
Richburg, Keith (1999) 'Seven Days in May that Toppled a Titan: Backroom Intrigue Led to Soeharto's Fall', in Edward Aspinall, Herb Feith and Gerry Van Klinken (eds) *The Last Days of President Soeharto*, Clayton: Monash Asia Institute, Monash University, pp. 75–80.

Rice, Robert and Hal Hill (1977) 'Survey of Recent Developments', *Bulletin of Indonesian Economic Studies* 5 (2).

Roberts, P.C. (1997) 'Asian Crisis Proves That Industrial Policy Doesn't Pay', *Australian Financial Review* 20–21 December: 47.

Robison, Richard (1986) *Indonesia: The Rise of Capital*, Sydney: Allen and Unwin.

—— (1987) 'After the Gold Rush: The Politics of Economic Restructuring in Indonesia in the 1980s', in Richard Robison *et al.* (eds), *Southeast Asia in the 1980s: The Politics of Economic Crisis*, Sydney: Allen and Unwin, pp. 16–51.

—— (1988) 'Authoritarian States, Capital Owning Classes, and the Politics of Newly Industrializing Countries: The Case of Indonesia', *World Politics* 41 (1): 52–74.

—— (1992) 'Industrialisation and the Economic and Political Development of Capital: The Case of Indonesia', in Ruth McVey (ed.) *Southeast Asian Capitalists*, Ithaca: Cornell University Press.

—— (1993) 'Indonesia: Tensions in State and Regime', in Kevin Hewison, Richard Robison, and Garry Rodan (eds) *Southeast Asia in the 1990s: Authoritarianism, Democracy and Capitalism*, Sydney: Allen and Unwin.

—— (1994) 'Organising the Transition: Indonesian Politics in 1993/94', *Indonesia Assessment 1994: Finance as a Key Sector in Indonesia's Development*, Singapore: Institute of Southeast Asian Studies; Canberra: Research School of Pacific Studies, Australian National University, pp. 49–74.

—— (1996) 'The Politics of Asian Values', *Pacific Review* 9 (3): 309–337.

—— (1997) 'Politics and Markets in Indonesia's Post-oil Era', in Garry Rodan, Kevin Hewison and Richard Robison (eds) *The Political Economy of Southeast Asia: An Introduction*, Melbourne: Oxford University Press.

—— (2001) 'Indonesia: Crisis, Oligarchy and Reform', in Garry Rodan, Kevin Hewison and Richard Robison (eds) *The Political Economy of Southeast Asia: Conflict, Crises, and Change*, Melbourne: Oxford University Press, pp. 104–137.

Robison, Richard and Vedi Hadiz (1993) 'Privatisation or the Reorganisation of Dirigism? Indonesian Economic Policy in the 1990s', *Canadian Journal of Development Studies* Special Issue (December): 13–31.

Robison, Richard, Mark Beeson, Kanishka Jayasuriya and Hyuk Rae Kim (eds) (2000) *Politics and Markets in the Wake of the Asian Crisis*, London: Routledge.

Robison, Richard, Kevin Hewison and Richard Higgott (eds) (1987) *Southeast Asia in the 1980s: The Politics of Economic Crisis*, Sydney: Allen and Unwin.

Robison, Richard and Andrew Rosser (1998) 'Contesting Reform: Indonesia's New Order and the IMF', *World Development* 26 (8).

—— (2000) '"Surviving the Meltdown": Liberal Reform and Political Oligarchy in Indonesia', in Richard Robison, Mark Beeson, Kanishka Jayasuriya and Hyuk Rae Kim (eds) *Politics and Markets in the Wake of the Asian Crisis*, London: Routledge, pp. 171–191.

Rocamora, J. (1974) *Nationalism in Search of an Ideology: The Indonesian Nationalist Party 1946–65*, PhD thesis, Cornell University, Ithaca.

Rodan, Garry (1996a) 'Theorising Political Opposition in East and Southeast Asia', in Garry Rodan (ed.) *Political Oppositions in Industrialising Asia*, London: Routledge, pp. 1–39.

—— (1996b) 'The Internationalisation of Ideological Conflict: Asia's New Significance', *Pacific Review* 9 (3): 328–351.

Rodan, Garry, Kevin Hewison and Richard Robison (eds) (2001) *The Political Economy of Southeast Asia: An Introduction*, Melbourne: Oxford University Press.

Rosser, Andrew (1999) *Creating Markets: The Politics of Economic Liberalisation in Indonesia Since the mid-1980s*, unpublished PhD thesis, Murdoch University, Perth.

—— (2002) *The Politics of Economic Liberalisation in Indonesia: State, Market and Power*, Richmond: Curzon.

—— (2003) 'Governance, Markets and Power: The Political Economy of Accounting Reform in Indonesia', in Geoffrey R.D. Underhill and Xiaoke Zhang (eds) *International Financial Governance Under Stress: Global Structures Versus National Imperatives*, Cambridge: Cambridge University Press, pp. 263–282.

Rowley, Anthony (1987) 'Economic Schizophrenia', *Far Eastern Economic Review* 10 November: 70–75.

Rudner, Martin (1976) 'The Indonesian Military and Economic Policy', *Modern Asian Studies* 10 (2).

—— (1991) 'Repelita V and the Indonesian Economy', *Review of Indonesian and Malay Affairs* 25 (2): 35–74.

Rueschemeyer, Dietrich and Peter B. Evans (1985) 'The State and Economic Transformation', in Peter B. Evans, Dietrich Rueschemeyer and Theda Skocpol (eds) *Bringing the State Back In*, Cambridge: Cambridge University Press, pp. 44–69.

Rueschemeyer, Dietrich, Evelyne Huber Stephens and John D. Stephens (1992) *Capitalist Development and Democracy*, Cambridge: Polity Press.

Ryter, Loren (1998) 'Pemuda Pancasila: The Last Loyalist Free Men of Soeharto's Order?', *Indonesia* 66: 45–73.

Sachs, Jeffrey (1997) 'IMF Orthodoxy Isn't What Southeast Asia Needs', *International Herald Tribune* 4 November: 8.

—— (1998a) 'The IMF and the Asian Flu', *The American Prospect* 37.

—— (1998b) 'Why the IMF Programs in Asia are Failing', *Project Syndicate* January, cited 9 August 2003 (http://www.project-syndicate.org/commentaries/commentary_text.php4?id=138&m=commentary).

Sadli, Mohammad (2000) 'Restoring Investor Confidence', *Jakarta Post* 2 September: 4.

Salim, Emil (1998) 'Reformasi Total', unpublished paper, 17 June.

Saludo, R and A. Shameen (1997) 'A Question of Openness', *Asiaweek* 3 October: 62, 63.

Samego, Indria *et al.* (1998) *Bila ABRI Berbisnis*, Bandung: Penerbit Mizan.

Sato, Yuri (1993) 'The Salim Group in Indonesia: The Development and Behaviour of the Largest Conglomerate in Indonesia', *The Developing Economies* 31 (4): 408–441.

Sacerdoti, Guy (1980) 'Acrobatic Technocrats Star in an Indonesian Balancing Act', *Far Eastern Economic Review* 16 May: 44–50

Schamis, Hector (1999) 'Distributional Coalitions and the Politics of Economic Reform in Latin America', *World Politics* 51 (January): 236–268.

—— (2002) *Re-forming the State: The Politics of Privatization in Latin America and Europe*, Ann Arbor: University of Michigan Press.

Schiller, Jim (1990) 'State Formation and Rural Transformation: Adapting to the "New Order" in Jepara', in Arief Budiman (ed.) *State and Civil Society in Indonesia*, Clayton: Monash University Centre of Southeast Asian Studies.

Schmitt, H.O. (1962) 'Foreign Capital and Social Conflict in Indonesia', *Economic Development and Cultural Change* 10 (3).

—— (1963) 'Post-Colonial Politics: A Suggested Interpretation of the Indonesian Experience', *The Australian Journal of Politics and History* 9 (2).

Schmitter, Philippe (1974) 'Still the Century of Corporatism?', *Review of Politics* 36: 85–101.

Schwarz, Adam (1990) 'Charity Begins at Home', *Far Eastern Economic Review* 4 October: 62–64.

—— (1994) *A Nation in Waiting: Indonesia in the 1990s*, Sydney: Allen and Unwin.

—— (1999) *A Nation in Waiting: Indonesia in the 1990s* (updated and expanded version), Sydney: Allen and Unwin.

Setiawan, Bonnie (1996) 'Organisasi Non-Pemerintah dan Masyarakat Sipil', *Prisma* 7 (July): 35–61.

Shin, Yoon Hwan (1989) 'Demystifying the Capitalist State: Political Patronage, Bureaucratic Interests and Capitalists in Formation in Soeharto's Indonesia', unpublished PhD thesis, Yale University, New Haven.

Shiraishi, Takashi (1990) *An Age in Motion: Popular Radicalism in Java, 1912–1926*, Ithaca: Cornell University Press.

—— (1994) 'Dukuh: A Golkar Village', in Takashi Shiraishi (ed.) *Approaching Suharto's New Order From the Margins*, Ithaca: Cornell University Southeast Asia Program.

Simandjuntak, Djisman S. (1989) 'Survey of Recent Developments', *Bulletin of Indonesian Economic Studies* 25 (1) April.

Simanjuntak, Marsillam (1994) *Pandangan Negara Integralistik: Sumber, Unsur dan Riwayatnya dalam Persiapan UUD 1945*, Jakarta: Grafiti.

Siregar, Reza Y. (2001) 'Survey of Recent Developments', *Bulletin of Indonesian Economic Studies* 37 (3): 277–303.

Sjahrir (1985) 'Privatisasi Menuju Efisiensi?', *Prisma* 7.

—— (1988) 'Ekonomi Politik Deregulasi', *Prisma* 9: 29–38.

—— (1991) 'Gebrakan Setelah Deregulasi Macet', *Tempo* 9 March: 87–90.

—— (1992) 'Demokrasi dan Pertumbuhan Ekonomi', *Warta Ekonomi* 22 June: 8.

Skocpol, Theda (1979) *States and Social Revolutions: A Comparative Analysis of France, Russia and China*, Cambridge: Cambridge University Press.

—— (1985) 'Bringing the State Back In: Strategies of Analysis in Current Research', in Peter B. Evans, Dietrich Rueschemeyer and Theda Skocpol (eds) *Bringing the State Back In*, Cambridge: Cambridge University Press, pp. 3–37.

Smith, Steve (2000) 'US Democracy Promotion: Critical Questions', in Michael Cox, John Ikenberry and Takashi Inoguchi (eds) *American Democracy Promotion: Impulses, Strategies and Impacts*, Oxford: Oxford University Press, pp. 63–84.

Soeharto (1988) *Pikiran, Ucapan dan Tindakan Saya: otobiografi seperti dipaparkan kepada G. Dwipayana and Ramadan K.H.*, Jakarta: PT Citra Lantoro Gung Persada. Available in English translation as *Soeharto, My Thoughts, Words and Deeds: An Autobiography as told to G. Dwipayana and Ramadan K.H.*, trans. Sumadi, Jakarta: Citra Lamtoro Gung Persada.

—— (1989) 'State Address'. Extracted and cited in David Bourchier and Vedi R. Hadiz (eds) (2003) *Indonesian Politics and Society: A Reader*, London: Routledge, pp. 192–195.

—— (1990) *Pidato Kenegaraan Presiden Republik Indonesia, Soeharto*, Jakarta: Departmen Kenegaraan Republik Indonesia, 16 August.

Soehoed, A.R. (1977) 'Commodities and Viable Economic Sectors: A Possible Basis for Development Planning', *Indonesian Quarterly* 5 (1).

—— (1982) 'Industrial Development During Pelita III', *Indonesian Quarterly* 10 (4).

Soekarno (1956) 'Let us Bury the Parties', in *Indonesia: Pilihlah Demokrasimu Jang Sedjati*, Jakarta: Ministry of Information. Cited in Herbert Feith and Lance Castles (eds)

(1970) *Indonesian Political Thinking 1945–1965*, Ithaca: Cornell University Press, pp. 81–83.
—— (1959) 'Returning to the Rails of Revolution', from a speech dated 17 August 1959. Cited in Herbert Feith and Lance Castles (eds) (1970) *Indonesian Political Thinking 1945–1965*, Ithaca: Cornell University Press, pp. 90–109.
Soekarnoputri, Megawati (1998) 'Pidato Ketua Umum DPP PDI, Megawati Soekarnoputri Menyambut HUT ke XXV PDI', speech given by Soekarnoputri, January.
Soemitro (1989) 'Aspiring to Normal Politics', *Far Eastern Economic Review* 6 April: 22–24.
—— (1990) 'Pembangunan Politik Sesudah Tahun 1993', *Analisis CSIS* 19 (1): 47–53.
Soesastro, Hadi (1989) 'The Political Economy of Deregulation', *Asian Survey* 29 (9): 853–869
—— (1991) 'Capital Formation and Equal Distribution', *Indonesian Quarterly* 19 (1): 62–73.
—— (1999) 'The 1999 Election and Beyond', *Bulletin of Indonesian Economic Studies* 35 (2): 139–146.
—— (2000) 'The Indonesian Economy Under Abdurrahman Wahid', in *Southeast Asian Affairs 2000*, Singapore: Institute of Southeast Asian Studies.
Soesastro, H. and M.C. Basri (1998) 'Survey of Recent Developments', *Bulletin of Indonesian Economic Studies* 34: 3–54.
Srinivasan, T.N. (1985) *Neo-classical Political Economy: The State and Economic Development*, New Haven: Economic Growth Center, Yale University.
Stiglitz, Joseph (1998a) 'Restoring the Asian Miracle', *Asian Wall Street Journal* 2 February: 8.
—— (1998b) 'Towards a New Paradigm for Development: Strategies, Policies and Processes', Prebisch Lecture at UNCTAD, Geneva, 19 October. Cited on 9 August 2003 (http://www.worldbank.org/html/extdr/extme/prebisch98.pdf).
Stone, J. (1997) 'Of Course Corrupt Asian Markets Will Crash', *Australian Financial Review* 30 October: 21.
Suh, Sangwon and Jose Manuel Tesoro (1999) 'Maneuvering to the Top Amid Chaos', *Asiaweek* 25 (43).
Suhartoyo (1981) 'Penanaman Modal dan Industrialisasi', paper given at a seminar on the Framework of National Development, Center of Strategic and International Studies, Jakarta.
Supriyanto, Enin (1989) 'Growth-oriented Strategy and Authoritarianism', in *Committee to Revatilise Student Activities, Bertarung Demi Demokrasi*, Bandung: Bandung Institute of Technology.
Suta, I Putu Gede Arya (2000a) 'Reforming the Indonesian Banking System', *Jakarta Post* 12 July: 5.
—— (2000b) 'Reforming the Indonesian Banking System', *Jakarta Post* 13 July: 5.
Sutter, J. (1959) *Indonesianisasi: Politics in a Changing Economy, 1940–55*, Data Paper No. 36 (4 vols), Ithaca: Department of Far Eastern Studies, Cornell University.
Tan, T.K. (ed.) (1967) *Sukarno's Guided Indonesia*, Brisbane: Jacaranda.
Taubert, A. (1991) 'Liberalism Under Pressure in Indonesia', in *Southeast Asian Affairs 1990*, Singapore: Institute of Southeast Asian Studies.
Therborn, Goran (1977) 'The Rule of Capital and the Rise of Democracy', *New Left Review* 103 (May–June): 3–41.
Thomas, K.D. and J. Panglaykim (1973) *Indonesia: The Effect of Past Policies and President Soeharto's Plans for the Future*, Melbourne: CEDA.
Toye, John (1987) *Dilemmas of Development*, Oxford: Basil Blackwell.

Trubek, David L. *et al.* (1993) 'Global Restructuring and the Law', Working Paper No. 1, Madison: Global Studies Research Program, University of Wisconsin.

Uhlin, Anders (1997) *Indonesia and the Third Wave of Democratization: The Indonesian Pro-Democracy Movement in a Changing World*, Richmond: Curzon.

Van Dijk, Kees (2001) *A Country in Despair: Indonesia Between 1997–2000*, Leiden: KITLV Press.

Van Klinken, Gerry (1999) 'How A Democratic Deal Might Be Struck', in *Reformasi: Crisis and Change in Indonesia*, Melbourne: Monash Asia Institute, Monash University, pp. 59–67.

Vatikiotis, M. (1993) *Indonesian Politics Under Soeharto: Order, Development and Pressure for Change*, London: Routledge.

Wade, Robert (1990) *Governing the Market: Economic Theory and the Role of Government in East Asian Industrialisation*, Princeton: Princeton University Press.

—— (1992) 'East Asia's Economic Success: Conflicting Perspectives; Partial Insights and Shaky Evidence', *World Politics* 44 (January): 270–320.

—— (1998) 'The Asian Debt and Development Crisis of 1997: Causes and Consequences', *World Development* 26 (8): 1535–1553.

Wade, Robert and Frank Venoroso (1998) 'The Asian Crisis: The High Debt Model Versus the Wall Street–Treasury–IMF Complex', *New Left Review* 228: 3–23.

Wagstaff, Jeremy (1999) 'Status Quo: Real Change Remains Elusive in Indonesia as Election Nears', *Asian Wall Street Journal* 3 May.

Wahid, Abdurrahman (1990) 'Kindling a Democratic Culture', interview in *Inside Indonesia* October: 4.

Walters, Patrick (1999) 'The Week of Living Dangerously', in Edward Aspinall, Herb Feith and Gerry Van Klinken (eds) *The Last Days of President Soeharto*, Clayton: Monash Asia Institute, Monash University, pp. 81–84

Ward, Ken (1974) *The 1971 Election in Indonesia: An Indonesian Case Study*, Monash Papers on Southeast Asia No. 2, Clayton: Monash University Centre of Southeast Asian Studies.

Wardhana, Ali (1994) 'Financial Reform: Achievements, Problems and Prospects', in Ross McLeod (ed.) *Indonesia Assessment 1994: Finance as a Key Sector in Indonesia's Development*, Singapore: Institute of Southeast Asian Studies; Canberra: Research School of Pacific Studies, Australian National University.

Weber, Max (1964) *The Theory of Social and Economic Organisation* (edited and with an introduction by Talcott Parsons), New York: Free Press.

—— (1978) *Economy and Society* vol. 2, Berkeley: University of California Press.

Weiss, Linda (1998) *The Myth of the Powerless State*, Cambridge: Polity Press.

Weiss, Linda and John M. Hobson (2000) 'State Power and Economic Strength Revisited: What's So Special about the Asian Crisis?', in Richard Robison *et al.* (eds) *Politics and Markets in the Wake of the Asian Crisis*, London: Routledge, pp. 53–74.

Whitehead, Laurence (1993) 'Introduction: Some Insights from Western Social Theory', *World Development* 21 (8): 1245–1261.

'Who's Who', *Jakarta Post.Com* (2003), cited on 7 August 2003 (http://www.thejakartapost.com/who/PeopleList.asp).

Wibisono, Christianto (1981) 'Saham Pri dan Non-Pri', *Tempo* 14 March: 70–72.

Williamson, J. (1990) 'What Washington Means by Policy Reform', in John Williamson (ed.) *Latin American Adjustment: How Much has Changed?*, Washington, DC: Washington Institute for International Economics, pp. 7–20.

Williamson, J. (1994) 'In Search of a Manual for Technopols', in J. Williamson (ed.) *The Political Economy of Policy Reform*, Washington, DC: Institute for International Economics, pp. 11–28.

Williamson, J. and S. Haggard (1994) 'The Political Conditions for Economic Reform', in J. Williamson (ed.) *The Political Economy of Policy Reform*, Washington, DC: Institute for International Economics.

Wilner, Ann Ruth (1973) 'The Neo-Traditional Accommodation to Political Independence: The Case of Indonesia', in John T. Mc Alister (ed.) *Southeast Asia: The Politics of National Integration*, New York: Random House, pp. 517–541.

Wingfield, Tom (2002) 'Democratization and Economic Crisis in Thailand: Political Business and the Changing Dynamic of the Thai State', in E.T. Gomez (ed.) *Political Business in East Asia*, London: Routledge, pp. 250–300.

Winters, Jeffrey (1996) *Power in Motion: Capital Mobility and the Indonesian State*, Ithaca: Cornell University Press.

—— (1997) 'The Dark Side of the Tigers', *Asian Wall Street Journal* 12–13 December: 10.

—— (2000) 'The Financial Crisis in Southeast Asia', in Richard Robison *et al.* (eds), *Politics and Markets in the Wake of the Asian Crisis*, London: Routledge, pp. 34–52.

Wolf C. (1998a) 'Blame Government for the Asian Meltdown', *Asian Wall Street Journal* 5 February: 14.

—— (1998b) 'Markets, Not Architects, Will Solve Economic Crisis', *Asian Wall Street Journal* 21 July: 8.

Wood, Ellen Meiksins (1990) 'The Uses and Abuses of Civil Society', *The Socialist Register* 1990: 60–84.

World Bank (1981) *Indonesia: Selected Issues of Industrial Development and Trade Strategy, Direct Private Foreign Investment in Indonesia*, Annex 5, Jakarta: World Bank Country Department III, East Asia and Pacific Region.

—— (1983) *World Development Report*, New York: Oxford University Press.

—— (1984) *Indonesia: Policies and Prospects for Economic Growth and Transformation*, Jakarta: World Bank.

—— (1985) *Indonesia: Policies for Growth and Employment*, Jakarta: World Bank.

—— (1991a) *Indonesia: Developing Private Enterprise*, Washington, DC: World Bank.

—— (1991b) *Managing Development: The Governance Dimension*, Discussion Paper, Washington, DC: World Bank.

—— (1992) *Strategy for Forest Sector Development in Asia*, Technical Paper 182 (Asia Technical Department Series), Washington, DC: World Bank

—— (1993a) *Indonesia: Sustaining Development*, Jakarta: World Bank Country Department III, East Asia and Pacific Regional Office, 25 May.

—— (1993b) *The East Asian Miracle: Economic Growth and Public Policy*, New York: World Bank.

—— (1994) *Indonesia: Stability, Growth and Equity in Repelita VI*, Jakarta: Country Department III, East Asia and Pacific Region, 27 May.

—— (1995) *Indonesia: Improving Efficiency and Equity – Changes in the Public Sector's Role*, Jakarta: World Bank Country Department III, East Asia and Pacific Region, 4 June.

—— (1996) *Indonesia: Dimensions of Growth*, Jakarta: World Bank Country Department III, East Asia and Pacific Region, 7 May.

—— (1997a) *World Development Report: The State in a Changing World*, Washington, DC: Oxford University Press.

—— (1997b) *Global Development Finance: Country Tables*, Washington, DC: World Bank.

—— (1997c) *Indonesia: Sustaining High Growth and Equity*, Jakarta: World Bank Country Department III, East Asia and Pacific Region.
—— (1998a) *East Asia: The Road to Recovery*, Washington, DC: World Bank.
—— (1998b) *Indonesia in Crisis: A Macroeconomic Update*, Washington DC: World Bank, 16 July.
—— (1999) *Indonesia: From Crisis to Opportunity*, Jakarta: World Bank, 21 July.
—— (2000) *Indonesia: Accelerating Recovery in Uncertain Times*, East Asia Poverty Reduction and Economic Management Unit, 13 October.
Yayasan Dharmais (1989–1990) *Dharmais*, Jakarta: Kantor Yayasan Dharmais.
Yayasan Supersemar (1989–1990) *Yayasan Supersemar*, Jakarta: Yayasan Supersemar.
YLBHI (Yayasan Lembaga Bantuan Hukum Indonesia) (1994) 'Laporan Pendahuluan Kasus Marsinah', report by fact-finding team, March.
Young, Ken (1999) 'Post-Suharto: A Change of Regime?', in Arief Budiman, Barbara Hatley and Damien Kingsbury (eds) *Reformasi: Crisis and Change in Indonesia*, Clayton: Monash Asia Institute, Monash University, pp. 69–104.
Zakaria, Fareed (1994) 'Culture is Destiny: A Conversation with Lee Kuan Yew', *Foreign Affairs* 73 (2): 109–126.
—— (1997) 'The Rise of Illiberal Democracy', *Foreign Affairs*, 76 (6): 22–43.
Zingales, Luigi and Robert McCormack (2003) 'A Choice Between the Rich and the Markets', *Financial Times* 3 March: 13.
Zysman, John (1994) "How Institutions Create Historically Rooted Trajectories of Change', *Industrial and Corporate Change* 3 (1): 243–283.

Journals, newspapers and weeklies

Adil
Agence France Presse
Antara
Australian Financial Review
Asia Inc
Asia Times
Asian Wall Street Journal (*AWSJ*)
Asiaweek
Australian
Bisnis Indonesia
Bulletin of Indonesian Economic Studies (*BIES*)
Business Week
Detik
D&R
Economic and Business Review Indonesia
Economist
Editor
Eksekutif
Expresso
Far Eastern Economic Review (*FEER*)
Forum Keadilan (*FK*)
Gatra
Independen

Indikator
Indonesia Business Weekly
Indonesian Commercial Newsletter
Indonesian Observer
Indonesian Quarterly
Infobank
Jakarta Jakarta
Jakarta Post (*JP*)
Jawa Pos
Kedaulatan Rakyat
Kompas
Matra
Media Indonesia
Merdeka
New York Times
Panji Masjarakat
Pelita
Prospek
Prospektif
Republika
Reuters
Sinar Harapan
South China Morning Post
Straits Times
Suara Karya
Suara Pembaruan
Surabaya Pos
Surya
Tempo
Tiara
Time
Tiras
Ummat
Warta Ekonomi

Online periodicals

Business Times Online (http://business-times.asia1.com.sg/)
Detik.Com (http://www.detik.com/)
Jakarta Post.Com (http://www.thejakartapost.com/)
Jawa Pos Online (http://www.jawapos.co.id/)
Kapital News Online (http://www.kapital.co.id/)
Kompas Cybermedia (http://www.kompas.com/index1.htm)
Kontan Online (http://www.kontan-online.com/)
Laksamana.net (http://laksamana.net/)
Pikiran Rakyat Online (http://www.pikiran-rakyat.com/)
Republika Online (http://www.republika.co.id/)
SiaR (http://apchr.murdoch.edu.au/minihub/siarlist/maillist.html)

Tempo Interactif (http://www.tempointeraktif.com/)
Tempo Online (http://www.tempo.co.id/)
Time.Com (http://www.time.com/time/)
Warta Ekonomi.Com (http://www.wartaekonomi.com/)

INDEX

Abda'oe, Faisal 209
Abdulrahim, Imaduddin 115
Abeng, Tanri 141, 159, 201, 205, 210
ABRI *see* military
Aceh 248, 257, 263
Aditjondro, George 58, 214, 248
administrative patrimonialism 42–3
Afiff, Saleh 90
agency 2, 41
Agung, Permana 207
Aidit, D.N. 45
Albright, Madeleine 182n
Alliance of Independent Journalists (AJI) 123
Allison, J.M. 8
Amang, Beddu 189, 211
Amsden, Alice 4
Anam, Subagyo 238
Anderson, Benedict 30, 34, 38n, 46, 47, 53, 126, 130, 134, 231–2, 247
Anspach, R. 44
Antlov, Hans 63–4
Anwar, Dewi Fortuna 211
Apkindo 87–8, 157–8, 200, 208
Aribowo, Tunky 91, 93
Arief, S. 47
Arifin, Bustanil 52, 57, 84, 209
Arndt, Heinz 8, 35n, 46
Arryman, Arief 236
Arthur Andersen 205
Asaat 45
Asian Crisis 141
Asian Development Bank 158
Asian economic crisis 4, 147–50, 258, 259; fracturing oligarchy 151–5; impact 150–1; neo-liberalism and convergence 6–10; and political capitalism 155–60; and political crisis 160–1, 164, 165–6, 174; public choice theory 21; significance 10–12
Aspinall, Edward 167
ASPRI 48
asset sales 197–9, 264
Association of Indonesian Economists (ISEI) 96, 133
Association of Indonesian Muslim Intellectuals *see* ICMI
Astra group 56, 75, 83, 91–2, 197–8
Atmadjaja, Usman 210
Attorney General's Office (AGO) 208, 210, 212
Australia 248
Australian, The 262
authoritarian corporatism 121–3, 225
authoritarianism 1, 13, 40–1, 120, 138–9; end of political openness 139–40; Guided Democracy 45; New Order 49; Soeharto's re-election 140–3; Soekarno 44
automobile industry 54, 56, 75, 90–2, 158, 200
azas tunggal 49, 104, 228

Baird, Mark 196
Baker, Gerard 33, 262
Bakorstanas 107
Bakrie, Aburizal 60, 85–6, 133, 141, 209, 218n
Bakrie family 79
Balak 204
Bali 248, 262
balkanisation thesis 247–8
Bandung group 111–12
Bandung Institute of Technology 169
Bank Andromeda 157

Bank Bali 189, 192, 208, 209, 210, 211, 214, 217n, 234–5, 242, 260
Bank Bumi Daya 81, 82, 92, 93
Bank Central Asia (BCA) 58, 153, 191–2, 198, 199, 220–1n, 260
Bank Dagang Negara 81, 93
Bank Duta 86–7
Bank Indonesia 81, 88, 155, 189; Liquidity Assistance 192–3; reform 204–5; rescue programme 83; SBI promissory notes 192; securities 94
Bank Niaga 198, 260
Bank Pacific 152
Bank Rakyat 81
Bank Summa 83, 95
Bank Umum Nasional 191
banks 52, 71–2, 97n, 99–100n, 187; economic crisis 148, 150, 152; and foundations 55; liberal criticism 125; and oligarchy 57, 58, 73, 74, 75; predatory raids 80–4, 154; recapitalisation 191–2; reform 156–7, 204–5; Sumarlin shocks 94–5
Banpres 59, 94
Banser 241
Bapindo 81–2, 99n, 204
Bappenas 59, 65, 76
Baramuli, Arnold 58–9, 173, 210, 242
Bardhan, Pranab 27
barisan sakit hati 120
Barro, Robert 35
Basri, Faisal 236
Basri, M.C. 153
Baswir, Revrisond 244
Batam 58
Bates, Robert 19
Batubara, Ridwan 246–7
Bawazier, Fuad 142, 159, 160, 189, 232, 237
BDNI 191
beer tax 76–7, 87
Beeson, Mark 26
Bellin, Eva 256
Benda, Harry J. 40–1
Bentoel 95
Berdikari 57
bersih lingkungan 107
Bhattacharya, A. 72
BI *see* Bank Indonesia
Bimantara 79, 92, 154
BKPM 59, 76
BLBI 192–3, 195, 204, 210, 218–19n
Bogasari 55, 56, 58, 74
Boileau, J. 49
Bourchier, David 120, 137, 242
bourgeoisie 31–3, 38n, 126, 130–4
BPIS (Board of Strategic Industries) 59, 87, 89
BPK 82, 189, 193, 205–6
BPKP 189, 201, 203, 205–6
BPPC 88, 200
BPPT (Board for Technology, Research and Development) 76, 87, 89, 237
Bresnan, John 21
Brown, Colin 107, 127
Brunei 244
Buchanan, James M. 19
Buchori, Mochtar 239, 240
Budianta, Melani 168
Budiman, Arief 179, 183n, 253–4, 256
Bulog 51–2, 54, 57, 58, 74, 205; Chinese business groups 55–6; corruption 208, 209, 221n, 222n; and political parties 206, 214, 234; reform 155, 159, 200, 203; Sekneg 59; Wahid 244
Busang goldmine scandal 118, 119n
Bustanil, Arifin 52, 57, 84, 209

Camdessus, Michel 6, 7, 158, 259
Cameron, Lisa 197
capital flight 95–6, 155, 160, 261
capitalism: civil society and bourgeoisie 31–3; collisions 26–7, 37–8n; and democracy 125, 134; institutions , collisions and social conflict 27–9; New Order 40–1
Castles, Lance 44, 46
Cendana group 78–9, 83, 154, 162n, 202
Cendana Palace 43
Central Executive Board 111
Centre for Information and Development Studies 89, 127
Centre for Policy and Development Studies 116–17, 127
CGI 72
Chalid, Idham 63
Chalmers, Ian 45, 50
Chamber of Commerce and Industry 131–2, 197
Chandra, Djoko S. 208, 210, 211
Chandra Asri 92–4, 196, 220n
Chaudhry, Kiren Aziz 5, 27
China 96
Chinese business groups 32, 33, 58, 60, 85,

131; asset sales 197; conflicts over rents 88; corruption 213; economic crisis 150; forestry industry 74; and military 226; and oligarchy 53, 54, 73; and political parties 232, 239–40, 250n; rise 51, 54–7; xenophobia towards 44, 132–3, 159, 169–70
CIDES 89, 127
Ciganjur group 178–9
Citra Lamtorogung group 79
civil society 31, 43, 104–5; disorganisation 103, 123, 138, 168, 253; militarisation 225; New Order 48–50
civilian militia 241
cloves 58, 88, 101n, 158, 200, 203
Cole, David 151–2, 162n
colonialism 43–4
COLT 93
Command for the Restoration of Stability and Order 48, 105, 107
Communication Forum of Jakarta Student Senates 181
Condon, Timothy 147
Consultative Group on Indonesia 72
convergence 7–8
corporatism 121–3, 225
corruption 10, 21, 125, 149, 151, 206–15, 216
CPDS 116–17, 127
Crescent and Star Party (PBB) 232, 237
Cribb, Robert 107, 127
cronyism 7, 10
Crouch, Harold 48, 51, 56, 164, 165, 175, 249, 255
CSIS 127
cultural relativism 136
Cumings, Bruce 4, 37–8n
currency 94, 147, 150, 156, 157, 158, 160, 174
Customs Office 72, 76

Dakab 55, 57–8, 112, 234
Dale, Reginald 7
Darusman, Marzuki 64, 113, 176, 212, 235, 239, 242, 243
debt *see* private sector debt; public debt
democracy 7–8, 35, 53, 103–5, 217, 226, 253–9; and capitalism 125; and globalisation 262–3; and middle class 123; Pancasila 138–9; and working class 134–5
deregulation 9, 72, 73, 94, 104; financial and banking systems 36n, 82–4; and industry policy 89, 90; and institutions 22; liberal reformers 124; resistance 74–5
Detik 139
Dewan Pembina 111, 114
Dhakidae, Daniel 129
Dhanutirto, Haryanto 116
Dharmala 96
Dharsono, General 120
Di Palma, Giuseppe 255
Dick, Howard 72, 192, 196
Disappointed Front 120
Djalil, Matori Abdul 240
Djarot, Eros 240
Djiwandono, Soedradjad 90, 147, 158
Djojohadikusumo, Hashim 62, 100n, 202, 210
Djojohadikusumo, Sumitro 62, 100n, 125, 143n
Djojonegoro, Wardiman 116
Dobb, Maurice 32
Dodsworth, John 196
Donor, Richard 25
Dorn, James A. 35, 262
Downer, Alexander 263
DPP 111
DPR 107, 176, 178, 224, 229, 230; 1999 elections 232, *233*, 250n; military 108, 249
Duta Ekonomi 57
dwifungsi 136

East Kalimantan 61
East Timor 242, 248, 251n, 257, 263
Easton, David 25
Ecip, Sinansari 167, 170, 171, 172
economic nationalism 50–1, 65, 86, 88–94, 198, 203, 216
economy: liberalisation 71–3, 103, 104; New Order 50–3, *see also* Asian economic crisis
Edwards, Richard C. 32
Eichengreen, B. 153
elections 223–4, 226, 227, 229–30, 232, *233*
Emmerson, Donald 41
Engels, Friedrich 32, 36–7n
Era Giat Prima 208
ethylene 92–3
Evans, Peter 23

Fatwa, A.M. 117, 120, 236, 245
Fauzi, Noer 135
Feith, Herbert 41, 45
fertiliser 206
Fidler, Stephen 33, 262
Financial Sector Policy Committee (FSCP) 195, 196, 199
Fine, Ben 25
Fischer, S. 158
FKPPI 61
FKSMJ (*Forum Komunikasi Senat Mahasiswa Jakarta*) 181
Ford, Michele 180, 228
foreign investment 34–5, 50, 51, 65–6, 66n, 67–8n, 71, 73, 92, 97n, 260–1
forestry industry 54, 56, 74–5, 80, 82, 97n; confiscation process 201; corruption 207, 209; military 106
Forum Demokrasi 128, 129, 171
Forum Kota 169, 177–8, 181
Fosko 120
foundations *see yayasan*
Frankel, Jefrey 7
Freeport 118
Front Pembela Islam (FPI) 247
FSPSI 134, 232

Gadjah Mada University 167, 171
Gafur, Abdul 170
Garnaut, Ross 10, 36n, 148
GARPRI 88
Garuda 78–9, 201, 221n
Gema Madani 174
Gerakan Pemuda Ka'bah 247
Ghalib, Andi M. 213
Gill, S. 153
Gills, Barry 35, 262
Ginandjar *see* Kartasasmita, Ginandjar
Gintings, Sutradara 235
globalisation 1, 13; ambiguous implications 34–5; and democracy 262–3; and market reforms 259–61
Golkar 122, 129–30, 166, 242, 265; 1999 elections 232, *233*; and Bank Bali 189, 214; and bureaucracy 230; funds 55, 214; Habibie 173, 175–6, 227, 229; and ICMI 116; internal conflict 61, 233–6; local elites 61, 62–3; middle class 127; and military 106, 107; and oligarchs 103, 110–15; and PDI-P 238, 239; *pribumi* 131–2, 133, 134; reform 126, 232; Soeharto 49, 141; and SPSI 137; Tutut 142; Wahid presidency 243
Gomez, Edmund 4
Gontha, Peter 93, 158
Gourevitch, Peter A. 35
government 17n
Government Financial and Development Comptroller 189, 201, 203, 205–6
GPK 247
Gramsci, Antonio 3, 16n
Greenspan, Alan 7, 259
Grindle, Merle S. 20
Guided Democracy 44–5, 46
Guided Economy 45–6
Gumelar, Agum 243

Habibie, B.J. 76, 110; business empire 58, 84; corruption scandals 206, 208; debt restructuring 193; Golkar 114; ICMI 115, 116, 117, 118; industry policy 51, 89, 90–1; loss of Presidency 242; and military 106; opposition 176–82; political reform 227–41; as President 172–6, 188–9; student activists 250n; as Vice-President 142–3, 159; warships 87, 139
Hadiz, Vedi R. 45, 50, 51, 104, 127, 134, 135, 137–8, 176, 180, 228, 229, 245, 256
Haggard, Stephen 4, 25
Halid, Nurdin 212
Hamid, Sjarwan 170
Hamilton, Gary 7
Hamilton-Hart, Natasha 76, 190, 204, 210
Hamzah Haz 242–243
HANKAM 59
Hanke, Steve 156, 158
Harberger, Arnold 20, 28
Hardjojudanto, Sigit 57, 58, 100n
Harmoko 58, 84, 96–7n, 114, 141, 170–1
Harris, Nigel 11, 33, 38n, 80
Harriss, John 25
Harseno, Kentot 110
Hartas, Harsudiono 110, 129
Hartarto 51, 89, 90–1, 93
Hartcher, Peter 28
Hartono, General 110, 116, 142
Hartono, Dimyati 239, 240
Hartono, Ibnu 232
Harun, Lukman 117
Hasan, Bob 54, 55, 58, 141; Apkindo 87–8, 200; Bank Umum Nasional 191;

in Cabinet 142, 159, 160; corruption prosecution 208, 209–10; Golkar 131
Hashim, Djojohadikusumo 62
Hasibuan, Albert 112
Hawkins, Everett D. 136
Haz, Hamza 242, 243
Hedijanti, Siti 100n, 142
Hefner, Robert W. 104, 139–40
Hendroprijono 110, 122
Hewison, Kevin 4, 53, 231, 232
Hikam, Muhammad 243
Hill, D. 139
Hill, Hal 21, 52, 72, 77, 89, 147, 148, 149, 152, 215, 216, 910
Hill, Robert 263
HIPMI 62, 132, 197
HIPPI 132
HKTI 134
Hobohn, Sarwar 93
Hoffman, Stanley 263
Holdiko 198
Hong Goei Siauw 198
Hong Kong 96
human rights 263
Humpuss group 91
Huntington, Samuel 41, 44, 49, 103, 254–5, 262
Hutasoit, Potsdam 238
Hutchcroft, Paul 42
Hutomo Mandala, Hutomo Putera *see* Soeharto, Tommy

IBRA 154, 155, 188, 189, 191, 206, 216; asset sales 197–9; debt restructuring 193, 194, 195, 196
Ibrahim, Marwah Daud 116, 173, 236
ICMI 61, 90, 104, 115–18, 127, 237, 241; and Golkar 126, 233–4, 235–6, 239; Habibie 227; *pribumi* 131–2, 134; Sasono 64
Idris, Fahmi 113, 118
Idris, Kemal 168
IGGI 50
Ikatan Keluarga Besar Arief Rachman Hakim 141
Ikatan Pemuda Karya 247
Imaduddin 117
IMF 19, 24, 142, 225, 264; asset sales 193, 197, 198, 199; banking system 154–5; capital flight 261; corruption 210–11; currency 156; debt 154, 191, 196; economic crisis 4, 6, 7, 147, 148, 149, 150–1, 165; gate-keeping institutions 203, 204, 205; Indonesia 8, 9, 10, 27, 29; reform programme 157–61, 166, 187, 188, 189, 259; Washington consensus 20
import substitution industrialisation 41, 51, 56
Indofood 95, 96, 153, 194, 219n
Indomobil group 56, 91–2, 199
Indonesian Centre for Working Class Struggle 123, 140
Indonesia Corruption Watch 213, 215, 217n
Indonesian Committee for Islamic World Solidarity 140, 228, 250n
Indonesian Communist Party (PKI) 8, 44, 45, 46, 63, 134
Indonesian Democratic Party *see* PDI
Indonesian Environmental Forum 75
Indonesian Journalists Association 123
Indonesian Muslim Students' Action Front 169, 181, 237
Indonesian Nationalist Party (PNI) 44, 49, 63, 238, 239
Indonesian Prosperity Trade Union 123, 137, 140
Indonesian Socialist Party (PSI) 236
Indonesian United Democratic Party 123
industrial action 105, 135, 144n, 168
industrial policy 51
industry policy 41, 51, 89–94, 136–7
Ingleson, John 136
institutions 25, 26, 43, 203–7, 256, 258; collisions and social conflict 27; neo-liberalism 21–3; and oligarchy 73, 76–7; and politico-business families 60; reform 199, 226–32; Soeharto versus the military 105–10
intelligentsia 103, 104, 124, 126–30, 139, 171, 238–9
Inter-Governmental Group on Indonesia 50
International Crisis Group 248
International Finance Corporation (IFC) 261
International Monetary Fund *see* IMF
Investment Co-ordinating Board 59, 76
IPCOS 245
IPTN 58, 89, 90, 156
Irian Jaya 248
ISEI 96, 133
Islam 49, 63, 115, 225, 241; KAMMI 169,

181; KISDI 140; Middle Axis 237; PAN 236; and Pancasila 122, 228; PBB 232; People's Sovereign Party 233; populism 44–5, 46, 173, 176, 227, 233–4; Soeharto 143; Sutrisno 175; xenophobia 132, *see also* ICMI; Nahdlatul Ulama
Ismail, Nurmahmudi 216, 243
ITB 169
Iwantono, Sutrisno 199

Jakarta Post 206, 211, 216–17
Jakarta Stock Exchange (JSX) 83–4, 95, 157
Japan 72, 90, 91, 218n
Jasa Marga 79
Java 46
Jayasuriya, Kanishka 4–5, 148, 262
Jenkins, David 172
Johnson, Chalmers 4
Joint Team to Eradicate Corruption 208, 209
Jones, Steven 58
judiciary 155, 194, 209, 210, 211–12, 213, 216, 222n, 244
Judohusodo, Siswono 61, 62, 82, 85, 133
Juoro, Umar 220n
Justice Party 232, 237
Justice and Unity Party 232, 235

Kadin 131–2, 197
Kalla, Yusuf 216, 243
KAMMI (*Kesatuan Aksi Mahasiswa Muslim Indonesia*) 169, 181, 237
Kanindotex 81
Kartasasmita, Agus 60
Kartasasmita, Ginandjar 62, 76, 84, 115, 188, 189; corruption 208–9, 210, 212; IMF 160; industry policy 51, 89, 90; and Panigoro 238; *pribumi* contractors 85, 86; Soeharto's fall 170, 172; Team Ten 59
Kaufman, Robert 25
Kedungombo 136
Kia 91
Kiemas, Taufik 214, 222n, 238, 240, 245, 257
King, Dwight Y. 45, 176
Kingsbury, Damien 256
KISDI 140, 228, 250n
KNPI 141
Kolognas 51–2
Komaruddin, Ade 232, 233
Kopassus 248
KOPKAMTIB 48, 105, 107
Korea 33, 90
KORPRI 112
korupsi, kolusi dan nepotisme (KKN) 167, 190, 208
Kosgoro Youth 141
Kostrad 170, 175, 209, 226
Krakatau Steel 52, 75, 89
Krueger, Anne O. 19
Kusumaatmadja, Sarwono 61, 62, 64, 111, 112, 113, 168, 235
Kutai aristocracy 61
Kwik Kian Gie 129, 130, 189, 199, 216, 240, 243; asset sales 96, 197; banks 80, 81, 125, 126; on bribery 213; capital flight 95; debt restructuring 195, 196; deregulation 124; Jakarta Stock Exchange 83; PDI-P 238–9

Lal, Deepak 19, 20
Lasykar Jihad 248
Latief, Abdul 82, 85, 133
Law, D. 153
Leftwitch, Adrian 25–6
Legal Aid Institute 110, 127, 136, 182n
Lehman Brothers 147
Lenin, V.I. 37n, 224
Lev, D. 44
Levine, D. 47
Leys, Colin 22
liberalism 5, 121–2, 123–30, 138
Liddle, Bill 37n, 41, 152, 165, 256
Lie, Alvin 237, 244–5
Liem, Anthony 198
Liem Sioe Liong 54, 55, 75, 92, 131; asset sales 198, 199; Bank Central Asia 58, 191–2; Bulog 200; debt restructuring 193, 194; economic crisis 153; food industry 56, 74; oil refinery 86–7, 88, 133; reorganising capital 95, 96
Lindsey, Tim 194, 211, 212, 256
Linz, Juan J. 258
Lippo conglomerate 240
Litaay, Alex 238
local power 63–4, 245–7
Lopa, Baharuddin 189, 212–13
Lubis, Todung Mulya 49, 128–9, 211

McBeth, John 234
McCormack, Robert 260
MacDougall, John 110, 115, 116
McFaul, Michael 257
MacIntyre, Andrew 10, 21, 22, 52, 148–9
McLeod, Ross 10, 21, 36n, 148, 149, 152, 204
McVey, Ruth 42, 44, 46
Madjid, Nurcholish 113
Magenda, Burhan 61
Mahendra, Yusril 117, 237, 242–3
Malaysia 89
Mallaby, Sebastian 262
Maluku 248
Mangkusubroto, Kuntoro 189, 201
Manning, Chris 256
Mantrust 95
Manulife Indonesia 265n
Mardiyanto, General 176
Mardjono, Ary 114
Marimutu, Sinivasan 213, 214
market capitalism 3–5
Marsinah 136
Marsudi, Djiteng 79–80, 202
Marubeni 196
Marx, Karl 31–2, 37n, 47, 258
Mas, Salim 58
Mas, Sinar 58
Masduki, Teten 213
Masjumi 45, 63, 236
Master Settlement Acquisitions Agreements 193–4, 195
Medan 139–40, 240, 246–7
Megawati *see* Soekarnoputri, Megawati
Meliala, Sembiring 114
Memed, Yogie 142
Merpati 78–9
Middle Axis 237, 242
middle class 31, 32, 40, 104, 161, 164, 182n, 253; New Order 46, 47, 48, 60–1, 63, 64, 65, 68n; reformism 123–30, 135; Soeharto's fall 168; student activism 180; and working class 137–8, *see also* bourgeoisie
Mietzner, Marcus 237
Miliband, Ralph 32
military: commercial interests 54–5; conflicts over rents 87; Forum Demokrasi 129; Golkar 111, 112–14, 175; ICMI 116–17; New Order 46, 47, 48; PDI-P 238; political role 103, 104, 225–6, 229, 230, 248–9, 257; reform 212; and Sekneg 59; and Soeharto 105–10, 120, 139, 141, 142–3; Soeharto's fall 169–70, 172; and United States 263; and working class 136, 137
Mindo 79
Mochdi, Gunarijah 115
Moerdani, Benny 59, 104, 106–7, 108, 110; Golkar 112; and Islam 117; and Sjafei 238; and Soemitro 105; and Sutrisno 175
Moerdiono 57, 59, 129
Moertopo, Ali 41, 48, 49, 59, 105, 117, 134, 239
Moon, Chung-in 4, 25
Mortimer, R. 44, 47
MPR 63, 107, 128, 174, 176, 224, 226, 229–30, 232; and military 226, 249; November 1998 Extraordinary Session 175, 178–9, 181; Pancasila 228; Soeharto's re-election 141–2, 164; Wahid presidency 244
MSAA 193–4, 195
Muhaimin, Yahya 243
Muhammad, Fadel 60, 116, 118, 133, 141
Muhammad, Mar'ie 77–8, 87, 90, 93, 139, 142, 152, 156–7
Muhammadiyah 63, 236, 237, 241
Mulyani, Sri 196
Mulyosudirdjo, Priyambodo 78
Munir, Rozy 214, 241

Nahdlatul Ulama (NU) 63, 128, 129, 240–1
Naipaul, V.S. 117
Nasakom 45
Nasution, Adnan Buyung 128–9
Nasution, Anwar 73, 76, 80, 81, 84, 152, 205
Nasution, Muslimin 201
National Awakening Party *see* PKB
National Economic Planning Board 59, 65, 76
National Labour Party 233
National Mandate Party *see* PAN
nationalism 225, 233–4; economic 50–1, 65, 86, 88–94, 198, 203, 216
Neiss, Hubert 193
neo-liberalism x, 1–2, 18–19; economic crisis 6–10, 148, 151–2; institutions 21–3; markets and democracy 35, 259, 262–4; problems 25–6; public choice and Washington consensus 19–21;

reform problems 215, 216–17; social capital 24–5; state capacity 23–4
New Order 21, 40–3; Chinese business groups 54–7; politico-business oligarchy 53–4, 57–61; social order 61–4; state capitalism 50–3; tensions within regime 64–6; triumph of the state 46–50, *see also* Soeharto
NGOs 25, 64, 124, 139
Nitisastro, Widjojo 216
North, Douglass 22, 36n
North Sumatra 245–7
Novanta, Setya 210
Nurmahmudi, Ismail 216, 243
Nursalim, Sjamsul 191, 194–5, 209, 210, 212–13, 214, 244
Nusamba group 57–8, 75

O'Donnell, Guillermo 164, 255
Oetojo, Bambang 115
oil 52, 65, 66, 67n, 71, 96n
oligarchy 5, 9, 11, 16–17n, 43, 265; asset sales 197–9; bank recapitalisation 191–2; confiscation 199–203; conflicts over rents 86–8; debt 94–5, 190–1, 192–6; diversity 84–6; economic crisis 149, 150, 151–5; economic nationalism 88–94; Golkar 110–15, 234; insurance policies 95–6; Pancasila 122; patrimonial 42; political regime 103–5; politico-business families 60–1; predatory raids on banks 80–4; reconstitution 13, 187–8, 224, 225; social order 61–4; and Soeharto's fall 166; triumph 73–80
Olson, Mancur 19
Ooetojo, Bambang 115
OPSUS 48–9, 105
organiasi masyarakat (ORMAS) 49

Pakpahan, Muchtar 123, 137, 139, 140, 183n, 233, 238
PAL 87, 89
Pam Swakarsa 179, 181, 183n
pamong praja 239
Pamungkas, Sri Bintang 116, 117, 123, 128
PAN (*Partai Amanat Nasional*) 228, 232, 239, 244–5; 1999 elections 232, 233, *233*; internal tensions 236–7; paramilitaries 247
Pancasila 49, 104, 122, 130, 138–9; Habibie 228; industrial relations 136; middle class critics 128; military 116; PDI-P 239; PKB 241
Pangaribuan, Robinson 48, 59, 60, 76, 106, 133
Pangestu, M. 72
Pangestu, Mari 52, 150, 151, 192, 216
Pangestu, Prajogo 85, 131; automobile industry 75, 83; Chandra Asri 92; corruption 209, 213, 214, 244; debt 81, 196; oil refinery 86–7, 88, 133; Taspen 82
Panglaykim, J. 44, 50
Panigoro, Arifin 60, 212, 214, 232, 238, 239, 240
Papua 248
paramilitaries 225, 247
parliament *see* DPR; MPR
Partai Keadilan dan Persatuan see PKP
Partai Kebangkitan Bangsa see PKB
Partai Kebangkitan Ummat 241
Partai Nahdlatul Ulama 241
Pasaribu, Bomer 243
patrimonial administrative state 5, 16n
patrimonial oligarchy 16n, 42
patronage 43, 61, 63–4, 188
Pauker, Guy 8
PBB 232, 237
PDBI 95
PDI 49, 105, 122, 126, 129, 130, 213, 228, 238; attack on headquarters 139–40
PDI-P (PDI-*Perjuangan*) 228, 231, 232, 237–9, 242, 243, 265; 1999 elections 232, 233, *233*; and Golkar 235; local power 246–7
PDR 233, 234, 250n
peasantry 134, 135–8, 231, 238
Pemuda Panca Marga 61
Pemuda Pancasila 61, 231, 247, 251n
penembakan misterius 127
PENI 92, 94
People's Democratic Party 123, 140, 169, 228
People's Sovereign Party 233, 234, 250n
Perlez, Jane 263
Permindo 201, 202–3
Perraton, J. 153
Perta 78, 201, 202–3
Pertamina 52, 54, 58, 59, 60, 62–3, 77, 155; confiscation process 201, 202–3;

corruption 208–9, 221n; feedstock 92; outsourcing 78; reform 203–4
Petisi 50 group 120
Petras, James 34, 127, 135
petrochemical industry 92–4
Philippines 42, 231–2
Pincus, Jonathan 57, 72, 94
PK 232, 237
PKB 214, 228, 231, 240–1, 242, 243; 1999 elections 232, 233, *233*
PKI 8, 44, 45, 46, 63, 134
PKP 232, 235
PKU 241
PLN 52, 77, 79–80, 155, 202, 261
Plunder, John 153
plywood 87–8, 157–8, 200
PNI 44, 49, 63, 238, 239
PNU 241
Polanyi, Karl 28
Poleko 59
political parties 8, 224, 226, 227–8, 229, 231, 232, 244–5, 259; Bulog funds 206; laws 104, 122; local power 245–6; New Order 49; post-colonial period 44, *see also individual parties*
politico-business families 53–4, 66, 84–5; become oligarchy 57–60; economic reform 72–3; as gate-keepers 54; as oligarchy 60–1
populism 21, 161, 163n, 225; and economy 50–1, 149, 216; Islamic 46, 117–18, 173, 176, 227, 233–4; post-colonial period 44–5
power generation industry 79–80, 202
PPBI 123, 140
PPMI 237
PPP 49, 63, 113, 122, 129, 130; 1999 elections 232, 233, *233*; Middle Axis 237; Nahdlatul Ulama 128; paramilitaries 247
Prabowo, Subianto
Prajogo *see* Pangestu, Prajogo
Prawiranegara, Sjafruddin 122
Prawiro, Radius 77, 151
Prabowo, Subianto 62, 100n, 110, 116, 170, 175, 182–3n, 237
PRD 123, 140, 169, 228
predatory capitalism 5, 13, 16n, 217; and military 226; rearrangement 223, 224, 225
predatory states 19–20, 23–4, 42
Presidency 224
presidium 171, 177
Pribadi, Henry 92
pribumi 60, 64, 77, 85, 237; conflicts over rents 88; protest 131–3; and Sekneg 106
Pribumi Business Association 132
PriceWaterhouse Coopers (PwC) 78
private sector debt 6, 10, 93, 102n, 151, 190–1, 264; economic crisis 150, 152–4, 155, 159–60, 161n, 162n; predatory raids 80–4; restructuring 192–6, 206, 260–1; Sumarlin shocks 86, 94–5
privatisation 9, 22, 77–80, 197–9
Probosutedjo 58, 92, 132, 141, 157, 208, 210
property rights 60, 201, 261
PSI 236
public choice theory 18–19, 41–2
public debt 6, 40, 50, 154, 161n
PUDI 123
Pupuk Sriwijaya (Pusri) 206
Pura, Raphael 58
Putnam, Robert 24–5
PWI 123

Radelet, S. 10, 84, 147, 151, 153
Rahardjo, Dawam 115, 118
Rais, Amien 28, 140, 141, 168, 171, 177, 178, 179–80, 237; Busang goldmine scandal 118, 182n; ICMI 115, 117; PAN 136, 228; Wahid presidency 242, 243
Ramage, Douglas 104, 115, 116, 118
Ramelan, Rahardi 209
Ramli, Rizal 57, 72, 75, 94, 96, 189, 195, 206
Ramli, Rudy 211
Reeve, David 45, 49, 111
reformasi total 167
regime 17n
regime change 33–4, 156
regulatory state 4–5, 43
rents 52, 53, 151, 199, 201; conflicts 86–8
Republic Party 231
Riau 248
Richburg, Keith 171, 172
Robison, Richard 4, 9, 22, 44, 46, 50, 51, 52, 55, 56, 57, 58, 62, 71, 72, 81, 91, 93, 104, 124, 142, 200
Rocamora, J. 44
Rodan, Garry 31

Rosser, Andrew 4, 34, 52, 72, 82, 83, 142, 148, 152
Rudner, Martin 9, 51
Rueschemeyer, Dietrich 127, 131, 134, 135, 231
Rukmana, Siti Hardijanti 58, 100n, 142, 159, 231; corruption 208, 209, 210, 222n; Golkar 115; toll roads 79
rupiah 94, 147, 150, 156, 157, 158, 160, 174
Russia 27–8
Rustam, Soepardjo 112
Ryter, Loren 247

Sabirin, Sjahril 205, 209, 216
Sachs, J. 10, 153
Sadikin, Ali 168
Saidi, Ridwan 113
Salim, Anthony 96, 141
Salim, Emil 62, 168, 174, 182n
Salim group 194, 195, 198
Saludo, R. 7
Sambuaga, Theo 113
Samego, Indria 116
Samsudin, General 108
Sang Young Ryu 4
Santoso, Amir 117
Sanusi, H.M. 120
Sasono, Adi 47, 64, 115, 118, 173, 197, 233
Sastrosatomo, Soedarpo 62
Sastrowardojo, Sanyoto 93
satgas 247
Sato, Yuri 85
SBSI 123, 137, 140
Scharmis, Hector E. 28
Schiller, Jim 63
Schmitt, H.O. 44
Schmitter, Philippe C. 66–7n, 255
Schwarz, Adam 55, 56, 74, 75, 79, 83, 85, 87, 88, 93, 106, 110, 114, 132, 148
secular nationalism 237, 239
Seda, Frans 62
Sekneg 48, 59–60, 76, 86, 106, 133
Semen Gresik 260
Sentral Organisiasi Buruh Seluruh Indonesia 45
separatism 247–8
Sepdalopbang 94
September 11 248, 262
Setiakawan 137
Setiawan, Bonnie 123
Shameen, A. 7
Shihab, Alwi 243
Shin, Yoon Hwan 57
Shiraishi, Takashi 63, 136
Sigit, Ari 76–7, 87
Sihaloho, Aberson Marle 238
Simandjuntak, Djisman 83
Simanjuntak, Marsillam 128–9
Sinar Mas 194
Sinivasan, Marimutu 206, 209, 234–5, 244
Sirait, Sabam 238
Siregar, Arifin 88
Siregar, Reza Y. 198
Sjafei, Theo 238
Sjahrir 72, 74, 95, 124, 125, 152, 254
Sjamsuddin, Din 117
Sjamsuddin, Sjafrie 110, 170
Skocpol, Theda 224
Slade, Betty 151–2, 162n
Smith, Steve 35, 262
SOBSI 45
social capital 24–5, 26
social justice 108
social order 61–6
social power 13, 14, 26, 165; New Order 42; reorganisation 27–8
Société Générale de Surveillance 72
Soedjono, Police Colonel Roekmini 108
Soegomo, Bambang 115
Soeharto 40; accession to power 46; authoritarianism 21, 49, 138; Chandra Asri 93; and Chinese business groups 54; Command for the Restoration of Stability and Order 48; corruption prosecution 208, 209, 210, 211; economic crisis 6–7, 27, 147, 155–6, 158–9, 161; economic nationalism 89; economy 50, 51–2; fall 164–7, 170–2; Golkar 49, 111, 112, 113, 114; ICMI 115, 117, 118; and military 104, 105–10; MPR 224; Pancasila 228; patronage 43; and political Islam 122; and *pribumi* contractors 133; rational calculations 41; re-election 140–3, 167; rents 86–7, 88; resisting reform 12; rise 5; and Soekarno business figures 199–200; sources of income 94; state capitalism 8–9; Timor decision 92; *yayasan* 55, 57, *see also* New Order
Soeharto, Tommy 57, 79, 100n; automobile industry 91, 196; cloves 88, 101n, 200; corruption prosecution 208,

209, 211, 212, 215; debt 154; loans 80, 81–2; taxes 87
Soeharto family 57, 58, 60, 71, 76–7, 84, 100n, 132, 152; automobile industry 75, 92; banks 156–7; and Chinese-owned business groups 85; confiscation process 200, 201, 202; conflicts over rents 87; debt 153–4; forestry industry 74; petrochemical industry 92; political parties 231; public procurement 78, 79, 80, *see also* Hardjojuntanto, Sigit; Rukmana, Siti Hardijanti; Trihatmodjo, Bambang
Soehoed, A.R. 51
Soekarno: authoritarianism 40, 44, 45; fall 8, 46, 127; state capitalism 40, 41
Soekarnoputri, Megawati 141, 177, 178, 179–80, 228, 229, 233; failed Presidential bid 241, 242–3; and military 248; nationalism 263; organicist vision 239; PDI 105, 129, 139–40; PDI-P 238, 240; and PKP 232; as President 217, 254; presidential funds 245, 257; secular nationalism 237; and student activism 171; as Vice-President 242, 243
Soemitro, General 105, 107, 141
Soedradjad, Djiwandono 90, 147
Soerjadjaya, Willem 56
Soesastro, Hadi 21, 72, 151, 152, 153, 216
Soewandi, Rini M.M. 206
Solidarity Free Trade Union 137
South Korea 4
Special Operations Command 48–9, 105
Special Presidential Assistants 48
Srinivasan, T.N. 20
Standard Chartered Bank 82, 192
state 17n; neo-liberalism 19–21; New Order 46–50
state capacity 23–4, 26
state capitalism 1, 9, 40; and economic crisis 155–60; New Order 50–3
state corporations 52, 77–80
state power 13, 14, 29–30, 120; Golkar 110–15; ICMI 115–18; Soeharto versus the military 105–10
state-qua-state 30, 47
State Secretariat 48, 59–60, 76, 86, 106, 133
Stiglitz, Joseph 22, 24
stock exchange 83–4, 95, 157
strikes 105, 135, 144n, 168
structure 2, 30
student activism: Habibie 177–9, 180–2, 250n; New Order 46, 48, 64–5, 124, 161; Soeharto's fall 166, 167–9, 171, 173, 183n
Subianto, Bambang 154
Sudarijanto, Cacuk 189, 195, 199
Sudarsono, Juwono 141
Sudharmono 57, 84; Golkar 111, 112, 113, 126; and military 106, 107
Sudharmono, Tantyo 115
Sudibyo, Bambang 211, 216, 243
Sudjana, Eggi 237
Sudjana, Ida Bagus 78
Sudrajat, Edi 109, 168, 175, 235
Sudwikatmono 57, 58, 141, 208, 210
Sugama, Bambang Yoga 81
Sugama, Yoga 81, 115
Sukardi, Laksamana 73, 125–6, 129, 130, 189, 216, 238–9, 240, 243
Sulbi, Endang 58
Sumargono, Ahmad 237
Sumarlin, J.B. 81, 90
Sumarlin shocks 94–5
Supreme Audit Agency 82, 189, 193, 205–6
Surasa 82
Suta, Putu Gede Arya 192
Sutomo, Yos 55
Sutowo, Ibnu 52, 58, 60, 62, 78, 85, 115
Sutowo, Ponco 58, 60, 62, 115, 133
Sutowo, Adiguna 141
Sutrisno, Try 109, 110, 142, 175
Sutter, J. 44

Tandjung, Akbar 113; corruption 206, 209, 212, 214, 215, 222n, 234; Golkar 175–6, 235, 239; Presidential challenge 242–3
Tandjung, Feisal 87, 110, 116, 141, 142
Tanjung Priok riots 120, 122
Tanoto, Sukanto 209
Tansil, Eddy 81
Taspen 82
Taufik, Iman 60
Team Ten 59–60, 63, 76, 86, 133
Team 39 93
technocrats 18, 19, 25, 30, 35–6n, 40, 48, 50, 86, 216–17; and debt 93, 94
telephone industry 79, 143–4n
Telkom 89
Tempo 123, 139, 231, 235

terrorism 248–9, 262, 263
Texmaco 196, 220n
Thailand 28, 34; bourgeoisie 126, 130; bureaucratic polity 42; democratisation 104, 125, 231; economic crisis 4; military coups 121–2; planes-for-rice deal 89
Therborn, Goran 134, 231
Thomas, K.D. 44, 50
Thomas, R. 22
Timor Putra Nasional 91–2, 200
Tirtosudiro, Achmad 52
Titiek 100n, 142
Tjahjadi, Robby 81
Tobing, Jakob 235, 238
Toemioen, Theo 239
Tolleng, Rachman 113, 129
Tommy *see* Soeharto, Tommy
Toye, John 19
Tri Polyta 92, 94
Tri Usaha Bhakti (TUB) 54
Trihadtmodjo, Bambang 57, 77, 82, 100n, 142, 202; automobile industry 92; banks 157; citrus monopoly 87; debt 154; Golkar 115; Knindotex 81; satellites 79
Trisakti University 169
Trubek, David L. 125
Tullock, Gordon 19
Tutut *see* Rukmana, Siti Hardijanti

United Development Party *see* PPP
United States 265; automobile industry 91; and democracy 35, 262; GSP status 137, 144n; human rights 263; regime change 33, 156; war on terror 248
University of Indonesia (UI) 167, 169
UP3DN 86, 100–1n

Van Diermen, Peter 256
Van Dijk, Kees 256
Van Klinken, Gerry 254
Vatikiotis, M. 21
Venoroso, Frank 26

Wade, Robert 4, 26
Wagstaff, Jeremy 235
Wahid, Abdurrahman 8, 63, 128, 129, 177, 178, 179–80; accountability 224; banks 105, 204; corruption scandals 206, 208, 214, 215; debt restructuring 196; Golkar 113; ICMI 115; judiciary 212; leadership style 215–16; money scandals 245; PKB 228, 229, 233, 240; and Rais 237; reforms 189, 241–4, 254, 257; and student activists 171, 173
Wahid, Hashim 214
Wahid, Salahuddin 214
Wahono, General 108, 109, 113
Walters, Patrick 170, 172
Wanandi, Jusuf 112
Wanandi, Sofjan 132, 195
Ward, Ken 49
Wardhana, Ali 90
Washington consensus 16n, 20, 21
Weber, Max 23, 42
Weiss, Linda 4
Whardana, Ali 209
Whitehead, Laurence 37n
Wibisono, Christianto 52, 95, 124, 125
Williamson, J. 4, 20
Wilner, Ann Ruth 41
Winata, Tommy 214
Wingfield, Tom 4, 53
Winters, Jeffrey 11, 34, 59, 60, 76, 148, 153
Wirahadikusumah, Agus 226, 249
Wiranto, General 110, 141, 142, 243, 251n; and Habibie 175–6, 179; Soeharto's fall 170, 172
Witoelar, Rachmat 64, 111–12, 113
Wolf, C. 7
Wolfowitz, Paul 263
Wood, Ellen Meiskens 31, 32
working class 127, 134–8, 164, 168, 181, 231, 238, 253
World Bank 6, 9, 19, 136, 147, 225, 259, 262, 263–5; asset sales 198; corruption 210, 212; debt 93, 195, 196; deregulation 36n, 51; economic crisis 12, 29, 150, 152, 153, 154, 155, 158; forestry industry 207; fund misappropriation 206; import controls 71; industry policy 50, 56, 65; institutions 22, 124–5, 189, 203, 204, 205; oligarchy 74, 75; Philippines 42; predatory raids 148, 151; private banks 83; privatisation 77, 78, 79; reform programme 72, 73, 76, 86, 215, 216; social capital 24, 25; state banks 81, 82; state capacity 23, 24; Washington consensus 20

xenophobia 44–5, 131, 132–3

Yamin, M. 238
yayasan 55, 57–8, 94, 208, 234
Yayasan Dakab 55, 57–8, 112, 234
Yayasan Dana Sejahtera Mandiri 133–4
Yayasan Dharma Putra (YDP) 54
Yayasan Dharmais 55, 57–8
Yayasan Eka Paksi 55
Yayasan Supersemar 55, 57–8
YLBHI 110, 127, 136, 182n
Yogyakarta 245–6, 247
Yogyakarta, Sultan 62, 178, 183n
Young Entrepreneurs Association 62, 132, 197
Yudhoyono, Susilo Bambang 243
Yusuf, Glen 193
Yusuf, Slamet Effendy 113

Zaire 66n
Zakaria, Fareed 257
Zingales, Luigi 260
Zon, Fadli 117, 237
Zysman, John 4, 26

Lightning Source UK Ltd.
Milton Keynes UK
UKOW05f1813280217
295569UK00014B/377/P

9 780415 332538